# Coalbed Methane and Coal Geology

Geological Society Special Publications
*Series Editor* A. J. FLEET

GEOLOGICAL SOCIETY SPECIAL PUBLICATION NO. 109

# Coalbed Methane and Coal Geology

EDITED BY

## R. GAYER & I. HARRIS

Department of Earth Sciences, University of Wales,
Cardiff, UK

1996

Published by

The Geological Society

London

# THE GEOLOGICAL SOCIETY

The Society was founded in 1807 as The Geological Society of London and is the oldest geological society in the world. It received its Royal Charter in 1825 for the purpose of 'investigating the mineral structure of the Earth'. The Society is Britain's national society for geology with a membership of around 8000. It has countrywide coverage and approximately 1000 members reside overseas. The Society is responsible for all aspects of the geological sciences including professional matters. The Society has its own publishing house, which produces the Society's international journals, books and maps, and which acts as the European distributor for publications of the American Association of Petroleum Geologists, SEPM and the Geological Society of America.

Fellowship is open to those holding a recognized honours degree in geology or cognate subject and who have at least two years' relevant postgraduate experience, or who have not less than six years' relevant experience in geology or a cognate subject. A Fellow who has not less than five years' relevant postgraduate experience in the practice of geology may apply for validation and, subject to approval, may be able to use the designatory letters C Geol (Chartered Geologist).

Further information about the Society is available from the Membership Manager, The Geological Society, Burlington House, Piccadilly, London W1V 0JU, UK. The Society is a Registered Charity, No. 210161.

Published by The Geological Society from:
The Geological Society Publishing House
Unit 7
Brassmill Enterprise Centre
Brassmill Lane
Bath BA1 3JN
UK
(*Orders*: Tel. 01225 445046
        Fax 01225 442836)

First published 1996

**British Library Cataloguing in Publication Data**
A catalogue record for this book is available from the British Library.

ISBN 1-897799-56-X

Typeset by Aarontype Ltd, Unit 47, Easton Business Centre, Felix Road, Bristol BS5 0HE, UK

Printed in Great Britain by
The Alden Press, Osney Mead
Oxford, UK

**Distributors**

USA
  AAPG Bookstore
  PO Box 979
  Tulsa
  OK 74101-0979
  USA
  (*Orders*: Tel. (918) 584-2555
          Fax (918) 560-2632)

Australia
  Australian Mineral Foundation
  63 Conyngham Street
  Glenside
  South Australia 5065
  Australia
  (*Orders*: Tel. (08) 379-0444
          Fax (08) 379-4634)

India
  Affiliated East-West Press PVT Ltd
  G-1/16 Ansari Road
  New Delhi 110 002
  India
  (*Orders*: Tel. (11) 327-9113
          Fax (11) 326-0538)

Japan
  Kanda Book Trading Co.
  Tanikawa Building
  3-2 Kanda Surugadai
  Chiyoda-Ku
  Tokyo 101
  Japan
  (*Orders*: Tel. (03) 3255-3497
          Fax (03) 3255-3495)

# Contents

# Preface

Resources of coalbed methane (CBM), i.e. methane trapped within the porous system of coal, may be as high as $250 \times 10^{12}\,m^3$ worldwide, many times greater than the collective reserves of all the known conventional gas fields. Yet only in the United States is this energy source produced commercially, and that largely due to enlightened federal tax concessions that encouraged the research into the exploration and production techniques necessary to establish a viable industry. Unlike the situation in conventional hydrocarbon plays, with CBM coal acts as both the source rock and the reservoir for the gas. This relationship leads to a major paradox whereby, in order for gas sourced by the coal not to have migrated, the coal must either be sealed or possess very low permeability. And yet for the coal bed to be an effective reservoir the gas must readily migrate into the well bore when the coal is penetrated by a production well. The solution to this paradox lies in a wide-ranging understanding of the geology of coal, and this volume aims to provide some answers.

The 24 papers in the volume are written by experts in CBM and coal geology from a range of countries in which CBM is either currently being produced commercially (USA) or in which there is active exploration (Europe and Australia). The first group of contributions covers CBM resources, from a coal geology perspective, in the United States and Europe, including the potential from the UK continental shelf and the exciting prospects of Russia and the Ukraine. The coverage also includes a geological assessment of the Appalachian and Variscan coal-bearing foreland basins in which CBM potential ranges from excellent to poor, and detailed appraisals of individual case studies that can be enhanced significantly by using a computerized database (e.g. the Ruhr).

The second and largest section contains papers that discuss aspects of coal as a gas reservoir, particularly in relation to permeability—the key to successful CBM production. Treatment ranges from in-depth studies of reservoir fracture systems that give the principal permeability pathway in coal, through the effects of coal microstructures on methane release, and models of matrix shrinkage during methane desorption, to adverse effects on reservoir permeability caused by mineralization. The theoretical treatments of the various contributions are supported by detailed case studies in South Wales, the Czech Upper Silesian coal basin, and the Bowen basin of Queensland.

The last section of the volume contains a range of papers covering aspects of traditional coal geology, relating key features to a fuller understanding of CBM. Thus the maceral content of a coal determines: (i) the amount of gas generated during its maturation, with liptinite and vitrinite being the greatest contributors; (ii) the amount of gas stored in the coal, with the micropore structure of inertinite providing the greatest storage capacity; and (iii) the speed of methane release to the well bore, facilitated by the meso-pore structure of fusain and the enhanced cleat development in vitrinite. Papers directly relevant to assessing the maceral composition of coals include discussion of a depositional model related to maceral content of the Lower Silesian coal basin in Poland, an automated method for microlithotype analysis of coal, and organic geochemical studies related to gas proneness. The section also contains papers investigating the mineral matter in coal, particularly pyrite, and the use of palynology in down-the-hole correlation of coal seams.

The CBM industry is relatively young and only now are exploratory wells being sunk in the UK, Europe and elsewhere in the World. There is a dearth of texts describing the geology of CBM and this volume will help to fill this gap. It will appeal to geologists involved with the CBM industry, and also to those connected with coal and conventional hydrocarbon resources, as well as to lecturers and students.

The editors would like to thank all those involved with the preparation of the volume, including the secretarial staff in the Department of Earth Sciences at Cardiff. We particularly thank the authors and referees of the 24 original papers.

Rod Gayer and Ian Harris
Department of Earth Sciences
Cardiff University

# Coalbed methane in the USA: analogues for worldwide development

## D. KEITH MURRAY

*D. Keith Murray & Associates, Inc., 200 Union Boulevard, Suite 215,
Lakewood, CO 80228-1830, USA*

**Abstract:** Preliminary estimates of coalbed gas in the major coal-bearing basins of the
world range from approximately 3000 trillion cubic feet (TCF) to more than 12 000 TCF
(85–340 trillion cubic metres, TCM). The successful development of this important energy
resource will largely depend on the judicious transfer of the coalbed methane (CBM)
technology developed in the USA during the past two decades. Innovative field research
has been conducted primarily in the San Juan and Black Warrior basins. Projects have
involved co-operative efforts between the Gas Research Institute, federal and state
geological surveys, academia, both large and small oil companies and private consultants.
As a result, significant improvements have been made in many aspects of CBM
technology, including: (1) dynamic openhole cavitation completions; (2) advanced
hydraulic stimulation procedures; (3) enhanced ultimate coalbed gas recovery using the
injection of nitrogen or carbon dioxide into the coalbed reservoir; (4) improved reservoir
simulation parametric techniques; and (5) seismic methods to predict areas of optimum
reservoir permeability. The appropriate applications of this rapidly developing technology
to the complex and varied conditions that exist in other coal-bearing basins throughout
the world is expected to lessen significantly the risks inherent in the successful
commercialization of coalbed gas in a number of countries.

The purpose of this paper is to present an up to
date summary of coalbed methane (CBM)
technology in the USA. Coalbed 'methane' is
more accurately termed coalbed *gas*, inasmuch
as heavier hydrocarbon components, including
high-gravity oil, and carbon dioxide sometimes
comprise the production stream. The implications of this development, the most advanced
and prolific in the world, to exploration of
the presumed very large undeveloped CBM
resources in the rest of the world, are alluded
to in this summary treatment.

## World coal resources and production

Coal is the most abundant mineral fuel in the
world, with reserves estimated at more than
$10^{12}$ t (Rice *et al.* 1993). The total coal
resources of the world have been variously
estimated from 9655 to 30 000 Mt; more than
90% is found in the Northern Hemisphere
(Landis & Weaver 1993). According to some
estimators, most of this resource is located in
Asia, North America and Europe, in order of
declining percentages. Most of the coal
resources occur in large present day sedimentary basins or sedimentary provinces (Landis &
Weaver, 1993).

According to Petroconsultants (1981), the
total known coal resources in the world have

been estimated at approximately $9155 \times 10^9$ SCE
(standard coal equivalent; one SCE = 1 t hard
coal with a heating value of 6879 kcal/kg,
equivalent to high-volatile bituminous coal by
ASTM standards). This is equal in energy
content to $4.3 \times 10^{13}$ BBLs oil-equivalent. Hard
coal comprises 79% of the total.

Coal resources are believed to be present in at
least 3000 coal basins or deposits located in
more than 70 political entities. More than 87%
of the world's coal resources occur in the
following countries: former USSR, 40.18%,
USA, 30.81%; and China, 16.38%. About 8%
of the resources are found in West Germany
(2.67%), Canada (2.18%), the UK (1.78%) and
Australia (1.45%) (Petroconsultants, 1981).

Matveev (1976) has published some of the
more optimistic (and realistic?) estimates of total
world coal resources—30 000 Mt. This estimate
appears to have been made on the basis of the
*geology* of the basins of coal deposition, rather
than on consideration of coalbed depth or
thickness, or on engineering factors related to
mineability and utilization.

In 1990, more than 4704 Mt of coal were
produced in the world, of which hard coal (i.e.
coal of bituminous or higher rank) comprised
more than 70%. In terms of all ranks, the
following countries produced 73% of the
world's coal in 1990 (in descending order):
China, USA, USSR, Germany, Poland and the
UK (Benkis 1992)

*From* Gayer, R. & Harris, I. (eds) 1996, *Coalbed Methane and Coal Geology,*
Geological Society Special Publication No 109, pp 1–12.

**Table 1.** *Major coal and coalbed methane resources in the world (modified from Kuuskraa et al. 1992)*

| Country | Coal resource ($10^9$ tonnes) | Methane resource (TCF, in-place) |
|---|---|---|
| Russia | 6500 | 600–4000 |
| China | 4000 | 1060–2800 |
| United States | 3970 | 275–650 |
| Canada | 7000 | 300–4260 |
| Australia | 1700 | 300–500 |
| Germany | 320 | 100 |
| United Kingdom | 190 | 60 |
| Kazakhstan | 170 | 40 |
| Poland | 160 | 100 |
| India | 160 | 30 |
| Southern Africa* | 150 | 40 |
| Ukraine | 140 | 60 |
| Total | 24 460 | 2976–12 640 |

\* Includes South Africa, Zimbabwe and Botswana.

## World coalbed methane resources

The immense amounts of coal known or suspected to be contained within the coal-bearing basins of the world have generated significant amounts of coalbed gas. Perhaps more than 90% of this gas resource is contained in coal deposits in the Northern Hemisphere. Hard coal deposits have generated more gas (generally thermogenic in origin) than have brown coals. However, the lower rank coals (subbitiminous and lower) in some instances have generated very large volumes of biogenic gas, some of which is being produced at commercial rates today. A notable example is the production of nearly pure methane from Tertiary Fort Union Formation coalbeds in the Powder River Basin, northeastern Wyoming, USA In the eastern part of this large Laramide basin, coalbed gas is being produced from upper Fort Union coals as much as 50–200 ft (15–61 m) in thickness at depths of approximately 300–1000 ft (152–305 m) and at rates in some instances exceeding 500 mega cubic feet (MCF) (14.2 Mm$^3$) per day (Tyler *et al.* 1991).

As shown in Table 1 (modified from Kuuskraa *et al.* 1992), most of the coal and CBM resources are concentrated in 12 countries. These estimates are, at best, tentative and based on an incomplete, but growing, database.

The estimated in-place CBM resources in the USA, as of 31 December 1992, range from 275 to 649 trillion cubic feet (TCF) (7.8–18.4 trillion cubic metres; TCM), the wide variation being dependent on assumed average *in situ* gas content and other variables (Potential Gas Committee 1993).

Most of the CBM resources in the USA occur in the Rocky Mountain basins, which contain coals of Cretaceous and Tertiary age and which

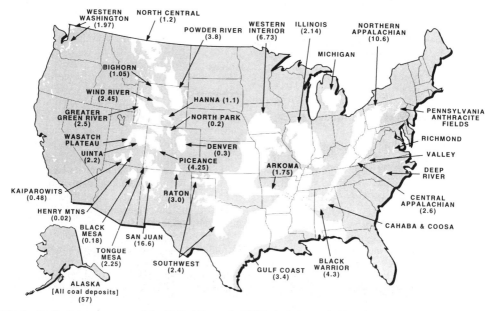

**Fig. 1.** Coalbed gas resources of the United States (in TCF); includes probable, possible and speculative potentially producible supply (Potential Gas Committee 1993).

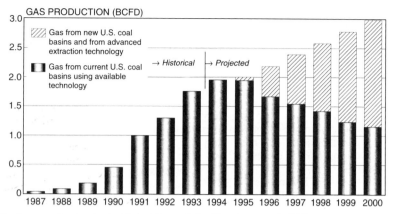

**Fig. 2.** Coalbed gas production in the United States (after Kuuskraa *et al.* 1992).

are largely undeformed. Lesser amounts of CBM resources are in the Midcontinent and Appalachian basins, where most of the coals are of Pennsylvanian (Namurian–Westphalian) age.

## Producible coalbed methane resources (supply) of the USA

Figure 1 (Potential Gas Committee 1993) shows the major coal-bearing regions in the USA (the entire state of Alaska is treated as one coal-bearing region due to the lack of specific data for the many basins in that state). For their 31 December 1992 report, the Potential Gas Committee considered 28 separate coal-bearing areas. The sum of their *most likely* estimates for the potentially recoverable supply of CBM in the conterminous USA and Alaska is 135 TCF (approximately 4 TCM), categorized as follows: 11.613 TCF (0.344 TCM) *probable*, 40.177 TCF (1.190 TCM) *possible* and 83.219 TCF (2.466 TCN) *speculative* (Potential Gas Committee 1993).

## Coalbed methane production in the USA

Figure 2 (after Kuuskraa *et al.* 1992) shows both historical and projected levels of CBM production in the USA. The *projected* part of this bar

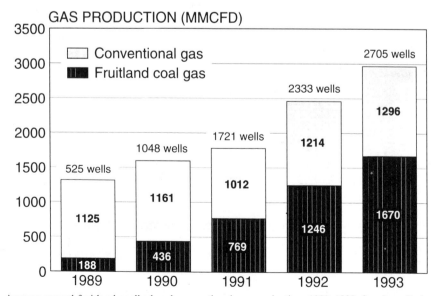

**Fig. 3.** Average annual fruitland coalbed and conventional gas production, 1989–1993, San Juan Basin,

graph is divided into two parts: (1) gas from currently producing coal-bearing areas using available technology; and (2) gas from newly developed coal-bearing areas plus advanced extraction technology used in all producing basins. On this basis, by the year 2000, daily production in the USA should reach 3 billion cubic feet (BCF) (0.08 billion cubic metres; BCM) per day. However, this may be a very conservative estimate, inasmuch as the San Juan Basin alone is predicted to reach a daily production rate during 1994 in excess of 2 BCF (0.06 BCM) per day.

Yearly production of CBM in the USA has increased from 4.77 BCF (0.13 BCM) from 160 wells in 1983 to 550 BCF (15.6 BCM) from 5743 wells in 1992, with an average daily rate of 1.5 BCF (0.04 BCM). Approximately 96% of the total CBM production in the USA came from the San Juan and Black Warrior basins in 1992.

Most of the production increase occurred in the San Juan Basin, which provided 80% (446.7 BCF; 12.6 BCM) of total CBM production in the USA from only 38% (2161) of all CBM wells. During 1992, CBM production in the San Juan Basin increased 62% over that of the previous year (see Fig. 3).

In the Black Warrior Basin, Alabama, CBM production during 1992 totalled 92 BCF (2.6 BCM), or 17% of CBM production in the USA. This was produced from 3047 (53%) of all the CBM wells in the USA.

Lesser, but increasing, amounts of CBM were produced from basins that include the Central Appalachian Basin (Virginia, West Virginia), the Western Interior Coal Region (Arkoma Basin, Oklahoma and Cherokee Basin, Kansas, Oklahoma and Missouri), the Powder River Basin (northeastern Wyoming), the Piceance Basin (western Colorado) and the Uinta Basin (eastern Utah) (see Fig. 1 for locations).

Table 2 represents an analysis of production rates of Fruitland CBM wells in the San Juan Basin, Colorado and New Mexico. Average rates are shown for five different categories, ranging from less than 100 MCF/d (2.8 MCM/d) to more than 5000 MCF/d (142 MCM/d). The weighted average for the 2257 CBM wells analyzed is 773 MCF/d (22 MCM/d) (pers. comm. from George Lippman, El Paso Natural Gas, El Paso, Texas, 1994).

## Selected examples of coalbed methane technology under development in North America

### Dynamic openhole cavitation completions

Dynamic openhole cavity completion techniques have been pioneered in the San Juan Basin, Colorado–New Mexico. This completion method for CBM wells has, in those areas where it has been successful, played an important part in making the San Juan Basin the most prolific CBM basin in the world.

In this type of openhole completion, the coalbed reservoirs are encouraged to slough into the well bore, creating numerous self-propped fractures of varying orientation and linking the well bore to the reservoir. Because the cavity is a by-product of the process and not the main objective of this completion technique, the term 'dynamic openhole completion' is more appropriate than 'openhole cavity completion' (see Logan *et al.* 1993).

In 1993, gas from coalbed wells amounted to 54% (more than 1.6 BCF/d; 0.05 BCM/d) of the total gas production from the basin (Fruitland Formation coalbed gas plus gas from older Cretaceous sandstones) (see Fig. 3). Openhole cavitated wells have produced more than 76% of the total (Kelso 1994). To date, more than 1000

**Table 2.** *Total producing Fruitland coalbed gas wells, San Juan Basin, Colorado and New Mexico (pers. comm., George Lippmann, El Paso Natural Gas Co., El Paso, Texas, 1994)*

| Rate (MCFD) | No. of wells | Average rate (MCFD) |
|---|---|---|
| <100 | 690 | 45 |
| 101–500 | 885 | 233 |
| 501–1000 | 235 | 704 |
| 1001–5000 | 384 | 2429 |
| >5001 | 63 | 6481 |
| Total | 2257 | — |
| Weighted average | — | 773 |

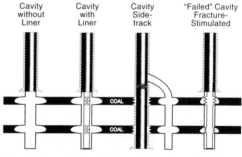

**Fig. 4.** Types of 'cavity' completions in coalbed gas well (after Kelso 1994).

Fruitland CBM wells have been completed, or recompleted, using dynamic cavity techniques (Fig. 4; Kelso 1994). Kelso (1994) believes that, in the San Juan Basin at least, a combination of the following criteria is required for successful cavity completions: (1) adequate reservoir permeability (at least 5 mD); (2) reservoir over-pressuring (say, greater than 0.50 PSI/ft); and (3) thermal maturity of the coalbed reservoir of at least high-volatile A bituminous in rank (at least 0.78% $R_o$).

*Background of development of cavitation technique.* From the 1970s through the mid- to late 1980s, the standard practice of coalbed gas well completions in the San Juan Basin was to run casing to total depth, cement the casing, perforate and hydraulically stimulate the target coal beds with various types of fracture treatments. In the mid-1980s, Meridian Oil Inc. began experimenting with openhole cavity techniques in its 30-6 Unit in Rio Arriba County, New Mexico. When the results of Meridian's highly successful completions became known to the petroleum industry, other operators (including Amoco, Devon Energy, Phillips Petroleum and ARCO Oil & Gas) followed suit on their acreage holdings in the northern San Juan Basin. Dramatic success in well performance often resulted from using this variation of openhole completion, which involves drilling through the Fruitland coal interval underbalanced in a controlled blowout situation, followed by cavitation of selected coal beds.

In 1990, Amoco Production Company, the Gas Research Institute and Resource Enterprises Inc. began a joint research effort designed to compare the performance of cased, hydraulically fractured wells with that of openhole cavity completions at a single location. This site, termed the Completion Optimization and Assessment Laboratory, is located on the Southern Ute Indian Reservation in La Plata County, Colorado. The primary focus of this co-operative research was '...to collect and interpret data concerning the performance of an open hole cavity well in a described, controlled environment' (Mavor 1991).

In the 'sweet spot,' or high-potential fairway, of the northern San Juan Basin, a well completed using the dynamic openhole cavity completion technique typically experiences a peak production rate of 1 MCF/d (0.03 MCM/d) or more at depths averaging about 3000 ft (914.4 m).

In the 'fairway region' of the northern San Juan Basin, cavity completions usually outperform hydraulically fractured wells by five to 10 times. Table 3 shows comparative performance of these two methods of completions. Of special note are two CBM wells completed in the 'sweet spot' of the basin: (1) Meridian Oil San Juan 30-6 No. 412, which to date has produced more than 21 BCF (0.6 BCM) and still (in 1993) was producing more than 10 000 MCF (283 MCM) per day; and (2) San Juan 30-6 No. 401, recompleted as an openhole cavity well, which experienced a 30-fold increase in rate of gas production (Kelso 1994).

**Table 3.** *Comparative performance of cased/hydraulically stimulated and openhole cavity completions in Northeast Blanco Unit (NEBU), northern San Juan Basin, New Mexico (Kuuskraa & Boyer 1993)*

| Originally cased and fractured wells | | Re-drilled openhole cavity wells | | |
|---|---|---|---|---|
| Well number | Short-term gas rate (MMCF/d) | Well number | Short-term gas rate (MMCF/d) | January 1992 gas rate (MMCF/d) |
| 400 | 0.11 | 400R | 2.6 | 3.6 |
| 403 | 1.00 | 403R | 10.9 | 4.2 |
| 404 | 0.15 | 404R | 22.6 | 4.6 |
| 413 | 0.40 | 413R | 10.8 | 4.5 |
| 421 | 0.80 | 421R | 22.9 | 5.5 |
| 423 | 0.24 | 423R | 10.0 | 4.3 |
| 425 | 0.28 | 425R | 13.2 | 3.7 |
| 427 | 0.86 | 427R | 8.9 | 3.0 |
| 429 | 0.47 | 429R | 2.0 | 1.5 |
| 441 | 0.72 | 441R | 0.7 | 1.7 |
| Total | 5.03 | Total | 104.6 | 36.6 |
| per well | 0.50 | per well | 10.5 | 3.7 |

The Fruitland coals in this area are high-volatile A bituminous or higher in rank; the basal Fruitland coal is usually thick (often 25–50 ft, 7.6–15.2 m or more), with relatively high permeability (10–25 mD) and characterized by higher than normal reservoir pressure (greater than 0.48 PSI/ft).

In a more general sense, this completion method has been successfully applied to coalbed reservoirs that (1) have absolute permeabilities greater than 5 mD, (2) are normally pressured to overpressured, (3) are friable and (4) are under low geological compressive stresses.

Reasons for commercial failure include: (1) depths greater than 3000 ft (914.4 m) in certain other coal basins; (2) failure to stabilize the cavity after it has been created; (3) excessive water production in areas of very high permeability; (4) non-cavitation of the coal; (5) inability to isolate effectively and complete all of the coal beds in a long openhole interval; and (6) limited understanding of the effects of this type of completion on the coalbed reservoir. In some instances where the cavitation was unsuccessful, operators have run and cemented liners and hydraulically stimulated the coals with some success (refer to Fig. 4).

*Summary.* Most of the production rate increase in a well completed using dynamic openhole cavitation techniques appears to be related chiefly to the improved permeability of the coalbed reservoir beyond the created cavity rather than the size of the cavity itself. However, additional research needs to be conducted to further verify the hypothesized permeability enhancement mechanisms for these types of completions.

## Advanced hydraulic fracture treatments

According to Conway (1994), many hydraulic fracture treatments of CBM wells are not optimized due to damage to the coalbed reservoir or to incomplete access to the available reservoir storage capacity. He maintains that hydraulic fracture treatments *can* be significantly improved, resulting in higher CBM well production rates. Frac fluids containing nitrogen foam, for example, can be designed to minimize reservoir damage. Conway (1994) maintains that only a relatively small percentage of coalbed reservoirs are amenable to cavity stress completions, which provides adequate incentives to improve fracture treatment designs. He presents (Conway 1994) some field-derived data from several areas in the San Juan Basin that point to new areas of

stimulation research that could lead to higher production rates from CMB wells completed using these types of treatments.

## Seismic characterization of fractured coalbed reservoirs

In 1984, the Reservoir Characterization Project was initiated by Professor Thomas L. Davis of the Department of Geophysics at the Colorado School of Mines. This project has been funded by a consortium of 30 companies, including both major and independent petroleum and industry service companies, domestic and international, together with the Gas Research Institute. In the early 1990s, multicomponent seismic surveys were run in Cedar Hill field, northern San Juan Basin, New Mexico, which produces gas from Cretaceous sandstones as well as Fruitland coalbed reservoirs (the CBM wells in the entire San Juan Basin are spaced one per 320 acres, or 130 ha, by state regulations).

*Summary of the technology.* The presence of open fractures in the subsurface can be detected and interpreted from multicomponent seismic information, provided that the fractures are sufficiently large in spatial extent. For open fractures in a reservoir at depth, which act as avenues of fluid flow, to remain open, a regime of stress relief must be present. Shear (S) waves are especially suited for detecting vertical fracture zones, whereas compressional (P) wave data are relatively insensitive to such detection. To quote from Davis & Benson (1992), 'Shear waves that propagate through a parallel, unidirectional, open fracture set will have components of displacement parallel and perpendicular to the fracture plane. The component of displacement parallel to the fracture direction will propagate faster than the perpendicular component, causing the polarized shear waves to split and be recorded at different times.' Permeability anisotropy in a reservoir is due in a large part to natural fracturing. This anisotropy induces velocity anisotropy between split shear waves. The time differentials between split shear wave arrivals will reveal the relationship between velocity anisotropy and reservoir permeability anisotropy. Three-dimensional multicomponent seismic data provide the means directly to map permeability anisotropy in a fractured reservoir, regardless of its lithology. The geophysical detection of the presence and orientation of open fractures in hydrocarbon-bearing reservoirs in the subsurface should have significant economic impact world-wide (Davis & Benson 1992). Ongoing research at the Colorado School of Mines is expected both to sharpen the resolution and to lower the cost of this developing technology.

*Case history, Cedar Hill CBM field.* Amoco Production Company completed the initial Fruitland coalbed gas well in this field in May 1977, producing from the thick (up to ≈25 ft, 7.6 m) basal coal of this formation. Public records show that, as of 31 October 1993, the field had produced nearly 54 BCF (1.5 BCM) of gas from 23 producing Fruitland wells. During the month of October 1993, average daily production per well was approximately 714 MCF (20 MCM). The most productive wells in this field are characterized by high fracture permeability.

A vertical seismic profile was acquired by the Colorado School of Mines for Mesa Petroleum's Hamilton No. 3 well, located on the west side of the field. Multicomponent seismic surveys run in the vicinity of the Hamilton well indicate a polarization direction of N60°W. This direction of open fractures corresponds to the orientation of the butt cleat in this part of the field, as measured in oriented cores. In the eastern part of Cedar Hill field, the direction of open fractures parallels that of the face cleat (a well-developed natural fracture in coal that is orthogonal to the lesser developed butt cleat).

The strongly reflective Fruitland coals in this area have both low density and low acoustic velocity; hence, they are distinctly imaged on seismic records.

The geophysical surveys run in Cedar Hill field demonstrated that the integration of multicomponent three-dimensional seismology and seismic amplitude versus offset technology characterizes changes in elastic properties of fractured reservoirs such as coal beds better than the more conventional seismic techniques. This developing technology is able to both predict reservoir properties and hydrocarbon presence.

## Enhanced recovery of coalbed methane by nitrogen injection

*Summary of the technology.* Puri & Yee (1990) schematically showed the mechanism for producing coalbed gas by reservoir pressure depletion (Fig. 5). This is the standard means by which coalbed gas is produced: (1) reservoir pressure is reduced by dewatering the coalbed reservoir; (2) gas then begins to desorb from the matrix and micropores of the coal by a process of diffusion; and (3) the desorbed gas then flows to the wellbore via the coal cleat and fracture system, along with any water that exists in the cleats and fractures (the usual situation), according to Darcy's law. Unfortunately, this method of production is not very efficient. Even after several decades of production, the ultimate recovery of coalbed methane is not expected to be greater than 50% of the original gas in place.

Figure 6 (Puri & Yee 1990) portrays schematically the mechanism of enhanced recovery of coalbed methane by nitrogen gas injection. Methane can be removed from a coalbed reservoir by lowering its partial pressure. By injecting an inert gas such as nitrogen into the coal, the molar fraction of methane is decreased, thus lowering its partial pressure. Laboratory tests have shown that the sorption isotherm is controlled by the methane partial pressure rather than by total reservoir system pressure. As a result, nearly all the gas (perhaps more than 85%, ultimately) can be stripped, or 'raked,' by nitrogen injection.

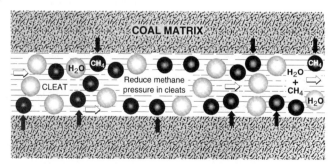

• Reduce cleat pressure by producing water
• Methane desorbs from matrix and diffuses to cleats
• Methane and water flow to wellbore

**Fig. 5.** Coalbed gas recovery by reservoir pressure depletion (after Puri & Yee 1990).

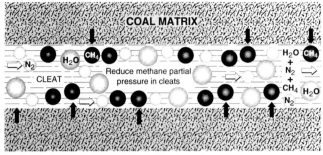

- Inject nitrogen into cleats
- Keep total cleat pressure high
- Reduce partial pressure of methane
- Methane desorbs from matrix and diffuses to cleats
- Methane, nitrogen and water flow to wellbore

**Fig. 6.** Enhance coalbed gas recovery by nitrogen injection (after Puri & Yee 1990).

*Description of the pilot test.* In February 1992, the US Environmental Protection Agency (EPA) approved Amoco's proposal to initiate a $2 million pilot project using their newly patented nitrogen injection process in La Plata County, Colorado, in the northern part of the San Juan Basin. In this test area, Amoco injected 250 MCF/d (7.1 MCM/d) nitrogen into each of four injection wells surrounding a single producing CBM well (i.e. a five-spot pattern). Nitrogen gas (95% $N_2$) was injected into the 20 ft (6.1 m) thick Fruitland coalbed reservoir at a depth of 2600 ft (792.5 m) and at a pressure of 1500 PSI. The injection process lasted from mid-1992 to mid-1993, followed by a one-year monitoring period. Preliminary tests of the pilot flood are 'encouraging' (*Western Oil and Gas World*, 1993) and the production well has followed the model production curve very closely. Amoco has high hopes that this innovative enhanced recovery method will significantly (1) increase the ultimate recoverable coalbed gas reserves, (2) increase production rates while decreasing well lives and (3) improve project economics.

The EPA contributed $250 000 to help finance the project, hoping that the nitrogen injection process may also assist in reducing methane emissions from coal mines, which would increase safety in underground coal mines and help to mitigate the reputed adverse environmental effect of methane emitted into the atmosphere and the waste of a valuable energy resource.

## Production of coalbed methane from 'deep' wells

An examination of the major coal-bearing basins in the world suggests that more than 50% of the in-place coalbed gas resources estimated for these basins is found in coals at depths below 5000 ft (1524 m). For example, in the Piceance Basin, western Colorado, it is believed that more than 70% of the gas in place in the basin, estimated at 84 TCF (2.4 TCM) or more, is found in coals below 5000 ft. These 'deep' coals are attractive targets due to their higher inherent gas contents and reservoir pressures. Assuming a constant permeability and pressure gradient and similar coal rank (i.e. degree of maturation), a coal bed found at 6000 ft (1829 m) could exhibit a peak gas production rate more than twice that for a coal bed at 3000 ft (914.4 m) (Kuuskraa & Wyman, 1993).

To understand the potential for deep coals to produce commercial amounts of coalbed gas, it is necessary to redefine the widely accepted theories concerning permeability versus depth relationships in view of the degree of stress present in a given basin, or in each part of a basin. Current research involving vertical, slant hole and horizontal drilling in deep low-permeability reservoirs—both coals and 'tight' sandstones—has provided hard data demonstrating that open fractures can exist in such reservoirs at present depths ranging from approximately 7000 to 10 000 ft (2134 to 3048 m) (Myal & Frohne 1991).

Production tests from Cretaceous coals at depths varying from approximately 5000 to 10000 ft (1524 to 3048 m) in Laramide basins of western North America have demonstrated the potential of 'deep' coalbed reservoirs to produce gas at 'commercial' rates. Kuuskraa & Wyman (1993), together with data uncovered in public records, provide information on deep CBM wells in the Piceance Basin, western

Colorado, Uinta Basin, eastern Utah and the Deep Basin, northwestern Alberta, Canada.

*Critique of initial depths versus permeability analysis.* According to Kuuskraa & Wyman (1993), in the case of vertical fractures the stress direction of interest is the *horizontal* component, which ordinarily is considerably *lower* than the vertical (lithostatic) stress. Much of the concern over low permeability in deep coals has been due to unfavourable depth versus permeability diagrams published in the 1980s. Some of these stress and permeability data were based on short-term slug tests in wells that had sustained high skin damage and had not been stimulated. However, once on production, these same wells exhibited permeabilities that were at least 500% higher than the original tests indicated.

These pessimistic depth versus permeability charts were based on the Kazeney–Carmen equation, which actually pertains to flow through conventional pores (as in a sandstone). This equation is *not* appropriate to use in predicting permeability in fractured reservoirs, whether they be coal seams, 'tight' sandstones or 'gas shales'.

In analysing the potential for deep coal beds to produce commercial volumes of coalbed gas, it is necessary to determine the presence of low ('relaxed') stress in a basin of interest. When such areas are located, depth (stress) versus permeability curves need to be developed. It has been demonstrated that coalbed gas can be produced at economic rates in coals below 5000 ft (1524 m) under low to moderate stress conditions (Table 4).

The gas and water saturation of coal are critical factors related to producibility. The following parameters offer the potential for a deep coalbed reservoir both to produce gas at high rates (say, above 1 MMCF/d; 0.03 MMCM/d) and to contain large recoverable reserves: 'Dry' coal (i.e. no mobile water production and free gas in the cleats and fractures); favorable (shallow) methane desorption isotherm (i.e. one that indicates rapid desorption of gas as the formation pressure is lowered); depth below 5000 ft (1524 m); and low stress conditions.

*Variation of coalbed permeability with time.* According to Harpalani (1991), three distinct phenomena occur simultaneously as a coalbed reservoir is dewatered and degassed: (1) desorption of gas as the reservoir pressure is lowered, enabling the rate of gas production to increase; (2) an increase in effective stress; and (3) two-phase flow with continuous reduction in water saturation.

In a conventional reservoir (such as a sandstone), increasing stress decreases permeability due to compaction. However, in a coalbed reservoir the observed reduction in permeability due to increased stress is considerably less than expected. It is postulated that other factors govern permeability. One possibility is that as methane desorbs, the coal matrix shrinks due to an increase in the free surface energy of the coal. As a result, cleats and fractures widen and the permeability increases. Harpalani (1991) has also found evidence that the effect of increased permeability is more pronounced at high stress, which, if demonstrated, should have important implications for gas production from deep coals.

## Case history of a long-lived coalbed methane well

Because of the relative infancy of the CBM industry, where well-documented production

**Table 4.** *Gas production and recovery from deep and shallow coals under three geological settings (Kuuskraa & Wyman 1993)*

| Stress condition | Deep coal | | Shallow coal | |
|---|---|---|---|---|
| | Permeability (md) | Rate (MCFD) | Permeability (md) | Rate (MCFD) |
| Depth versus peak gas rate | | | | |
| Low stress | 3 | 2730 | 30 | 2130 |
| Moderate stress | 0.03 | 100 | 7 | 900 |
| High stress | — | — | 2 | 430 |
| Depth versus estimated ultimate recovery | | | | |
| Low stress | 3 | 2800 | 30 | 2700 |
| Moderate stress | 0.03 | 200 | 7 | 1700 |
| High stress | — | — | 2 | 700 |

histories seldom exceed 10 years, a detailed case study by Hale & Firth (1988) provides some important clues regarding the potential productive life of such wells.

The San Juan 32-7 Unit No. 6 well, San Juan County, New Mexico, is located in the Los Piños Fruitland, South, pool, in T3N, R7W, NMPM, in the northern San Juan Basin. This well, originally designated the 6-17 San Juan 32-7 Unit, was completed by Phillips Petroleum Co. in August 1953. According to Hale & Firth (1988), this Fruitland Formation (Upper Cretaceous) well has produced more than 1.364 MMCF (0.04 MMCM) of gas over a period of 34 years (i.e. through 1987) with no decline in rates or reservoir pressure.

The original Phillips well was drilled to 3054 ft (931 m), at which point 7 inches (17.8 cm) casing was set. The drilling mud was removed and the well was drilled with gas to a total depth of 3240 ft (988 m) to provide a safe means of detecting any water and to protect against a large gas flow. At 3232 ft (985 m) a large flow of gas was noted, measuring as much as 2420 MCF/d (68.5 MCM/d) at 3236 ft (986 m). Drilling continued to 3240 ft total depth, leaving a 187 ft (57 m) openhole section, which was not logged. The well registered an initial potential of 1790 MCF/d (51 MCM/d). The producing reservoir is moderately overpressured, with a pressure gradient of 0.47 PSI/ft.

Although this well was not logged, correlation with nearby logged wells leaves little doubt that the producing reservoir is a lower Fruitland coal bed approximately 40 ft (12.2 m) thick. Analyses of the produced gas demonstrate that it is typical pipeline quality coalbed gas: 94.24% methane,

4.97% carbon dioxide, 0.36% ethane and traces of nitrogen and heavier hydrocarbon gases, with an average BTU content of 969 BTU/SCF.

Figure 7 (after Hale & Firth 1988) shows that the production rate of the 32-7 Unit No. 6 well has remained remarkably constant for about 30 years, between 150 and 180 MCF/d (4.2 and 5.1 MCM/d), except for a few anomalous months.

In the autumn of 1978, a 32-day pressure build-up test was conducted. The extrapolated reservoir pressure of 1502 PSIG was extremely close to the pressure measured 25 years earlier, in 1953, of 1504 PSIG, indicating no measurable pressure depletion.

The San Juan 32-7 Unit No. 6 well is one of the oldest continuously producing coalbed gas wells in the USA, if not the world. It was completed long before the significance of this unique reservoir had been realized. It lies just south of the synclinal axis of the San Juan Basin and appears not to exhibit any structural enhancement. The well has never been stimulated, nor has it required any appreciable maintenance. No significant water production has ever been reported. The production rate has remained essentially constant since 1953. Of the most significance is the fact that this well demonstrates that coalbed gas, under conditions of favourable geology and reservoir characteristics, can be produced economically for a period of at least 30–40 years.

## Summary and conclusions

Total world coal resources may be as large as 30 000 Mt, based on *geologic*, not *economic*, considerations (Matveev 1976). Most of this

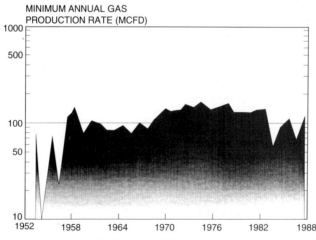

**Fig. 7.** Production history, San Juan 32-7 Unit No. 6 Well, San Juan County, New Mexico (after Hale & Firth 1988).

resource is in Asia, North America and Europe—i.e. in the Northern Hemisphere (Landis & Weaver 1993).

Estimates of in-place CBM resources are as high as 12 640 TCF (340 TCM), based on an incomplete, but growing, database. The most prospective coals, in terms of gas-generating capacity, are the 'hard coals'—i.e. those of bituminous and higher ranks.

Coalbed methane production in the USA currently exceeds 550 BCF (15.66 BCM) per year from more than 5700 wells. By the year 2000, this figure should more than double, based on current increasing rates of production. The San Juan Basin alone is expected to produce an average of approximately 2 BCF (0.06 BCM) per day during 1994.

The Rocky Mountain region in the USA contains the most prolific CBM wells in the world (in the San Juan Basin), some of which produce from 5 to more than 20 MMCF (142 to 566 MMCM) per day; the deepest sustained production (in the Piceance Basin, Colorado), from a depth of approximately 8800 ft (2682 m); and the longest recorded production (in the San Juan Basin) of more than 34 years.

The Laramide basins in western North America, together with the Black Warrior Basin, Alabama, have been the sites of innovative government and industry co-operative field research in CBM technology development since the late 1970s. These ongoing technological advances include the following: (1) dynamic openhole completions; (2) advanced hydraulic fracture stimulations; (3) multicomponent seismic characterization of coalbed reservoirs; (4) enhanced recovery of CBM by inert gas injection; (5) requirements for production of CBM from 'deep' wells; and (6) analyses of data from long-lived CBM wells.

The growing world-wide demand for natural gas, especially in those countries lacking large reserves of conventional natural gas, but blessed with significant resources of hard coal, requires that the methane known or suspected to be contained in these coals be developed as rapidly as possible.

It is imperative that the new technology being developed, especially that in the USA, is transferred to the exploitation of this indigenous, efficient, clean-burning fuel. Utilization of this new energy resource should both improve the economies and reduce atmospheric pollution in these countries, as well as in their neighbours' lands.

Most of the CBM production data presented herein was obtained from publications of agencies of the USA Government, the several states involved and the Gas Research Institute (GRI). In addition, the references deposited in the GRI-sponsored Western Natural Gas Resource Center at the Arthur Lakes Library, Colorado School of Mines, were extensively utilized.

# References

BENKIS, R. A. 1992. An overview of some selected coal resources of the world, *In*: RYAN, B. & CUNNINGHAM, J. (eds) *In*: Proceedings, Canadian *Coal and Coalbed Methane Geoscience Forum, Parksville, BC, 2–5 February 1992.* Victoria, British Columbia, Geological Survey Branch, Ministry of Energy, Mines, and Petroleum Resources, 256–289.

BOWMAN, K. C. 1978. Los Piños Fruitland, South. *In*: FASSETT J. E., THOMAIDIS, N. D. & MATHENY, M. L. (eds) *Oil and Gas Fields of the Four Corners Area,* Vol. II. Four Corners Geological Society, 393–394.

CONWAY, M. W. 1994. Is it time to put the final nail in the fraccing coffin? *Quarterly Review of Methane from Coal Seams Technology*, **11**(3 & 4) 7–12.

DAVIS, T. L. & BENSON, R. D. 1992. Fractured reservoir characterization using multicomponent seismic surveys. *In*: SCHMOKER, J. W., COALSON, E. B. & BROWN, C. A. (eds) *Geological Studies Relevant to Horizontal Drilling: Examples from Western North America.* Rocky Mountain Association of Geologists, Denver, 89–94.

HALE, B. W. & FIRTH, C. H. 1988. Production history of the San Juan 32-7 Unit No. 6 well, northern San Juan Basin, New Mexico, *In*: FASSETT, J. E. (ed.) *Geology and Coalbed Methane Resources of the Northern San Juan Basin, Colorado and New Mexico.* Rocky Mountain Association of Geologists, Denver, 199–204.

HARPALANI, S. 1991. Production of methane from coalbeds. *Quarterly Review of Methane from Coal Seams Technology*, **8**(4), 43.

KELSO, B. S. 1994. Geologic controls on open-hole cavity completions in the San Juan Basin. *Quarterly Review of Methane from Coal Seams Technology*, **11**(3 & 4), 1–6.

KUUSKRAA, V. A. & BOYER, C. M. II 1993. Economic and parametric analysis of coalbed methane. *In*: LAW, B. E. & RICE, D. D. (eds) *Hydrocarbons from Coal.* American Association of Petroleum Geologists, Studies in Geology, **38**, 373–394.

—— & WYMAN, R. E. 1993. *Deep Coal Seams: an Overlooked Source for Long-term Natural Gas Supplies.* Society of Petroleum Engineers Paper, **26196**.

——, BOYER, C. M. II & KELAFANT, J. A. 1992. Coalbed Gas 1—hunt for quality basins goes abroad. *Oil and Gas Journal, OGJ Special— Unconventional Gas Development*, **Oct. 5**, 49–54.

LANDIS, E. R. & WEAVER, J. N. 1993. Global coal occurrence, *In*: LAW, B. E. & RICE, D. D. (eds) *Hydrocarbons from Coal.* American Association of Petroleum Geologists, Studies in Geology, **38**, 1–12.

LOGAN, T. L., MAVOR, M. J. & KHODAVERDIAN, M. 1993. Optimizing and evaluation of open-hole cavity completion techniques for coal gas wells. *In*: *Proceedings, 1993 International Coalbed Methane Symposium, Birmingham, Alabama,* 609–622.

MATVEEV, A. K. 1976. Distribution and resources of world coal. *In*: MUIR, W. L. G. (ed.) *Coal Exploration—Proceedings of the First International Coal Exploration Symposium, London, UK, 18–21 May.* Miller Freeman, San Francisco, 77–88.

MAVOR, M. J. 1991. *Evaluation of Coalbed Natural Gas Openhole Cavity Completion Productivity.* Report to Gas Research Institute, GRI-91/0374.

MYAL, F. R. & FROHNE, K.-H. 1991. *Slant-hole Completion Test in the Piceance Basin, Colorado.* Society of Petroleum Engineers Paper, **21866**.

PETROCONSULTANTS 1981. *Coal Worldwide*, 5 vols. Petroconsultants, Geneva.

POTENTIAL GAS COMMITTEE 1993. *Potential Supply of Natural Gas in the USA (December 31, 1992).* Potential Gas Agency, Colorado School of Mines, Golden.

PURI, R. & YEE, D. 1990. *Enhanced Coalbed Methane Recovery.* Society of Petroleum Engineers Paper, **SPE 20732**.

RICE, D. D., LAW, B. E. & CLAYTON, J. L. 1993. Coalbed gas—an undeveloped resource. *In*: HOWELL, D. G. (ed.) *The Future of Energy Gases.* US Geological Survey, Professional Paper, **1570**, 389–404.

TYLER, R., AMBROSE, W. A., SCOTT, A. R. & KAISER, W. R. 1991. *Coalbed Methane Potential of the Greater Green River, Piceance, Powder River, and Raton Basins.* Gas Research Institute, Chicago, Topical Report (Jan. 1991–July 1991), GRI 91/0315, Powder River Basin, 107–144.

*Western Oil and Gas World* 1993. Best of the Rockies—nitrogen work in coalbeds wins award for Amoco. *Western Oil and Gas World*, June, 26.

# Coalbed methane potential of some Variscan foredeep basins

## THOMAS G. FAILS

*4101 East Louisiana Avenue, Suite 412, Denver, CO 80222, USA*

**Abstract:** Five Variscan foredeep coal basins in Germany and Great Britain are compared for coalbed methane (CBM) potential with the productive Black Warrior coal basin in Alabama. Although generally geologically similar, compared with the Black Warrior the European foredeep basins usually have much thicker coal measures, with more numerous coalbeds and thicker net coal thicknesses. Also compared are coal ranks, the present methane contents of the coals, potential depths of production and post-coalification histories. Despite basin to basin similarities, the CBM productive potential depends on characteristics unique to each coal basin. Only the unmined portions of the eastern Ruhr and western South Wales basins appear to be prospective for CBM; both are characterized by high exploration risks for potentially high economic returns. The Oxford and Kent basins appear to be non-prospective for CBM. The CBM potential of the Bristol–Somerset basin is speculative, given the lack of methane content data and severe tectonic compression. Subaerial exposure and erosion during Stephanian–Permian–early Triassic times may have destroyed the CBM potential of the Oxford, Kent and Bristol–Somerset basins. The Belgian and French foredeep basins are not evaluated here.

Development of coalbed methane (CBM) production on a commercial scale in two USA coal basins—the Black Warrior and San Juan—since 1986 has been so successful that CBM now constitutes about 5% of the 167 trillion cubic feet (TCF) (4675 billion cubic metres; BCM) natural gas reserve in the USA. The CBM production potential of hard coal basins in Africa, Asia, Australia and Europe are now in the preliminary stages of assessment. In Europe, Germany, Poland, the Ukraine and the UK appear to be most prospective. Discussion of the CBM potential of the many coal basins in these countries is beyond the scope of this paper. Instead, several Variscan foredeep coal basins in Germany and the UK will be compared with the Black Warrior coal basin in Alabama, birthplace of the CBM industry.

Basin analysis for CBM prospectivity is an important preliminary part of the CBM exploration process, although rarely discussed in published work. CBM-oriented basin analysis parallels that for conventional hydrocarbons in many ways. A preliminary analysis, undertaken to determine whether a portion or all of a coal basin should be the subject of an application for CBM exploration rights, is essential. Pre-application analyses must sometimes be completed in haste in competitive situations; however, they must be as accurate and complete as the available data allow. More thorough analyses must be undertaken after an exploration licence or permit is obtained before a work obligation to drill one or more exploration boreholes can be proposed or accepted; see Table 1.

Evaluation of coal basins for CBM exploration and production potential utilizes both the basin geological factors important to conventional basin analysis and numerous coal data factors. As with all sedimentary basins, geologic structure plays a key part, particularly with respect to the type and time(s) of tectonism relative to the coalbeds, and whether compression, extension or compression followed by extension have affected the coal measures. The present state and direction of stress, if known, may be very important. Insufficient permeabilities of Carboniferous coals often occur deeper than 1500 m; the depth range of the coalbeds is thus of critical importance. A modest amount of thermal activity may enhance prospectivity by elevating coal ranks. Excessive heat and/or volcanism may have adverse effects, reducing permeabilities or destroying *in situ* methane. Stratigraphy of the coal measures and post-coal measures strata must also be considered, especially with respect to drilling costs, water disposal and the occurrence of conventional hydrocarbons. Formation pressures of coalbed reservoirs are determined by hydrogeological factors; the aquifer–coalbed relationship, if known, is useful. In previously mined areas, the resulting hydrogeological changes may adversely affect the CBM prospectivity of adjacent unmined coalbeds. Artesian conditions may alter the distribution of methane within the basin.

When a basin is analysed for CBM prospectivity, coal data of varying types must also be considered. Coal rank data are usually available and are useful for preliminary estimates of both

*From* Gayer, R. & Harris, I. (eds) 1996, *Coalbed Methane and Coal Geology*,
Geological Society Special Publication No 109, pp 13–26.

**Table 1.** *Important factors affecting coalbed methane exploration and production*

Basin geology
  Geological structure
    Periodicity and type of tectonism
    Compression versus extension
    Depth range of coalbeds
    Thermal activity
  Stratigraphy
    Coal measures
    Post-coal measures
    Presence, conventional hydrocarbons
  Hydrogeology
    Hydrology controls formation pressures
    Aquifer relationship to coalbeds
    Effects of mining on aquifer
    Recharge and artesian areas
    Water quality

Coal data
  Coal rank
    For original methane generation estimate
    For original methane retention estimate
  Coalbed thicknesses
    Thickness, individual coalbeds
    Number of coalbeds
    Vertical distribution within objective
      coal measures
    Net objective coal thickness
    Net coal percentage of objective coal measures
  Present *in situ* methane content of coals
    Gas saturation versus water saturation
    Free gas in cleats and fractures
    Thickness of shallow degassed zone
  Coalbed reservoir permeability
    Syndepositional fracturing and faulting
    Cleat development
    Post-coalification tectonism and fracturing
    Compression versus extension
    Stress regimes, past and present
    Secondary mineralization of fractures
    Reservoir pressure and hydrology

Present *in situ* methane content and coalbed premeability are the most critical factors for CBM production

Premeability can be determined only by drilling and testing

the original methane volumes generated and the capacity of the coalbeds to retain this methane (Fig. 1). Varying volumes of the original methane generated escape for a variety of reasons. Determination of the current methane contents of the objective coalbeds is thus of vital importance. Uplift and erosion may have occurred. Water may have displaced some, perhaps most, of the methane originally generated. Shallow coalbeds in most coal basins have been degassed to various depths, sometimes exceeding 1000 m. Most coal basins contain

deep coalbeds with higher gas contents, prospective for CBM production, but a few appear to be totally degassed. Gas contents vary laterally in some instances, sometimes from fault block to fault block.

The commerciality of a CBM producing well will depend in part on the gross thickness of the methane-bearing coalbeds penetrated (i.e. the 'net pay'). Thin coalbeds, less than 0.3 to 0.6 m in thickness, do not warrant the cost of completion unless they are closely associated with additional coalbeds of greater thicknesses. Therefore the number, thickness and vertical distribution of the objective coalbeds are very important. Two net coal quantities necessary for effective CBM basin analysis are the net coalbed thickness within the methane-bearing objective section and that objective section's net coal percentage. For instance, in the tax-subsidized Black Warrior basin, a total of only 3–6 m of coalbeds occur in a 2.5% net coal objective section. In the European coal basins, a net coal thickness less than 20 m is of questionable commerciality at present. Further, an objective coal measures section that is only 2% or less net coal is equally questionable, unless the net coal percentages of the underlying objective coal measures exceed 4–5%.

Coalbed permeabilities exceeding about 0.4–0.5 md should be present to allow the stimulation treatments (hydrofracturing or cavitation) of the coalbeds necessary for CBM production at commercial rates. The fracture and cleat permeabilities are determined by injection testing after the test hole has been drilled, cored and wireline-logged. The original cleat fracture system produced during coalification may be enhanced by syn-depositional or post-coalification tectonism and fracturing, especially of an extensional nature.

Among the coal data factors just discussed, the permeabilities and current *in situ* methane contents of the objective coalbeds are the most important. They are the primary risk factors in CBM exploration. Unfortunately, permeabilities can be determined only by drilling and testing a borehole. Current *in situ* methane contents are sometimes determined during coal mining and coal exploration activities in some actively mined coal basins, but rarely before the 1950s. Further, where methane content data do exist, they are usually proprietary and cannot always be obtained easily.

## Zone of shallow, degassed coalbeds

The shallow 'zone of degassed coalbeds' has not been discussed often in publications on CBM,

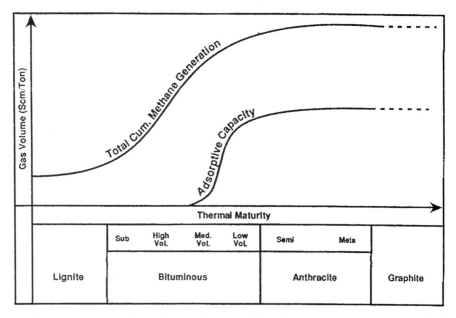

- Coals Generate More Gas Than Can be Adsorbed
- Gas and Water Expelled Into Surrounding Strata

**Fig. 1.** Theoretical relationships between original methane volume generation by coal and adsorptive capacity volume of coal for methane compared with coal rank. In nature, the present methane contents of coals are typically much below these theoretical volumes. Published with permission of Resource Enterprises, Inc.

but is an important consideration in comparison and evaluation for CBM production potential of the subject basins. 'Zone of degassed coal seams' is defined in Kotas (1991) as 'a zone of degassed coal seams which is located in uppermost, outcropping parts of Carboniferous strata, which, in turn is followed by a zone of gassy coal. The transition to the gassy coal zone is in most cases rapid . . .'. The discussion in Kotas (1991) continues, (1) to include Carboniferous strata subcropping unconformably against Miocene cover strata and (2) to further define the 'zone of degassed coal seams' by the methane contents of the coals therein being in the $0–4.5\,m^3$/Mt (daf) range. Kotas (1991) deals with the Upper Silesian Coal Basin of Poland, where extensive coal exploration borehole programmes with 1 km spacing provide abundant data from which a structure contour map on the 'top surface of gassy coal zone' has been made for the unmined area and most of the mined areas of the basin. This surface ranges from $-250$ to $-1250\,m$, relative to sea level. The concept of a shallow degassed zone occurring above a 'top of gassy coals zone' has been adopted for use in this paper, as have the

defining methane contents of $0–4.5\,m^3$/Mt. Coals within the shallow zone of degassed coal beds (1) are less hazardous for underground mining operations and (2) probably will not produce CBM in commercial volumes.

## Variscan and Appalachian foredeep basins

The thicknesses, drilling depths, net coal thicknesses, coal ranks and present methane contents of the coal measures in six Variscan/ Appalachian foredeep basins ranging from Alabama through Great Britain to Germany will be discussed and are compared graphically in Fig. 2. A number of 'anomalies' exists, and some of the basins reviewed do not appear to be prospective for commercial CBM production. The post-coalification history of a coal basin may have a profound effect on its CBM production potential. Unfortunately, as many questions as answers characterize this paper, which is an indication of the 'state of the art' of CBM basin analysis in 1994.

All of the six subject basins—Black Warrior, South Wales, Bristol–Somerset, Oxford, Kent

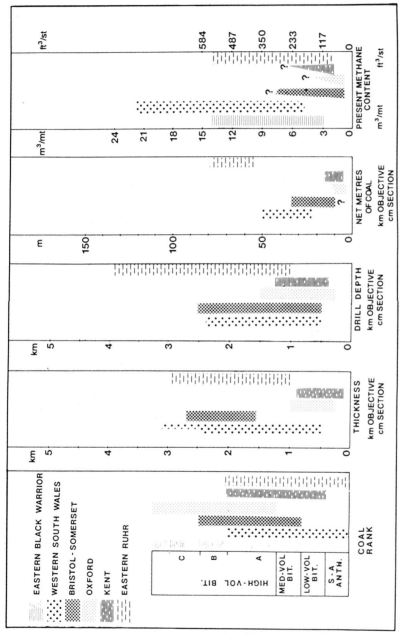

**Fig. 2.** Comparison of factors important in determination of coalbed methane production potential for six Variscan/Appalachian foredeep basins. Coal measures of Westphalian A through D ages occur in the South Wales, Bristol–Somerset, Kent and Ruhr basins, only Westphalian C and D in the Oxford and only Westphalian A in the Black Warrior Basin. Methane contents for the Bristol–Somerset Basin are speculative as no actual measurement is known. Ruhr Basin methane contents are based on limited information. References for this figure are: Black Warrior, McFall *et al.* (1986), Montgomery 1986; Hill (1993), South Wales: Moore (1954), National Coal Board (1960), Woodland & Evans (1964), Adams (1967), Archer (1968), Squirrell & Downing (1969), Gill *et al.* (1979), Creedy (1988), Ayers *et al.* (1993), Cornelius *et al.* (1993); Bristol–Somerset, National Coal Board (1960); Kellaway & Welch (1993), Oxford, Dunham & Poole (1974), Foster *et al.* (1989); Kent, Stubblefield (1954), National Coal Board (1960); Ruhr, Deutloff (1976), Petroconsultants (1977), Drozdzewski *et al.* (1981), Fiebig & Groscurth (1984), Strack & Freudenberg (1984), Steuerwald & Wolff (1985), Littke (1987), Juch *et al.* (1988), Anon. (1989), Littke & LoTenHaven (1989), Strehlau (1990), Juch (1991, this volume), Lommerzheim (1991), Freudenberg *et al.* (this volume).

and Ruhr—are present day erosional remnants of a series of originally larger coal measures depositional basins. These basins developed during Westphalian times within the Variscan/Appalachian foredeep system before the culmination of the orogeny in Stephanian and lower Permian times and the opening of the Atlantic (Fig. 3). Only the Black Warrior and Ruhr basins were paralic in nature. The other four basins were bounded on the then-north by the Wales–Brabant massif (Fig. 4).

The coal basins of Belgium and northern France are excluded from this discussion due to limited available data. Further, excluding the Campine basin on the northeast flank of the Wales–Brabant massif, these basins have been heavily mined and are mainly autochthonous, lying beneath the Midi thrust. Reduced coalbed reservoir permeabilities are probably a result. The CBM production potential of these sub-Midi basins may thus be low. Judgement is reserved concerning the prospectivity of the Campine basin. Coalbed methane prospectivity of the small, heavily mined Aachen basin in Germany, at the eastern end of the sub-Midi basins, is considered to be limited.

Development and tectonism of the subject foredeep coal basins and the Variscan/Appalachian orogeny are discussed in Wills (1951), Moore (1954), Woodland & Evans (1964), Archer (1968), Squirrell & Downing (1969), Deutloff (1976), Drozdzewski, et al. (1980; 1981), Steuerwald & Wolff (1985), Wolf (1985), Montgomery (1986a), Brix et al. (1988), Juch et al. (1988), Gayer & Jones (1989), Strehlau (1990), Cole et al. (1991), Jones (1991), Lommerzheim (1991), Pashin (1991), Cameron et al. (1992), Cope et al. (1992), Ayers et al. (1993), Cornelius et al. (1993), Gayer et al. (1993) and Kellaway & Welch (1993).

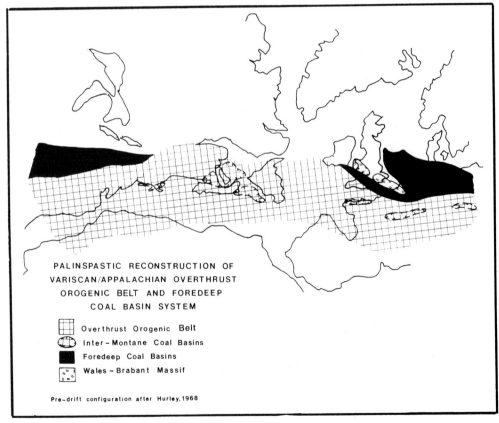

**Fig. 3.** Palinspastic reconstruction showing Variscan/Appalachian overthrust orogenic belt including intermontane coal basins, foredeep coal basins and Wales–Brabant massif. Time is about mid-Westphalian, before opening of the Atlantic. Intermontane basins are Westphalian A–B and Westphalian C–D in age. References for this figure are: Wills (1951), Hurley (1968), Strehlau (1990), Cameron (1992), Cope et al. (1992), Gayer et al. (1993), Hill (1993), Ryan and Boehner (1994), Geologic Map of the USA.

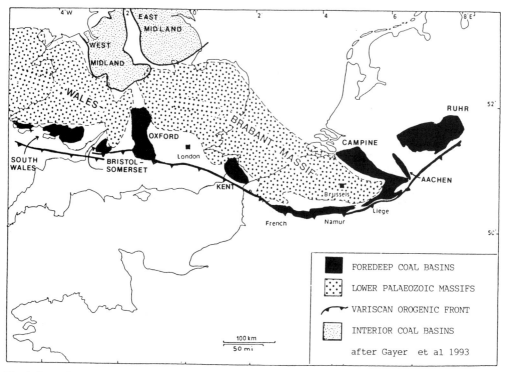

**Fig. 4.** Part of northwest Europe showing Variscan foreland thrust belt, lower Palaeozoic massifs, Westphalian foredeep coal basins and portions of Westphalian interior coal basins. Coal basins are shown in their present configurations as sub-basin remnants of originally more widespread depositional systems. The Campine, Aachen and Ruhr coal basins shown represent portions at mining and CBM prospective depths of a much larger paralic foredeep coal basin extending at depth to the north and northwest (see Fig. 3). After Gayer *et al.* (1993), modified with data from Ayers *et al.* (1993).

## North American upper carboniferous/ Pennsylvanian coal basins

The Black Warrior Basin, northwestern Alabama, lies at the southwestern end of the large Appalachian foredeep basin. It is the largest and most important of several CBM-productive areas in the Appalachian Basin (Fig. 5). A larger, deeper portion (about 90% of the total Black Warrior Basin area) lies to the west, mainly in Mississippi. It is productive of conventional hydrocarbons, but not of CBM. Only limited exploration for CBM has taken place due to the depth and low coal thicknesses and ranks. The Cincinnati arch, which separates the Appalachian basin from the Michigan, Illinois, Mid-Continent and Arkoma basins to the northwest, was a more passive feature than the Wales–Brabant massif. Although the Cincinnati arch affected coal measures distribution and development, especially during Westphalian A and B times, due to the great width of the Appalachian Basin it played little part in tectonic deformation of the coal measures during the Appalachian

orogeny. The CBM production potential of the Appalachian and Mid-Continent Palaeozoic coal basins in the USA appears to be relatively modest, considering their vast areas. With only about 20 m net of coal present at best, and 10–20 m being more common, little incentive exists for exploration under current gas market conditions; see Hill 1993*a*; 1993*b*; McFall *et al.* 1986; Montgomery 1986*b*; Pashin 1991.

In the Canadian Maritime region, the Appalachian/Variscan orogenic belt closely approaches the Canadian Shield (Fig. 5). There is no obvious upper Palaeozoic foredeep basin preserved in this region. The Westphalian coal basins of Maritime Canada may have CBM potential; however, little, if any, methane content data are available for these coals. The Maritime coal basins appear to be of the intermontane type (Ryan & Boehner 1994), as are the Westphalian Saar–Lorraine and Lower Silesian coal basins of Europe, among others.

When the Black Warrior and other Appalachian foredeep basin CBM-producing areas are compared with the European Variscan foredeep

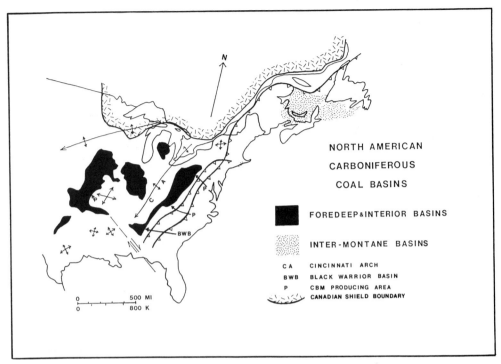

**Fig. 5.** Eastern North America showing Appalachian overthrust orogenic front, foredeep and interior coal basins in the US, the Cincinnati arch, the principal CBM producing areas of the Appalachian foredeep basin and the intermontane coal basins of Maritime Canada. References for this figure: Geologic Map of the USA; Hill (1993); Ryan and Boehner (1994).

basins (Fig. 2), the thickness of the coal measures, number of coal seams and net coal thicknesses of the latter are generally much greater than in the USA. So the question arises: do the European foredeep coal basins have greater potential for CBM production as well?

## Coal basin comparisons for CBM production potential

The Upper Carboniferous/Pennsylvanian coal measures of the foredeep basins are generally similar stratigraphically but vary in thickness. However, the strong lower argillaceous/upper arenaceous nature of the generally much thicker European coal measures is absent in the USA. There the thinner coal measures typically display classic cyclothem development.

The readily available publications about coal, coal rank and methane content, coal measures stratigraphy and coal field structural geology are substantial in some coal basins and scant in others. References, by basin, include: Black Warrior, McFall *et al.* (1986), Montgomery (1986a; 1986b), Pashin (1991); South Wales, Levine (1992), Hill (1993), Moore (1954), National Coal Board (1960), Woodland & Evans (1964), Adams (1967), Archer (1968), Squirrell & Downing (1969), Petroconsultants (1977), Jones (1991), Cope *et al.* (1992), Gayer (1992), Cornelius *et al.* (1993), Gayer *et al.* (1993); Bristol–Somerset, Kellaway & Welch (1948; 1993); Oxford, Dunham & Poole (1974), Foster *et al.* (1989); Kent, Stubblefield (1954); and Ruhr, Deutloff (1976), Petroconsultants (1977), Strack & Freudenberg (1984), Teichmuller *et al.* (1984), Steuerwald & Wolff (1985), Littke (1987), Juch *et al.* (1988), Anon. (1989), Littke & LoTenHaven (1989), Strehlau (1990), Juch (1991), Lommerzheim (1991), Freudenberg *et al.* (this volume), Juch (this volume).

Among the European foredeep basins, varying subsidence histories are discussed in Gayer *et al.* (1993). The modest subsidence history curves of the relatively thin Oxford and Kent basin coal measures shown in Gayer *et al.* (1993) have been averaged, as shown on Fig. 6. Limited

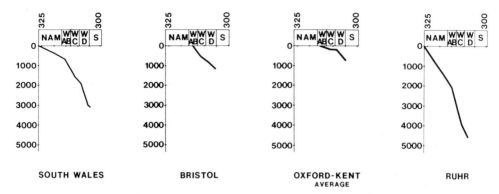

**Fig. 6.** Subsidence histories, five European foredeep coal basins. Compare the Oxford–Kent average subsidence history with those of the South Wales and Ruhr basins. The numbers at the tops of the individual curves refer to the number of millions of years before present. After Gayer *et al.* (1993). NAM, Namurian; W, Westphalian; and S, Stephanian.

subsidence and stratigraphic development may be factors in the low CBM potential of these basins. However, although the Bristol–Somerset Basin exhibits a subsidence history curve commencing later than, but more similar to, those of the South Wales and Ruhr basins, its CBM production potential is questionable. The South Wales and Ruhr basins, in contrast, have strong, early established subsidence histories and appear to be the most prospective among the non-producing basins considered here.

All of the coal basins compared (Fig. 2), except the Oxford, contain coals of the medium-volatile to high-volatile A bituminous ranks, which are generally considered to be the most prospective for CBM production (Fig. 1). Low-volatile bituminous and semi-anthracite/anthracite coals, with higher methane generation and retention characteristics than lower rank bituminous coals, occur in all the basins discussed herein, except Oxford and Kent. However, these higher rank coals are possibly less prospective for CBM production due to greater densities and hence potentially lower permeabilities; see National Coal Board (1960), Dunham & Poole (1974), Petroconsultants (1977), Juch *et al.* (1988), Juch (1991; this volume) and White (1991) for coal rank data for the European basins discussed herein.

Coal measures thicknesses appear to be adequate for potential CBM prospectivity in all six basins, providing other factors are favourable. Three—the South Wales, Bristol–Somerset and Ruhr basins—have coal-bearing sections exceeding 1500 m in thickness.

Relatively thick objective sections with multiple coalbeds can be penetrated above a drilling depth of 1500 m in all of the subject basins except the Ruhr (Fig. 2; Juch *et al.* 1988). Exploration boreholes for CBM are often drilled to depths of between 1300 and 1500 m. Although permeable coalbeds occur in the Cretaceous coalbeds of the CBM productive Piceance Basin to drilling depths of approximately 2400–2800 m, it is widely believed at present that favourable permeabilities in Palaeozoic coals may occur down to drilling depths of only about 1500 m. However, the net metres of coalbeds thicker than 30 cm present in the Black Warrior, Oxford and Kent coal basins, regardless of depth, are so limited (Fig. 2) that the CBM commerciality of these three basins is limited as well. (Among these three basins, only the Black Warrior Basin coals appear to have adequate methane contents.) In contrast, note the thick net coal thicknesses in the Bristol–Somerset, South Wales and Ruhr basins (Fig. 2).

*Low methane content basins*

The present *in situ* methane content of the objective section coalbeds is a most critical analysis factor (Fig. 2). Hypothetical methane contents of the coalbeds in the Oxford and Kent basins appear to be so low (0.5 and 1.7 $m^3$/Mt, respectively; Ayers *et al.* 1993) that these basins are probably within the shallow zone of degassed coalbeds at most levels. Regarding Bristol–Somerset Basin mining, Kellaway & Welch (1948) comment on 'the almost complete absence of firedamp in the mines and with a few exceptions, naked light working is universal'. Kellaway & Welch (1993) report that 'with the exception of the Nettle-bridge area, firedamp was normally absent and naked lights could be used.' The lower and middle coal measures of the Nettlebridge area, at the southern extreme of the Bristol–Somerset Basin, are heavily tectonized, with multiple overthrusts and overturned beds. In the much larger remaining portion of the basin, tecton-ism is far less severe, except on the overthrust Kingswood anticline. Mines up to 550 m deep (Easton and Pennywell Road) appear to have been worked without methane problems (Kell-away & Welch 1993). It thus appears that the shallow degassed zone of coalbeds may be 550 m or more thick in most of the Bristol–Somerset Basin. Methane contents are unknown at greater depths and could be subcommercial. Methane content and perme-ability are the major risk factors for the Bristol–Somerset Basin.

## High methane content basins

Only the Black Warrior, South Wales and the eastern Ruhr basins are known to have consistent, multi-seam *in situ* methane contents exceeding $7 \, m^3/Mt$. Established productivity confirms that acceptable methane contents occur in the Black Warrior Basin. Reliable *in situ* methane contents in the South Wales Basin have been measured by British Coal Corporation (Creedy 1988). Methane content data are difficult to obtain in the Ruhr Basin. Nevertheless, it has been established that *in situ* methane contents in the eastern Ruhr basin are higher than in the western part of the basin; see Freudenberg *et al.* (this volume), Lommerzheim (1991) and Teichmuller *et al.* (1984). Further, the 'top of the gassy coals' occurs immediately below the unconformity at the top of the coal measures in the eastern Ruhr Basin, versus several hundred metres below this unconformity

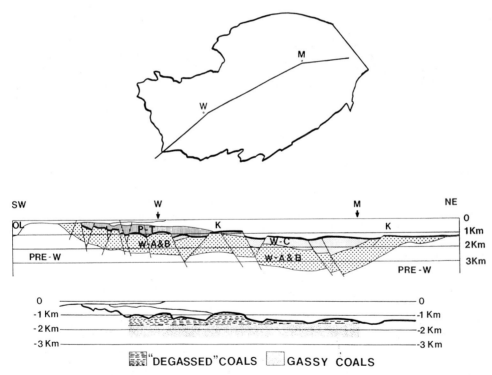

**Fig. 7.** Cross-sections, Rhur coal basin. Upper cross-section shows generalized geology of this basin along the line of section shown above; note the occurrence of post-Westphalian strata unconformably overlying truncated Westphalian A B and C coal measures. The relationship of the 'degassed' coals in the Westphalian to the underlying gassy coals and the unconformably overlying strata of various ages is illustrated in the lower cross-section. Note that the 'degassed' coal zone is thicker in the western area overlain by Permo-Triassic strata. After, in part, map and cross-section in Deutloff (1976) and a map in Juch *et al.* (1988). W, Westpahlian; P-T, Permo-Triassic; K, Cretaceous.

in the west (Fig. 7). As a result, the eastern Ruhr Basin appears to be more prospective for CBM production, providing acceptable permeabilities exist, than the western Ruhr Basin.

## Anomalous relationships

In 1992, Levine commented on several simplistic assumptions commonly used in the CBM industry, cautioning that specific issues were more complex than generally accepted. Similarly, when the data discussed here and displayed in Fig. 2 are compared, some relationships which might be considered as anomalous to commonly held CBM assumptions emerge. Several significant anomalies are discussed in the following.

### Coal rank versus present in situ methane content

Based on coal rank alone, the Kent, Oxford and western Ruhr basins should have higher methane contents than now exist. This also is true of the shallow portion (550 m) of the Bristol–Somerset Basin, where medium-volatile and high-volatile A bituminous coals occur. All four basins appear to be degassed to varying degrees, probably in response to post-coalification erosion and other problems.

### Exposure to erosion versus present in situ methane content

Coarse clastic sediments shed from the advancing Variscan orogenic belt characterize the Westphalian C and D of all five European foredeep basins considered here. These basins were probably buried beneath youngest Carboniferous (Stephanian) and lower Permian syn-orogenic clastic sediments, as the underlying coal measures were contemporaneously deformed to varying degrees. Normal faulting, both syn-orogenic and post-orogenic, is ubiquitous in all except the Oxford and Kent basins. In contrast, cyclothemic coal deposition persisted into the lowermost Permian in the Appalachian Basin, affected later by orogeny than were the basins in Europe. Compressional deformation was severe in the South Wales and Bristol–Somerset and moderate in the Black Warrior and Ruhr basins, whereas the Oxford and Kent basins were only broadly folded. In all six, the coal measures were subsequently truncated by erosion and are unconformably overlain by

younger strata of widely varying ages, except in South Wales. The period of post-orogenic exposure and erosion of the coal measures in the eastern Ruhr, Black Warrior and South Wales basins cannot be determined with the available data. In the western Ruhr, remnants of Permo-Triassic post-orogenic cover strata are known, as are similar Triassic (Bunter?) rocks in the Bristol–Somerset and Oxford basins (Dunham & Poole 1974; Foster et al. 1989; Kellaway & Welch 1993). Cover strata in the Kent Basin are younger, being of Jurassic and lower Cretaceous ages (Stubblefield 1954). Sharply folded and truncated coal measures of Namurian and Westphalian A ages crop out along the southern edge of the Ruhr basin. Truncated and more broadly folded coal measures, predominantly of Westphalian A B and C ages, are overlain by Cretaceous strata in most of the eastern and central parts of the Ruhr basin (Fig. 7; Brix et al. 1988; Drozdzewski et al. 1980; 1981). In the west, remnants of Stephanian, Permian (Zechstein) and Triassic (Bunter) sediments supplement the Cretaceous and are overlain in turn in some areas by Oligocene rocks (Deutloff 1976; Wolf 1985). Zechstein evaporites lie directly on truncated Westphalian A and B coal measures in part of this area. The coal measures in the South Wales basin are exposed at the surface. Except close to the basin margins, the exposed South Wales coal measures are thick, tight lithic arenites and conglomerates with thin coals of the Westphalian C and D Pennant series. Around the basin margins the lower and middle coal measures crop out, consisting of mudrocks of Westphalian A and B with numerous thick coals.

Relatively brief exposure to truncation and erosion of the coal measures before reburial could be interpreted as being favourable to the preservation of the methane resource. Longer periods of exposure might promote degassing of the coalbed reservoirs. However, in the six basins discussed, this does not appear to be the case with respect to present in situ methane contents. The basins with the shortest period of exposure and erosion before reburial by the oldest cover strata—Bristol–Somerset, Oxford, Kent and the western Ruhr (Fig. 7)—are the least prospective for CBM production. Oxford and Kent appear to be essentially degassed, as is Bristol–Somerset down to at least 550 m. The thickness of the degassed coal measures section in the western Ruhr, often occurring unconformably beneath Permo-Triassic strata, is about 500–700 m (Fig. 7). In contrast, the CBM-productive coal measures in the Black Warrior Basin are either exposed or lie beneath

Cretaceous alluvium of varying thicknesses. The shallow degassed coal measures there are usually about 300–400 m thick, but only 150 m in some areas. Further, the prospective eastern Ruhr Basin has Cretaceous cover strata unconformably overlying a very thin degassed section in the Westphalian that thickens to the west. In the prospective South Wales Basin, coal measures are exposed at the surface, with shallow degassed zone thicknesses similar to those common in the Black Warrior. In South Wales, the highest methane contents occur in basin centre coalbeds overlain by Pennant sandstones; truncated Westphalian A and B coalbeds exposed along the basin margins have lower methane contents.

Foredeep coal basins with the older cover strata—Bristol–Somerset, Oxford, Kent and western Ruhr—appear to be the least prospective and those with the youngest cover strata the most prospective for CBM production. Thus the length of the possible post-coalification exposure period may not affect CBM prospectivity as much as the geological time and environment of exposure.

In the western Ruhr Basin, exposure and erosion of the coal measures may have occurred during Stephanian and/or lowest Permian times (Fig. 7) and in the Bristol–Somerset and Oxford basins during the Stephanian, Permian and/or earliest Triassic. Five periods of post-Variscan tectonism in the western Ruhr are described in Wolf (1985); the first three—pre-Zechstein, mid-Zechstein and upper Triassic Keuper—are extensional. The higher coal ranks of the Kent Basin suggest deeper and/or longer burial; however, the lack of methane suggests earlier exposure, possibly during late Permian and/or early Triassic times. In the European foredeep basin area, the Stephanian through Triassic period was apparently characterized by extremely hot, dry desert conditions. As a result, the water-table may have been very deep, with the exposed coal measures effectively dewatered to considerable depths. Dewatering would create conditions whereby the efficient natural diffusion of methane in large volumes from the coals into fractures could occur. Escape of methane to the surface would be possible, especially in areas subject to extension (i.e. the western Ruhr) (Fig. 7). If this explanation is valid, it suggests that the eastern Ruhr, South Wales and Black Warrior coal measures remained buried beneath younger strata during this period, and were exposed later during the Mesozoic or, in the case of South Wales, during the Tertiary.

## Compression versus present coalbed permeability

In general, compression is deemed to be a destroyer of coalbed permeabilities and extension to be a creator/preserver of the same. At present, the Black Warrior Basin is the only one of the six basins discussed known to contain acceptable coalbed permeabilities. The other five are untested. Nevertheless, the Black Warrior Basin was moderately compressed during the Appalachian/Variscan orogeny. Normal faults with strikes perpendicular to the major fold trends occur in extensional patterns in the Black Warrior, a structural pattern also present in the Ruhr and South Wales basins.

## Lateral variation in coalbed methane contents

Classic concepts involving increasing coal ranks and methane contents with increasing depth as a dominant factor in CBM basins have become increasingly irrelevant. It is now known that both coal rank and methane content vary laterally, often independently, in the Black Warrior and San Juan basins. Coal rank and methane content vary laterally, as well as vertically, in a systematic manner in both the Ruhr and South Wales foredeep basins (National Coal Board 1960; White 1991; Juch 1991).

### Conclusions

Each of the six foredeep basins compared has unique characteristics, some pre-coalification, but more post-coalification, which determine whether that basin is prospective for CBM production. Coalbed methane basin analysis and productivity prediction are not easy processes. Analogues do not appear to be particularly useful. Attempts to use 'rules of thumb' may do more harm than good. Each basin must be evaluated on its own merits, using all available relevant data.

Of the six Variscan/Appalachian foredeep basins compared, only three appear to have potential for commercial CBM production: Black Warrior—productive since the late 1970s; South Wales (unmined portion)—high methane content coals, high estimated methane in place, but questionable permeability characteristics due to compressive history; and the Ruhr (unmined eastern portion)—high estimated methane in place, but questionable permeability characteristics due to the depths of the most prospective coalbeds.

Of these three, only the Black Warrior coalbeds have proved permeability. Acceptable permeabilities may exist in the other two, but will not be proved until a number of holes has been drilled and tested.

The Oxford and Kent basins appear to be non-prospective for commercial CBM production, based on available data. Potential commerciality of the Bristol–Somerset Basin is questionable, given the lack of methane content data, severe compression and the environmental opposition to production operations that will probably develop in this area.

Exposure and erosion of Upper Carboniferous coal measures in Europe during the Stephanian through Triassic period may have had serious negative effects on the methane contents and thickness of the shallow degassed coal measures.

The high methane contents and relatively thin (300–400 m), shallow, degassed zone of the South Wales coal basin suggest that inversion of this basin and removal of the post-Westphalian D cover strata may be geologically recent events.

# References

ADAMS, H. K. 1967, *The Seams of the South Wales Coalfield*. Institution of Mining Engineers, London.

ANON. 1989. *Steinkohle—sichere Energie fur die Zukunft*. Geologisches Landesamt Nordrhein-Westfalen, Krefeld.

ARCHER, A. A. 1968. *Geology of the South Wales Coalfield, The Upper Carboniferous and Later Formations of the Gwendraeth Valley and Adjoining Areas*. British Geological Survey Special Memoir.

AYERS, W. B., JR., TISDALE, R. M., LITZINGER, L. A. & STEIDL, P. F. 1993. Coalbed methane potential of Carboniferous strata in Great Britain. *In: Proceedings, 1993 International Coalbed Methane Symposium, Birmingham, Alabama*, 1, pp. 1–14.

BRIX, M. R., DROZDEWSKI, G., GREILING, R. O. & WREDEN, V. 1988. The N. Variscan Margin of the Ruhr Coal District (Western Germany): structural style of a buried thrust front? *Geologische Rundschau*, 77, 115–126.

CAMERON, T. D. J., CROSBY, A., BALSON, P. S., JEFFERY, D. H., LOTT, G. K., BULAT, L. & HARRISON, D. J. 1992, *The Geology of the Southern North Sea*. United Kingdom Offshore Regional Report, British Geological Survey, HMSO, London.

COLE, J. E., MILIORIZOS, M., FRODSHAM, K., GAYER, R. A., GILLESPIE, P. A., HARTLEY, A. J. & WHITE S. C. 1991. Variscan structures in the opencast coal sites of the South Wales Coalfield. *Proceedings of the Ussher Society*, 7.

COPE, J. C. W., INGHAM J. K. & RAWSON, P. F. (eds) 1992. *Atlas of Paleogeography and Lithofacies*. Geological Society, London, Memoir, 13.

CORNELIUS, C. T., HARTLEY, A., GAYER R. & ROSS C. 1993. Coal deposition and tectonic history of the South Wales Coalfield, UK—implications for coalbed methane resource development. *In: Proceedings, 1993 International Coalbed Methane Symposium, Birmingham, Alabama*, 1, 161–172.

CREEDY, D. P. 1988. Geological controls on the formation and distribution of gas in British Coal Measures strata. *International Journal of Coal Geology*, 10, 1–31.

DEUTLOFF, O. (ed.) 1976. Geologie, Lieferung 8. *In: Deutscher Planungs Atlas, Band 1—Nordrhein—Westfalen*. Veroffentlichungen der Akademie für Raumforschung und Landesplanung, Herman Schroedel, Hanover.

DROZDEWSKI, G., BORNEMANN, O., KUNZ, E. & WREDE, V. 1980. *Beitrage zur Tiefentektonik des Ruhrkarbons*. Geologisches Landesamt Nordrhein-Westfalen, Krefeld.

——, JANSEN, F., KUNZ, E., PIEPER, B., A. RABITZ, STEHN, O. & WREDE V. 1981. *Geologische Karte des Ruhrkarbons 1 : 100 000*. Geologisches Landesamt Nordrhein-Westfalen, Krefeld.

DUNHAM, K. C. & POOLE, E. G. 1974. The Oxfordshire Coalfield. *Journal of the Geological Society, London*, 130, 387–391.

FIEBIG, H. & GROSCURTH, J. 1984. Das Westfal C im nordlichen Ruhrgebiet. *In: HILDEN, H. D. (ed.) Nordwestdeutsches Oberkarbon, Band 33, Teil 1, Fortschritte in der Geologie von Rheinland und Westfalen*. Geologisches Landesamt Nordrhein-Westfalen, Krefeld, 257–268.

FOSTER, D., HOLLIDAY, D. W., JONES, C. M., OWENS, B. & WELSH, A. 1989. The concealed Upper Paleozoic rocks of Berkshire and South Oxfordshire. *Proceedings of the Geologists' Association*, 100, 395–407.

GAYER, R. A. 1992. *The Evolution of the South Wales Coalfield Foreland Basin*. Basin Dynamics Special Topic Research Grant, National Environmental Research Council Ref. GST/02/350.

—— & JONES, J. 1989. The Variscan foreland in South Wales. *Proceedings of the Ussher Society*, 1989, 177–179.

——, COLE, J. E., GREILING, R. O., HECHT, C. & JONES, J. A. 1993. Comparative evolution of coal bearing foreland basins along the Variscan northern margin of Europe. *In: GAYER, R. A., GREILING, R. O. & VOGEL, A. (eds) Rhenohercynian and SubVariscan Fold Belts*. Verlag Vieweg, Wiesbaden.

GILL, W. D., KHALAF F. I. & MASSOUD, M. S. 1979. Organic matter as indicator of the degree of metamorphism of the Carboniferous rocks in the South Wales coalfields. *Journal of Petroleum Geology*, 1(4), 39–62.

HILL, D. G. (ed.) 1993a. Coalbed methane—state of the industry; Appalachian Basin. *Quarterly Review of Methane from Coal Seams Technology*, **11**, 6–9.

—— (ed.) 1993b. Coalbed methane—state of the industry; Black Warrior Basin, Alabama. *Quarterly Review of Methane from Coal Seams Technology*, **11**, 10–13.

HURLEY, P. M. 1968. The confirmation of continental drift. *In*: ANON. (ed.) *Continents Adrift: Readings from Scientific American*. Freeman, San Francisco.

JONES, J. A. 1991. A mountain front model for the Variscan deformation of the South Wales Coalfield. *Journal of the Geological Society, London*, **148**, 881–891.

JUCH, D. 1991. Das Inkohlungsbild des Ruhrkarbons-Ergebnisse einer Ubersichtsauswertung. *Gluckauf Forschungs-hefte*, **52**, 37–46.

—— & GIS ARBEITSGRUPPE 1988. *Aufbau eines geologischen Information-systems fur die Steinkohlenlager-statten Nordrhein-Westfalens und im Saarland*. Geologisches Landesamt Nordrhein-Westfalens, Krefeld.

KELLAWAY, G. A. & WELCH, F. B. A. 1948. *Bristol and Gloucester District*, 2nd edn. British Regional Geology Series, British Geological Survey, HMSO, London.

—— & —— 1993. *Geology of the Bristol District*. Memoir for 1:63360 Geological Special Sheet, British Geological Survey, HMSO, London, 199pp.

KOTAS, A. (ed.) 1991. *Coal-bed Methane Potential of the Upper Silesian Coal Basin (Poland)*. State Geological Institute, Upper Silesian Branch, Sosnowiec.

LEVINE, J. R. 1992. Oversimplifications can lead to faulty coalbed gas reservoir analysis. *Oil and Gas Journal*, **90**, 63–69.

LITTKE, R. 1987. Petrology and genesis of Upper Carboniferous seams from the Ruhr region, West Germany. *International Journal of Coal Geology*, **7**, 147–148.

—— & LOTENHAVEN, H. 1989. Paleoecologic trends and petroleum potential of Upper Carboniferous coal seams of Western Germany as revealed by their petrographic and organic geochemical characteristics. *International Journal of Coal Geology*, **13**, 529–574.

LOMMERZHEIM, A. 1991. *Die Geothermische Entwicklung des Munsterlunder Beckens (NW-Deutschland) und ihre Bedeutung für die Kohlenwasserstottgense in Diesem Raum*. DGMK-Berichte 468, 319–372.

MCFALL, K. S, WICKS. D. E. & KUUSKRAA., V. A. 1986. *A Geological Assessment of Natural Gas from Coal Seams in the Warrior Basin, Alabama*. Topical Report, Final Geological Report (September 1985–September 1986), Gas Research Institute, Chicago.

MONTGOMERY, S. L. (ed.) 1986a. *The Black Warrior Basin: Proving the Potential of the Southeast*. Petroleum Frontiers, Vol. 3, No. 3, Petroleum Information Corporation, Denver.

—— 1986b. *Coalbed Methane: an Old Hazard Becomes a New Resource*. Petroleum Frontiers, Vol. 3, No. 4, Petroleum Information Corporation, Denver.

MOORE, L. R. 1954. The South Wales Coalfield. *In*: TRUEMAN, A. E. (ed.) *Coalfields of Great Britain*, Edward Arnold, London, 93–125.

NATIONAL COAL BOARD, SCIENTIFIC DEPARTMENT, COAL SURVEY 1960. *The Coalfields of Great Britain: Variation in Coal Rank*. National Coal Board, London.

PASHIN, J. C. (ed.) 1991. *Structures, Sedimentology, Coal Quality and Hydrology of Black Warrior Basin in Alabama: Controls on Occurence and Produceability of Coalbed Methane*. Topical Report (August 1987–December 1990) GRI 91/0034, Gas Research Institute, Chicago.

PETROCONSULTANTS 1977. *Coal Worldwide*. Petroconsultants, Geneva.

RYAN, R. J. & BOEHNER, R. C. 1994. *Geology of the Cumberland Basin—Nova Scotia*. Mines and Energy Branches, Memoir, **10**, Nova Scotia Department of Natural Resources.

SQUIRRELL, H. C. & DOWNING, R. A. 1969. *Geology of the South Wales Coalfield, Part I*, 3rd edn, *The Country Around Newport*. British Geological Survey, Memoir. New Series Map Sheet 249.

STEUERWALD, K. & WOLFF, M. 1985. Des tektonische Bau des Lippe-und Ludinghausener Hauptmulde zwischen Marl und Ludinghausen (Westfalen). *In*: HILDEN, H. D. (ed.) *Nordwestdeutsches Oberkarbon*, Band 33, Teil 2, *Fortschritte in der Geologie von Rheinland und Westfalen*. Geologisches Landesamt Nordrhein-Westfalen, Krefeld, 33–49.

STRACK, A. &. FREUDENBERG, U. 1984. Schichtenmachtigkeiten und Kohleninhalte im Westfal des Niederrheinisch-Westfalischen Steinkohlen reviers. *In*: HILDEN, H. D. (ed.) *Nordwestdeutsches Oberkarbon*, Band 33, Teil 1, *Fortschritte in der Geologie von Rheinland und Westfalen*. Geologisches Landesamt Nordrhein-Westfalen, Krefeld, 243–255.

STREHLAU, K. 1990. Facies and genesis of Carboniferous coal seams of Northwest Germany. *International Journal of Coal Geology*, **15**, 245–292.

STUBBLEFIELD, C. J. 1954. The Kent Coalfield and other possible concealed coalfields south of the English Midlands. *In*: *Coalfields of Great Britain*, 154–162.

TEICHMULLER, M., TEICHMULLER, R. & BARTENSTEIN, H. 1984. Inkohlung und Erdgas—eine neue Inkohlungskarte der Karbon-Oberflache in Nordwest-deutschland. *In*: HILDEN, H. D. (ed.) *Nordwestdeutsches Oberkarbon*, Teil 1. Geologisches Landesamt Nordrhein-Westfalen, Krefeld, 11–34.

WHITE, S. 1991. Palaeo-geothermal profiling across the South Wales Coalfield. *Proceedings of the Ussher Society*, **7**.

WILLS, L. J. 1951. *A Palaeogeographical Atlas of the British Isles and Adjacent Parts of Europe*. Blackie, Glasgow.

WOLF, R. 1985. Tiefentektonik des linksniederrheinischen Steinkohlengebietes. *In*: DROZDZEWSKI, G., ENGEL, H., WOLF, R. & WREDE, V. *Beitrage zur Tiefentektonik Westdeutscher Steinkohlenlagerstatten.* Geologisches Landesamt Nordrhein-Westfalen, Krefeld.

WOODLAND, A. W. & EVANS, W. B. 1964. *The Geology of the South Wales Coalfield, Part IV*, 3rd edn, *The Country Around Pontypridd and Maesteg.* British Geological Survey, Memoir. New Series Map Sheet 248.

# Geological controls on coalbed prospectivity in part of the North Staffordshire Coalfield, UK

F. J. MacCARTHY[1], R. M. TISDALE[2] & W. B. AYERS, JR[2]

[1] *Kinetica Limited, 20 Bedfordbury, Covent Garden, London WC2N 4BL, UK*
[2] *Taurus Exploration Inc., 2101 Sixth Avenue North, Birmingham, Alabama 35203, USA*

**Abstract:** Westphalian coal seams in the Kinetica/Taurus exploration licence EXL 282, which covers part of the North Staffordshire Coalfield, are estimated to contain over $5.6 \times 10^{10}$ m$^3$ of initial coalbed gas in place. In the evaluation of coalbed gas prospectivity in the licence area, data from the coal and the oil and gas industries were used to map coal occurrence, coal rank, the gas content of coal seams, gas in-place, structure, face cleat and *in situ* stress.

The North Staffordshire Coalfield is structurally and stratigraphically complex. Westphalian strata occur in a SSW plunging synclinorium that is highly segmented by faults with displacements as great as 500 m. These faults may either be barriers or conduits for fluid migration. The maximum *in situ* horizontal compressive stress is generally oriented NNW, subparallel to face cleats and some faults, which may favour openness and transmissivity of those fractures. In the southern part of the licence area, intraformational unconformities occur in the Westphalian. Total coal thickness is greatest (more than 40 m) and coal rank is highest (medium-volatile bituminous) in the northern part of the licence area, coincident with the greatest thickness of Westphalian strata. Predictably then, gas content and initial gas in-place per unit area are greatest in the northern part of the licence area, coincident with the occurrence of thick, high rank coals. In this area, initial gas in-place is estimated to exceed $4.0 \times 10^8$ m$^3$ per km$^2$ in some areas. Although gas resource trends have been identified, production rates of coalbed gas can only be established through well testing.

Exploration Licence EXL 282 was awarded to Kinetica Limited and Taurus Exploration UK Limited in 1993. This licence area is situated in the west-central UK (Fig. 1) and covers 600 km$^2$ in the counties of Staffordshire, Shropshire and Cheshire. The Westphalian coal-bearing sequence in the area was part of the Late Carboniferous Pennine Basin (Guion & Fielding 1988) within which the coalfields of central and northern England were initially contiguous. The Pennine Basin has been interpreted on a regional scale (Leeder 1982) as a thermal sag basin, infilling an earlier (Early Carboniferous–Namurian) rifting stage, with a basin depocentre in the south Lancashire region. Within the Pennine Basin, coal-bearing strata are of Namurian to Westphalian D age, with most of the coal resource occurring in the Westphalian A to C interval. Coals crop out in the North Staffordshire area and in the Coalbrookdale Coalfield in the southwest of the licence area (Fig. 2); coal-bearing strata are overlain by Permo-Triassic sediments in the Stafford Basin between these two exposed coalfields.

The Westphalian Coals in the licence area were deposited in fluvio-deltaic environments with occasional marine influence and were subject to a multi-phase deformation during the Variscan Orogeny. At that time some coal and gas resource were lost due to the erosion of coals and the removal of overburden required

for a gas seal. In addition the Westphalian, strata comprising the coalbed reservoir interval were segmented by faulting. The subsequent Permian and Mesozoic history of the area is poorly understood due to the incomplete stratigraphic record of these ages.

This paper focuses on the northern part of the licence area (North Staffordshire Coalfield Area; Fig. 3), where there is a large volume of data in the public domain, in contrast with the relative scarcity of data elsewhere in EXL 282.

## Stratigraphy

An outline of Westphalian and Stephanian stratigraphy in north EXL 282 is given in Table 1. The main coal-bearing sequence is of Westphalian A to Westphalian C age and is divided into the stratigraphic units of Lower, Middle and Upper Coal Measures (British Geological Survey 1993), but minor coals also occur in the overlying Newcastle Formation (Westphalian D). The Coal Measures strata were deposited in fluvio-deltaic environments within which two broad facies associations were recognized in a study of the Westphalian A and B (Guion & Fielding 1988); an upper delta plain facies (no marine influence) and a lower delta plain facies (some marine influence). There are considerable variations in the thickness of the

*From* Gayer, R. & Harris, I. (eds) 1996, *Coalbed Methane and Coal Geology*,
Geological Society Special Publication No 109, pp 27–42.

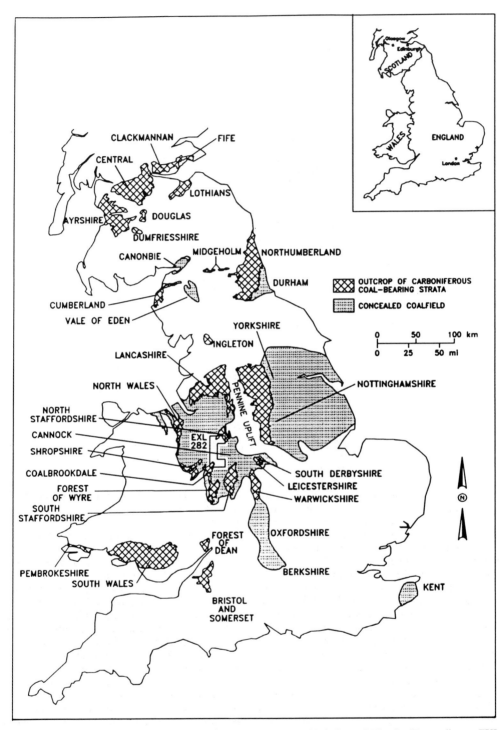

**Fig. 1.** Location of the principal Upper Carboniferous Coalfields of Britain, and Kinetica/Taurus licence EXL 282. Modified from National Coal Board (1974), British Geological Survey (1985) and Institute of Geological Sciences (1979).

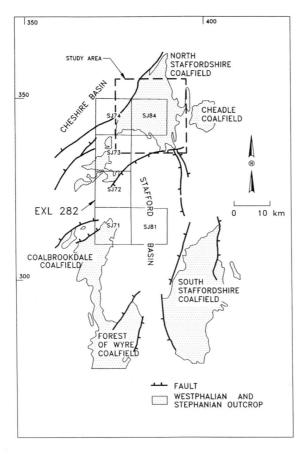

**Fig. 2.** Location of the EXL 282 licence area relative to coalfields in the west-central UK.

Coal Measures in north EXL 282 that can be attributed to tectonic controls on deposition and, in some instances, erosion. The thickest Coal Measures sequence occurs in the core of the Potteries Syncline west of Newcastle-under-Lyme (Fig. 4), and the Coal Measures are progressively eroded towards the flanks of the North Staffordshire synclinorium. On part of the axis of the Western Anticline (Fig. 4) the Upper and Middle Coal Measures have been completely removed by erosion. The preserved strata of the Coal Measures show that, within north EXL 282, a depocentre existed west of the Newcastle-under-Lyme area, and to the east and west the sequence becomes condensed, particularly towards the Western Anticline, over which there was less subsidence during most of the Westphalian (Corfield 1991).

The coal-bearing section (Fig. 5) can also be viewed as two discrete target intervals, separated by strata with an impoverished or absent coal resource. The lower interval comprises strata from the base of the Coal Measures to the Birchenwood Coal and the upper interval is from the Twist seam to the Blackband seam.

The Upper Coal Measures are diachronously overlain by barren red beds of the Etruria Formation. This formation comprises well-drained alluvial plain facies with subordinate alluvial fan facies, which interdigitate with coal-bearing rocks in the North Staffordshire Coalfield (Besly 1988). The red beds appear on the south margins of the Pennine Basin in the Westphalian A and B and spread diachronously northwards from Coalbrookdale towards North Staffordshire. Therefore, the lateral equivalents of much of the Coal Measures to the south are the red beds of the Etruria Formation. In general, the red beds of the Etruria Formation occurred earlier in areas which had low rates of subsidence during the Westphalian A and B (Besly 1988).

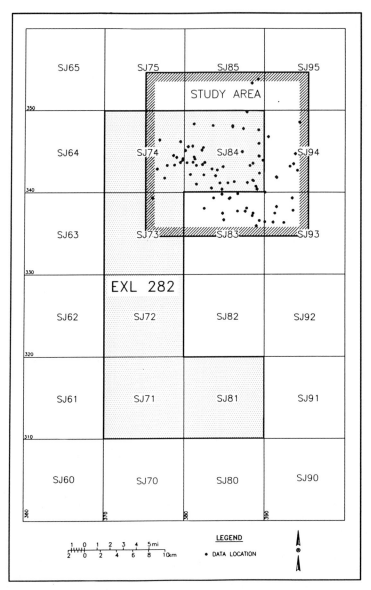

**Fig. 3.** Locations of the boreholes and mine shafts from which data was taken in north EXL 282.

## Structural setting

The Westphalian strata of the North Stafford-shire Coalfield occur in a SSW plunging synclinorium that is segmented by several fault groups. The structure of the coalfield has been analysed (Corfield 1991) and the principal structural events in north EXL 282 are summarized in Table 2. The primary structural imprint in the coalfield dates from the late Carbonifer-ous to early Permian Variscan Orogeny, with many of the Variscan structural trends inherited

from reactivated faults in underlying basement rocks. Corfield showed that three basement fault trends occur in the North Staffordshire Coalfield area: northeast–southwest, north–south and northwest–southeast, the latter being the least well developed in the area. The Red Rock Fault is an example of the northeast–southwest trend and the north–south trend is represented by the Lask Edge Fault, which is not exposed and is situated close to the Blackwood Anticline (Fig. 4). The Western Anticline can be related to the northeast–southwest trend, and the

**Table 1.** *Summary of Westphalian and Stephanian stratigraphy in north EXL 282. Data from borehole logs and mine shaft sections, with additional information from Besly (1988), Guion & Fielding (1988), Crofts (1990), Rees (1990) and Wilson (1991)*

| Name of stratigraphic unit | Age of stratigraphic unit | Thickness of stratigraphic unit (m) | Primary lithologies | Secondary lithologies | Environment of deposition |
|---|---|---|---|---|---|
| Keele Formation | Westphalian D –Stephanian | Max. recorded 620 | Mudstone, siltstone, sandstone | Conglomerate, limestone, coal | Alluvial, alluvial fan and lacustrine |
| Newcastle Formation | Westphalian D | 80–150 | Mudstone, sandstone | Coal, limestone | Alluvial and lacustrine |
| Etruria Formation | Westphalian C–D | Max. recorded 430 | Mudstone | Sandstone, limestone and conglomerate | Well drained alluvial plain |
| Upper Coal Measures | Westphalian C | 0–350 | Mudstone and siltstone | Sandstone, coal, ironstone and limestone | Upper and lower delta plain |
| Middle Coal Measures | Westphalian B–C | 0–650 | Mudstone and siltstone | Sandstone, coal, ironstone and limestone | Upper and lower delta plain |
| Lower Coal Measures | Westphalian A | 380–635 | Mudstone, siltstone and sandstone | Coal, limestone and ironstone | Upper and lower delta plain |

Werrington and Blackwood Anticlines are parallel to the north–south trend of the Lask Edge Fault (Corfield 1991). Corfield also identified outcrop-scale thrusts that trend northeast–southwest on the limbs of the Western Anticline.

Differential subsidence during the Westphalian exerted an important control on coal formation in north EXL 282. Corfield (1991) showed that coals amalgamate and thicken over the Western Anticline, indicating that folding was contemporaneous with sedimentation and that this folding was a result of local compression (northwest to southeast) in an overall post-rift thermal subsidence regime. Several stages of Variscan deformation are recognized in the North Staffordshire Coalfield (Corfield 1991; Table 2). Syn-depositional folds in the coalfield developed into major structures, which were transected by faults that formed at a later folding stage. These faults are generally oriented west–northwest to east–southeast, with displacements in some instances of several hundred metres. Two other Variscan fault sets are recognized in EXL 282: east–west trending normal faults, which dip both north and south, and NNW–SSE normal faults that dip both east and west. These faults are typically high-angle, planar structures which post-date coal deposition (Corfield 1991). The timing of the faulting is poorly constrained and is estimated by Corfield to be of end Westphalian D to early Triassic age. During the Variscan Orogeny, the coal reservoir was segmented by faulting, and in some instances

uplift and erosion removed the overburden required for an effective hydrostatic seal, in addition to the removal of some of the coals.

The post-Variscan history of north EXL 282 is largely unknown. No Permian stratum is known from north EXL 282, and the Triassic section is incomplete. Research by Lewis *et al.* (1992) indicates that the Mesozoic sediment cover was of kilometre-scale thickness in north EXL 282. Most of these strata were eroded as a result of regional uplift in the Tertiary. As in the Variscan event, part of the gas resource was lost by the erosion of coals and the loss of hydrostatic seal above the coal reservoir.

## Number of coals/net coal thickness

Net coal thickness and total number of coals were mapped in north EXL 282 in a stratigraphic interval (Fig. 5) from the base of the Lower Coal Measures to the Blackband seam (Upper Coal Measures). In these calculations, derived primarily from borehole data (Fig. 3), coals less than or equal to 0.3 m were not considered.

The net coals in the Westphalian A to C interval in north EXL 282 occur in a northwest–southeast trend with a maximum net thickness exceeding 40 m and a maximum number of coals projected to exceed 35 in block SJ84 (Figs 6 and 7). A broad area where the net coal thickness exceeds 20 m, representing a depocentre in Westphalian A to C times, extends southeastward into SJ93 (Fig. 7).

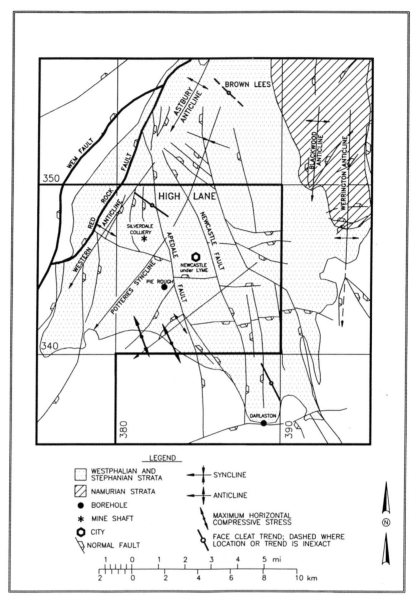

**Fig. 4.** Major structural features in north EXL 282, with face cleat orientation and maximum *in situ* horizontal stress orientation. Data from Brereton & Evans (1987), Corfield (1991), and British Geological Survey (1993).

This area coincides with the occurrence of a pervasive trend of 35 or more coal seams (Fig. 6). The effects of erosion on net coal thickness and the total number of coals in the Westphalian A to C interval are seen in the east of the coalfield, where the coal-bearing strata ascend to outcrop (Figs 2, 4 and 6). The eastern line of zero coal ('0' line) has been projected in the western half of SJ94. On the northwestern side of the licence, the coals have not been mapped west of the Red Rock Fault, due to the absence of publicly available coal data in this area.

The largest proportion of the coal in the licence area occurs within the Middle Coal Measures. In the northern part of the licence area, the Middle Coal Measures account for over 45% of the total coal thickness. Individual coal seams within the Lower, Middle and Upper Coal Measures vary in thickness from the minimum thickness cut-off of 0.3 m to over 4.0 m.

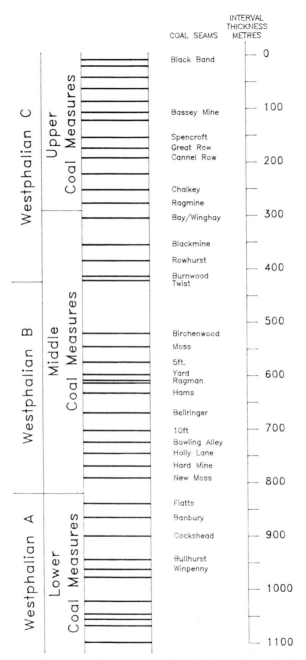

**Fig. 5.** Generalized stratigraphic column of the primary coal-bearing interval in the North Staffordshire Coalfield.

## Coal rank

The capacity of coal to store gas generally increases with increasing coal rank. The rank of coal in the North Staffordshire Coalfield has been the subject of several studies, including those by Wandless (1960), Millott (1942), Crofts (1953) and Millott *et al.* (1946) that used parameters including the calorific value of coal and the percentage of volatile matter to describe the nature and variation of coal rank in the coalfield.

The highest recorded coal rank in the coalfield occurs in the core of the Potteries Syncline west

**Table 2.** *Summary of the principal Westphalian to Tertiary structural stages in the North Staffordshire Coalfield, modified from Corfield (1991)*

| Event age | Major structures | Orientation of principal compresive stress | Comments |
|---|---|---|---|
| Tertiary | Reactivation of faults, uplift and erosion | | Erosion of coals and reservoir seal, some degassing of coals. Removal by erosion of much of Mesozoic cover |
| Triassic to Cretaceous | Possible Jurassic extension across Red Rock Fault | | Research (Lewis *et al.* 1992) indicates kilometre-scale sediment cover in the Mesozoic |
| Westphalian D–Early Triassic | Normal faulting, including reactivation of Red Rock Fault | NNW–SSE | NNW–SSE trending faults that dip to east and west. Initiation of Cheshire Basin |
| Westphalian D–Early Triassic | Normal faulting | E–W | E–W trending faults that dip to north and south |
| Westphalian D–Early Triassic | Normal faulting | WNW–ESE | WNW–ESE trending faults, formed at late folding stage. Some face cleat formation associated with this stage |
| Westphalian D–Early Triassic | Major folds | WNW–ESE to NW–SE | Local NNE–SSW extensional faults in crest of Western Anticline |
| Westphalian A–?Westphalian D | Precursors to major folds | NW - SE | Local compression within an setting of post-rift thermal subsidence. |

of Newcastle-under-Lyme (Figs 4 and 8), from where the rank diminishes towards the flanks of the synclinorium. In this high rank area, coals of the Moss seam (Westphalian B) have volatile matter contents of less than 37% (high-volatile A bituminous), increasing to more than 40% (high-volatile B bituminous) at the margins of the coalfield.

The variation in coal rank with stratigraphic position in an early petroleum exploration borehole (Pie Rough) is summarized in Fig. 9. The coal rank in the Pie Rough borehole generally increases with depth; in the Upper Coal Measures, the sampled coals have volatile matter levels of 39–46% (high-volatile B to C bituminous), the Middle Coal Measures have a range of 31–42% (high-volatile A to C bituminous) and in the Lower Coal Measures, the coals have between 28 and 33% volatile matter (medium-volatile to high-volatile A bituminous) (Millott *et al.* 1946).

## Coal quality

The ash content of coals is an important element in coalbed methane exploration because ash effectively reduces the gas storage capacity of coal seams. The ash content of a coal sample is the inorganic residue remaining after combustion and is expressed as the percentage weight of air-dried coal. In the licence area, the ash content can in some instances vary significantly within a coal seam section, from one coal seam to the next, and over kilometre-scale lateral distances.

In the Pie Rough borehole, ash contents were measured from most of the seams encountered during drilling (Millott *et al.* 1946). The recorded ash contents range from 2 to 11.8%, and the average ash content is 5.9%. The ash contents were measured from drill bit cuttings, which had been separated from non-coal fragments, and the total seam ash content may have been understated as a result.

Samples from another borehole in north EXL 282 (Creedy 1986) have an average ash value of 6.6%, based on 24 samples from 21 seams covering the Lower to Upper Coal Measures interval, with values that range from 1.1 to 17.2% in the Banbury and Bellringer seams, respectively. In this instance the values may also be understated, as a small sample size (30–70 g) was used and sample selection was biased away from dirty coal and rock partings (Creedy 1986; 1991).

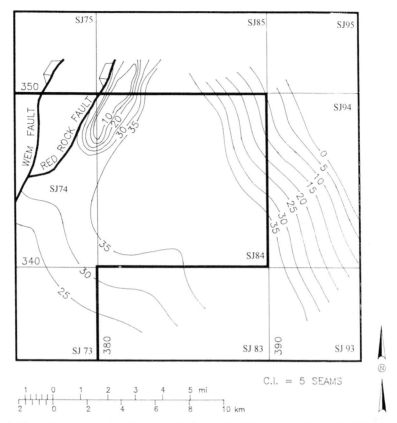

**Fig. 6.** Distribution of the number of coal seams in the Lower Coal Measures to Upper Coal Measures interval in north EXL 282. Seams under 0.3 m thick were excluded. C.I., Contour interval

In contrast, ash content values for three full-seam samples of the Cannel Row and Rowhurst seams ranged from 13.2 to 21.3% and averaged 17.8% (Bolton *et al.* 1988). The ash content of the Winghay seam, in Silverdale Colliery (Fig. 4), is as great as 22% (Belcher & Hill 1981).

The ash content within a single seam may also show significant variation. At one locality in a North Staffordshire colliery, ash contents in a section through the Cannel Row seam ranged from 9.7 to 22.2% (average 18.9%). A section through the Rowhurst seam displayed an ash range of 4.6–35.3% (average 13.2%) (Bolton *et al.* 1988). In both these sections, the highest ash content value, excluding dirt partings, occurs at the base of the coal seam.

Ash contents in North EXL 282 may show significant lateral changes over relatively short distances (Bolton *et al*, 1988). The ash content in the Cannel Row seam varies in some instances from more than 25% to less than 15% within a distance of 1 km.

## Coalbed gas

Published gas content data for coals (Creedy 1985; 1991) and proprietary analyses from the British Coal Corporation (BCC) were used to map gas content and composition in EXL 282. Most of the coal samples analysed for gas content are from surface exploration boreholes drilled by BCC. The reported gas content values comprise methane and ethane, expressed in cubic metres per tonne at 20°C and 101.3 kPa, and corrected for ash content. The actual in-place gas will depend in part on the ash content of the coal and this will be less than the mapped, ash-free values.

The methods used in coal sampling and gas measurements are outlined in Creedy (1986). Gas content values were averaged over the productive section and then contoured (Fig. 10). In the North Staffordshire Coalfield, the average gas content of Coal Measures seams increases from 6 m³/t near the eastern margin of the coalfield to more than

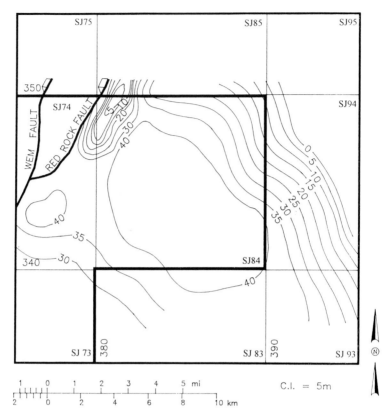

**Fig. 7.** Distribution of the total thickness (in metres) of net coals in the Lower Coal Measures to Upper Coal Measures interval in north EXL 282. Seams under 0.3 m thick were excluded.

$9 \, \text{m}^3/\text{t}$ in high- to medium-volatile bituminous coal in an area around the centre of the Potteries Syncline (Fig. 10).

Both methane and ethane generally increase with depth and coal rank in north EXL 282. Some reasons for deviations from predicted values may include differences in coal quality, gas migration, gas lost before the samples were placed in canisters, small size of test samples, testing inaccuracy, proximity to mining and prior degasification by geological processes. It has been shown elsewhere (Creedy 1991) that coals which occur within a distance of approximately 150 m below the top Carboniferous unconformity have lost much of their stored gas.

The percentage of the combined hydrocarbon component (methane and ethane) in coalbed gas that is composed of methane $[CH_4/(CH_4 + C_2H_6) \times 100]$ ranges between 80 and 95% in the Hobgoblin borehole (Creedy 1986). Generally, methane percentage decreases with depth in Hobgoblin; it averages 94.2% in Upper Coal Measures seams, 88.7% in the

Middle Coal Measures and 81.1% in Lower Coal Measures seams (based on data in Creedy 1983; 1986).

Unfortunately, only partial analyses are available for British coalbed gases. Most published gas analyses report only methane and ethane, even though significant other components, including nitrogen and carbon dioxide, are commonly present (Creedy & Pritchard 1983; Creedy 1985; 1991). Nitrogen is an inert gas that makes up 2–8% of UK coalbed gas, and carbon dioxide is a non-combustible, corrosive gas that may comprise 0.2–6%, on average, of UK coalbed gas (Creedy & Pritchard 1983; Creedy 1988; 1991).

The Darlaston Borehole in the North Staffordshire Coalfield (Fig. 4) is the only available UK borehole with published analyses of nitrogen and carbon dioxide reported by seam and depth. In this borehole, nitrogen averaged $0.5 \, \text{m}^3/\text{t}$, the same as the average for all 73 samples measured in the UK (Creedy & Pritchard 1983, Creedy 1985; 1988). At Darlaston, nitrogen comprises

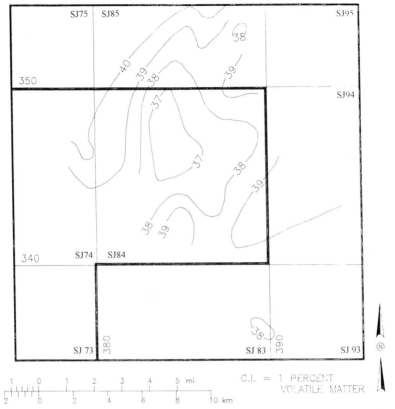

**Fig. 8.** Variations in the percentage of volatile matter content in coal from the Moss seam (Middle Coal Measures) in north EXL 282. Values are for dry, ash-free coal. Data from National Coal Board (1960).

3–14% of the coalbed gas. Because the nitrogen content does not vary with depth at Darlaston (whereas the total of all other gases increases), the average percentage of nitrogen in the coalbed gas decreases from 8.9% in Westphalian C coals to 5.2% in Westphalian B coals; Westphalian A coals were not tested.

Carbon dioxide content decreases with depth at Darlaston (Creedy & Pritchard 1983). It averages 4.2% in Westphalian C coals (highest value is 10.6 percent) and 1.0% in Westphalian B coals.

*Initial gas in-place resource*

The initial gas in-place resource was calculated and mapped (Fig. 11) using the following formula:

Gas resource = coal density × coal thickness

× gas content × area

The gas resource is reported in units of billions of cubic feet/square kilometre (BCF/km$^2$) where

1 BCF = 10$^9$ ft$^3$, following oil industry convention (Fig. 11). All calculated values are resources or in-place gas, rather than reserves or recoverable gas. Part of the initial coalbed gas resource in EXL 282 has been removed as a consequence of coal mining activities in the area. The initial in-place gas resource was calculated for coals at depths less than 1500 m with over 200 m of Carboniferous overburden and seams greater than 0.3 m thick in the licence area. A constant density value was used in the resource calculations. Because of the paucity of in-place ash content data, gas contents are reported on an ash-free basis and, consequently, the in-place gas values are inflated by the percentage of ash in the coal.

The initial in-place coalbed gas resource in the total licence area is estimated to exceed 2.0 × 10$^3$ BCF (5.6 × 10$^{10}$ m$^3$). Within north EXL 282 the highest volume of initial gas in-place per unit area is coincident with the area of thick, high rank coals described in the preceding sections (Fig. 10). In some places within this

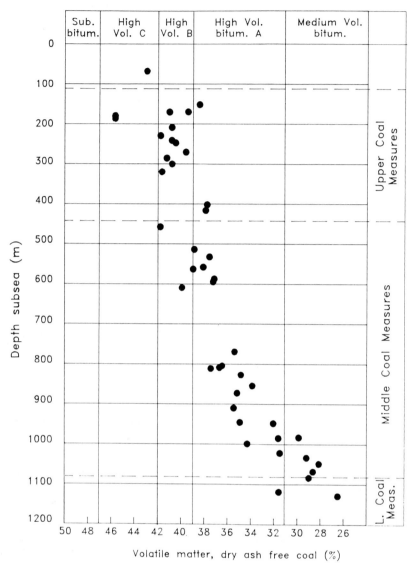

**Fig. 9.** Volatile matter versus depth in the Pie Rough Borehole. Derived from data in Millott *et al.* (1946).

area, estimates of initial gas in-place per square kilometre exceed 14 BCF ($4.0 \times 10^8 \, m^3$). The volumes of economically recoverable gas (reserves) cannot be calculated until additional reservoir engineering parameters are determined by *in situ* testing of coal seams.

In addition to the coal seams, the Upper Carboniferous sandstones in the study area are possible hydrocarbon reservoirs. The structural and stratigraphic setting in the area has some potential for forming hydrocarbon accumulations in anticlinal, fault-related and possibly stratigraphic traps. The possible hydrocarbon

resource in non-coal reservoirs has not been included in the above resource estimate.

## Cleat and permeability

In the UK, there is no available analysis of coalbed permeability from measurements taken in surface boreholes. Fluid (gas or water) flow in coal seams is primarily through the cleats, which are natural fractures formed during coalification. Because the face (dominant) cleats are more extensive than the butt (subordinate)

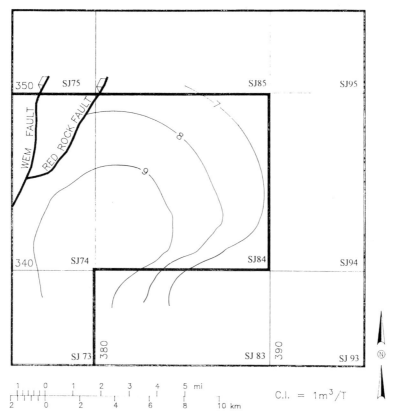

**Fig. 10.** Variations in coalbed gas content in the Coal Measures in north EXL 282. Modified from proprietary British Coal data.

cleats, permeability is commonly anisotropic and is greater in the face cleat direction. Cleat intensity, defined as the number of fractures per unit distance perpendicular to the cleat, is important in determining the magnitude of flow in a coal seam. Cleat intensity varies with coal type and bed thickness (Spears & Caswell 1986; Tremain *et al.* 1991) and, commonly, it increases with coal rank from the high-volatile to low-volatile bituminous stage of coalification (Ting 1977), above which cleat intensity decreases (Ammosov & Eremin 1963) or remains constant (Law 1993).

The high-volatile C to medium-volatile bituminous rank of coals in the north of the licence area (National Coal Board 1974) suggest that these coals should be well cleated, but data are meagre. Coals of similar rank in the Yorkshire Coalfield have from 0.3 to 1.0 cleats/cm and average 0.6 cleats/cm (Macrae & Lawson 1954). In the Yorkshire area, the dull coal bands commonly have a wider cleat spacing (lower intensity) and a blocky character, whereas the

bright coal is more highly cleated and should have better permeability. This relation between coal type and cleat intensity also occurs in the Cannock Coalfield, where face cleat intensity is 3.0–6.0 cleats/cm in 'bright' coals, but the average intensity for all coals, 'bright' and 'dull', is 2.2 cleats/cm (Spears & Caswell 1986).

The orientation of the face cleat commonly parallels the direction of maximum horizontal compressive stress at the time of coalification. At Brown Lees opencast mine, near the axis of the Potteries Syncline, and at High Lane opencast mine on the northwest flank of the syncline, face cleats are well developed and trend northwestward, perpendicular to the fold axis (Fig. 4). Corfield (1991) described three cleat sets at High Lane: 125°, the face cleat, with calcite infill; and two subordinate cleat sets at 155° and 105° which lack mineralization. Butt cleat is well developed and strikes perpendicular to dip at High Lane Colliery, but rather than being perpendicular to bedding it is inclined towards the vertical, indicating that folding, in part,

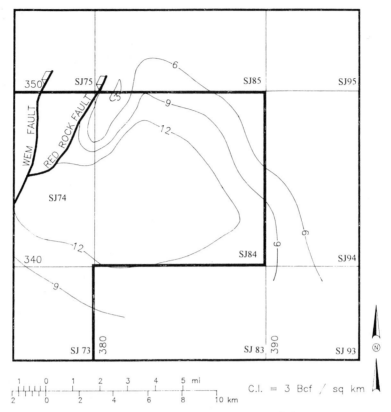

**Fig. 11.** Initial in-place coalbed gas resource in north EXL 282. 1 BEF $= 10^9$ ft$^3$.

pre-dates coalification. This conclusion is supported by the observation that coal isoranks cross bedding planes (Crofts 1953). In workings at Barlaston, in the deep, southern part of the North Staffordshire Coalfield, face cleat strike is 163° and butt cleat strike is 045° (Fig. 4) (Corfield 1991). Moseley & Ahmed (1967) indicate that the face cleat in the coalfields of the South Midlands may trend northward.

The extent to which cleats are occluded by minerals (which decrease permeability) is unknown in EXL 282. At High Lane Colliery, calcite is common in the face cleat only, and pyrite and ankerite occur locally (Corfield 1991).

Coalbed permeability may be affected by both the orientation and the magnitude of *in situ* stress. Laboratory and field studies show that coalbed permeability decreases with increasing stress (Durucan & Edwards 1986; McKee *et al.* 1987; Decker *et al.* 1991). Moreover, Sparks *et al.* (1993) convincingly demonstrate that coalbed methane production is inversely related to stress in the Cedar Cove field of Alabama, USA. In many coal basins, permeability in coal

seams decreases primarily as a function of increased pressure associated with depth; the pressure closes the cleat of the highly compressible coal. In some areas unusually high horizontal stress is present, and in these areas the collective overburden and horizontal stresses cause low permeability in comparatively shallow coal seams. However, the magnitude of horizontal stress in central England does not appear to be unusually high (Evans 1987) and, thus, depth will most likely exercise the primary control on stress reduction of permeability in the coals of the North Staffordshire Coalfield.

Both cleat and tectonic fractures (joints and faults) are more likely to be open if they are oriented parallel to the direction of maximum horizontal compressive stress than if they are perpendicular to this stress direction. Previous studies (Klein & Barr 1986; Evans 1987; Brereton & Evans 1987; Evans and Brereton 1990), indicate that the maximum horizontal stress is northwest (325°–340°) in the area of north EXL 282 (Fig. 4). These trends are parallel or subparallel to cleats and to some major faults in the area, which may favour openness of both types of fractures.

The magnitude and orientation of principal stress affect the orientation of hydraulic fractures induced during the stimulation of coal seams. In isotropic rocks, hydraulically induced fractures are thought to open in a plane perpendicular to the minimum principal stress direction. Recent stress studies confirm that the minimum horizontal stress is less than the theoretical overburden pressure in most regions of central England (Evans 1987), but data were not reported for the North Staffordshire Coalfield. These studies suggest that, at typical target depths, induced fractures will be vertical and they will generally trend NNW, which is subparallel or parallel to observed face cleat trends in North Staffordshire.

## Conclusions

In the North Staffordshire Coalfield, the net coal thickness is estimated in some instances to exceed 40 m in more than 35 coal seams. The present day occurrence of coal varies according to the initial sedimentary controls in the area, including marked differential subsidence, and also to the effects of erosion. The coal reservoirs in the coalfield have been variously segmented by several fault systems, most of which formed during the Variscan Orogeny. Two stages of partial degassing of the coal resource have been identified, one associated with the Variscan Orogeny and one with a second degassing event during Tertiary uplift.

The coal rank varies from high-volatile C bituminous to medium-volatile bituminous. Average gas contents in the Coal Measures in north EXL 282 vary from $6 \, \text{m}^3/\text{t}$ to over $9 \, \text{m}^3/\text{t}$, with the highest values of gas content and initial gas in-place per unit area coincident with an area of thick, high rank coals. Initial gas in-place is estimated to exceed $4.0 \times 10^8 \, \text{m}^3$ per square kilometre in some areas.

Cleat is likely to be well developed throughout the North Staffordshire Coalfield and may be subparallel or parallel to the maximum *in situ* horizontal compressive stress, which generally trends NNW in the area, promoting openness of cleats; however, the magnitude of the *in situ* stresses in the area is unknown. Data on many other parameters, for example reservoir engineering, are unknown in the licence area and can only be determined effectively by drilling and flow-testing the coal seams.

We thank M. J. Allen, formerly Chief Geologist, British Coal, for permission to use proprietary British Coal gas content data in this paper. We also thank P. A. Ferguson, D. M. McSwain, G. W. Murrie and R. P. Roark for their help with this paper. We acknowledge the improvements to the paper suggested by the referees. We are grateful to Kinetica and Taurus management for permission to present this paper, but the views expressed here are the authors' own and not necessarily those of Kinetica and Taurus.

## References

AMMOSOV, I. I., & EREMIN, I. V. 1963. *Fracturing in Coal* (translated from Russian). Israel Program for Scientific Translations.

BELCHER, J. A. & HILL, A. J. 1981. Drivages at Silverdale Colliery. *Mining Engineer*, **140**, 707–715.

BESLY, B. M. 1988. Palaeogeographic implications of late Westphalian to early Permian red-beds, Central England. *In*: BESLY, B. M. & KELLING, G. (eds) *Sedimentation in a Synorogenic Basin Complex: the Upper Carboniferous of Northwest Europe*. Blackie, Glasgow, 200–221.

BOLTON, T., ATKINS, A. S. & PROUDLOVE, P. 1988. The sulphur, ash relationship in coal seams and its implications in mine planning in the United Kingdom. *Mining Science and Technology*, **7**, 265–275

BRERETON, N. R. & EVANS, C. J. 1987. *Rock Stress Orientations in the United Kingdom from Borehole Breakouts*. Report of the Regional Geophysics Research Group, British Geological Survey, **RG 87/14**.

BRITISH GEOLOGICAL SURVEY 1985. *Map 1: Pre-Permian Geology of the United Kingdom (South). Map 2: Contours on Top of the pre-Permian Surface of the United Kingdom (South)* 1 : 1 000 000.

——1993. *Stoke-on-Trent. England and Wales Sheet 123. Solid Geology.* 1 : 50 000. British Geological Survey, Nottingham.

CORFIELD, S. M. 1991. *The Upper Palaeozoic to Mesozoic Structural Evolution of the North Staffordshire Coalfield and adjoining areas*. PhD Thesis, University of Keele.

CREEDY, D. P. 1983. Seam gas content database aids firedamp prediction. *The Mining Engineer*, **143**, 79–82.

——1985. *The origin and distribution of firedamp in some British coalfields*. PhD Thesis, University College, Cardiff.

——1986. Methods for the evaluation of seam gas content from measurement on coal samples. *Mining Science and Technology*, **3**, 141–160.

——1988. Geological controls on the formation and distribution of gas in British Coal Measure strata. *International Journal of Coal Geology*, **10**, 1–31.

——1991. An introduction to geological aspects of methane occurrence and control in British deep coal mines. *Quarterly Journal of Engineering Geology*, **24**, 209–220.

—— & PRITCHARD, F. W. 1983. Nitrogen and carbon dioxide occurrence in UK coal seams. *International Journal of Mining Engineering*, **1**, 71–77.

CROFTS, H. J. 1953. Coking coals of North Stafford-shire. *Transactions of the Institute of Mining Engineers*, **112**, 719–732.

CROFTS, R. G. 1990. *Geology of the Trentham District*. British Geological Survey Technical Report, **WA/90/06**.

DECKER, A. D., WHITE, J. & REEVES, S. R. 1991. Coalbed methane strategies successfully applied in the Bowen Basin, Queensland, Australia. *In*: *Proceedings of the 1991 International Coalbed Methane Symposium*. The University of Alabama School of Mines and Energy Development, Paper, **9114**, 317–330.

DURUCAN, S. & EDWARDS, J. S. 1986. The effects of stress and fracturing on permeability of coal. *Mining Science and Technology*, **3**, 205–216.

EVANS, C. J. 1987. *Crustal Stress in the United Kingdom*. Investigations of the Geothermal Potential of the UK. Technical Report, British Geological Survey, Nottingham.

—— & BRERETON, N. R. 1990. In situ crustal stress in the United Kingdom from borehole breakouts. *In*: HURST, A., LOVELL, M. A. & MORTON, A. C. (eds) *Geological Application of Wireline Logs*. Geological Society, London, Special Publication, **48**, 327–338.

GUION, P. D. & FIELDING, C. R. 1988. Westphalian A and B sedimentation in the Pennine Basin, U. K. *In*: BESLY, B. M. & KELLING, G. (eds) *Sedimentation in a Synorogenic Basin Complex: the Upper Carboniferous of Northwest Europe*. Blackie, Glasgow, 153–177.

INSTITUTE OF GEOLOGICAL SCIENCES 1979. *Geological Map of the United Kingdom (South)*, 3rd edn, solid, 1 : 625 000.

KLEIN, R. J. & BARR, M. V. 1986. Regional state of stress in western Europe. *In*: STEPHANSSON, O. (ed.) *Proceedings of International Symposium on Rock Stress and Rock Stress Measurements*. Stockholm, 33–44.

LAW, B. E. 1993. The relationship between coal rank and cleat spacing: implications for the prediction of permeability in coal. *In*: *Proceedings of the 1993 International Coalbed Methane Symposium*. The University of Alabama School of Mines and Energy Development Paper, **9341**, 435–441.

LEEDER, M. R. 1982. Upper Palaeozoic basins of the British Isles—Caledonide inheritance versus Hercynian plate margin processes. *Journal of the Geological Society, London*, **139**, 479–491.

LEWIS, C. L. E., GREEN, P. F., CARTER, A. & HURFORD, A. J. 1992. Elevated K/T palaeotemperatures throughout Northwest England: three kilometres of Tertiary erosion? *Earth and Planetary Science Letters*, **112**, 131–145.

MACRAE, J. C. & LAWSON, W. 1954. The incidence of cleat fractures in some Yorkshire coal seams. *Transactions of the Leeds Geological Association*, **6**, 227–242.

MCKEE, C. R., BUMB, A. C. & KOENIG, R. A. 1987. Stress-dependent permeability and porosity of coal. *In*: FASSETT, J. E. (ed.) *Geology and Coal-Bed Methane Resources of the Northern San Juan Basin, Colorado and New Mexico*. Rocky Mountain Association of Geologists, 143–153.

MILLOTT, J. O'N. 1942. Regional variations in properties of the Eight Foot Banbury or Cockshead seam in the North Staffordshire Coalfield. *Transactions of the Institute of Mining Engineers*, **101**, 2–24.

——, WOLVERSON COPE, F. & BERRY, H. 1946. The seams encountered in a deep boring at Pie Rough, near Keele, North Staffordshire. *Transactions of the Institute of Mining Engineers*, **105**, 528–586.

MOSELEY, F. & AHMED, S. M. 1967. Carboniferous joints in the north of England and their relation to earlier and later structures. *Proceedings of the Yorkshire Geological Society*, **36**, 61–90.

NATIONAL COAL BOARD 1960. *North Staffordshire Coalfield Seam Maps*. National Coal Board, London.

——1974. *The Coalfields of Great Britain—Location of the Main Classes of Coal. 1 : 2 000 000*. National Coal Board, London.

REES, J. G. 1990. *Geology of the Hanley District*. British Geological Survey Technical Report, **WA/90/05**.

SPARKS, D. P., LAMBERT, S. W. & MCLENDON, T. H. 1993. Coalbed Gas well flow performance controls, Cedar Cove area, Warrior Basin, USA. *In*: *Proceedings of the 1993 International Coalbed Methane Symposium*. The University of Alabama School of Mines and Energy Development Paper, **9376**, 529–548.

SPEARS, D. A. & CASWELL, S. A. 1986. Mineral matter in coals: cleat minerals and their origin in some coals from the English Midlands. *International Journal of Coal Geology*, **6**, 107–125.

TING, F. T. C. 1977. Origin and spacing of cleats in coal beds. *Journal of Pressure Vessel Technology*, **99**, 624–626.

TREMAIN, C. M., LAUBACH, S. E. & WHITEHEAD, N. H. III 1991. Coal fracture (cleat) patterns in Upper Cretaceous Fruitland Formation, San Juan Basin, Colorado and New Mexico: implications for coalbed methane exploration and development. *In*: AYERS, W. B. Jr. *et al.* (eds) *Geologic and Hydrologic Controls on the Occurrence and Producibility of Coalbed Methane, Fruitland Formation, San Juan Basin*. Gas Research Institute Topical Report, **GRI-91/0072**, 97–117.

WANDLESS, A. M. 1960. *The Coalfields of Great Britain, Variation in Rank of Coal*. Scientific Department, Coal Survey, National Coal Board, London.

WILSON, A. A. 1991. *Geology of the Silverdale District*. British Geological Survey Technical Report, **WA/90/04**.

# Coal thickness distributions on the UK continental shelf

J. L. KNIGHT[1,4], B. J. SHEVLIN[2], D. C. EDGAR[3] & P. DOLAN[3]

[1] The Barn, Bell Lane, Collingham, Nottinghamshire NG23 7LR, UK
[2] The Surrey Technology Centre, 40 Occam Road, The Surrey Research Park,
Guildford, Surrey GU2 5YH, UK
[3] IKODA Limited, 5 Old Lodge Place, St. Margarets,
Twickenham, Middlesex TW1 1RQ, UK
[4] Present Address: UK Nirex Limited, Curie Avenue, Harwell,
Didcot, Oxfordshire OX1 0RH, UK

**Abstract:** A review of the coals on the UK continental shelf as conventional dry gas source rocks and coalbed methane reservoirs has been undertaken using wireline log data from 1747 oil/gas wells. Coals ranging from Late Devonian to Pleistocene age were recorded, with important coal-bearing sequences in the Lower and Upper Carboniferous, Middle Jurassic and Palaeogene. The coals range from lignite to anthracite rank. Coal thicknesses were interpreted from a combination of gamma, sonic and density wireline logs. However, it is apparent that relatively large source/detector spacings and high logging speeds used in deep oil/gas wells fail to completely resolve thin coals. Frequency distributions of coal thickness for the offshore data set cover about 4720 observations in subsets based on geological age and indicate a high positive skewness with evidence of truncation at thicknesses less than approximately 2 ft (0.6 m). A sample of coal thicknesses based on 1227 observations from 66 cored boreholes in the onshore Westphalian displays similar frequency characteristics, but with the truncation confined to coals less than 1 in (0.03 m) thick. Data truncation seen in the offshore wells is attributed to incomplete wireline log resolution. At the maximum end of the distributions, the data generally display an asymptotic shape for larger thicknesses, normally somewhere in excess of 10 ft (3.0 m). To estimate the unresolved thin coals as well as test the statistical validity of the distributions, comparisons were made of the observed frequency distributions with a number of theoretical distributions of the exponential family, including the generalized Pareto distribution. In addition to this frequential approach, a series of thickness-weighted plots, indicative of coal volume, are introduced. Subsequently, conclusions are drawn regarding the estimation of the total coal resource base from this data. In particular, the effects of the truncation can be quantified and are of the order of 20–30%. These are important adjustments to the estimates of overall coal volumes which are used to assess gas-generating capacity.

As of 31 December 1993, the British Government's published estimate for total UK continental shelf (UKCS) initially recoverable gas reserves discovered to date is 100 trillion cubic feet (TCF) (2835 billion cubic metres; BCM), of which 50% is attributed to the southern North Sea (Department of Trade and Industry 1994). In broad terms, these figures imply an equal split between Westphalian coal-derived dry gas in conventional reservoirs of the southern North Sea and what has generally been regarded as 'wet' gas generated from over-mature oil-prone marine source rocks (mostly Jurassic) in the central and northern North Sea areas.

A study of composite logs from 1747 released wells on the UKCS has demonted that coals, ranging from anthracite to lignites, occur throughout the stratigraphic column of the UKCS, from the Upper Devonian upwards. In certain sequences other than those of Westphalian age, notably the Lower Carboniferous Firth

Coal Formation and the Middle Jurassic Pentland Formation, the burial history and abundance of coal is sufficient to have generated significant amounts of gas that may have migrated from the coals into conventional reservoirs.

Coals can, of course, act as both sources and reservoirs for gases. In their role as a source rock, it is obviously of paramount importance to estimate the most likely volume of coal present that can act as a gas source. Even the very thinnest coal leaves, as long as they represent a significant volume of coal relative to the volume of coal in seams of more 'usual' thickness, are of interest from the standpoint of their 'conventional' gas-generating capacity.

Coalbed methane (CBM) is, by its very nature, an accompaniment to the gas that is liberated from coals to migrate into conventional reservoirs. Hence an understanding of the distribution of coal-sourced gas in conventional reservoirs can also be used in an obverse

*From* Gayer, R. & Harris, I. (eds) 1996, *Coalbed Methane and Coal Geology*,
Geological Society Special Publication No 109, pp 43–57.

sense to predict where CBM may be of interest on the UKCS. When considering coalbeds as reservoirs for CBM, there is a need to identify the thicker seams which will allow economic well completions. In this instance, their actual proportion of the overall volume of coal is of less importance.

This paper considers some of the problems of obtaining reliable estimates of coalbed thicknesses and volumes in coalbeds on the UKCS. It summarizes what we can reliably infer from, as well as the various limitations of, observed thickness distributions (in both frequency and volume terms). It also speculates on possible relationships between the time factors involved in coal accumulation and the gross statistical patterns of the distributions. As a result, various inferences can be drawn as to the suitability of the coal-bearing formations considered as either conventional gas sources and/or possible CBM objectives.

## Stratigraphic and geographical distribution

The stratigraphic and geographical distributions of the main coal-bearing units of the UKCS have been outlined previously by Knight *et al.* (1994). This paper provides a more detailed account of the actual coalbed thickness distributions within selected coal-bearing stratigraphic units. With the sole exception of the Westphalian of the southern North Sea, the lithostratigraphic nomenclature of Cameron (1993*a, b*), Knox & Holloway (1992) and Richards *et al.* (1993) has been used.

The general stratigraphic and geographical distribution of coal-bearing units on the UKCS is summarized in Table 1 and Figs 1–3. Although coals of Carboniferous, Jurassic and Tertiary age dominate volumetrically and economically, it is worth noting that coals have been found at least locally in sequences of Late Devonian, Permo-Triassic and Cretaceous age. Only coal-bearing units with more than 148 coal penetrations in the released well database have been considered here and include the Tertiary Moray and Montrose groups, the Jurassic Pentland and Ness formations and the Carboniferous Westphalian and Firth Coal Formation. It should be noted that the number of coal penetrations in our database

**Table 1.** *Maximum cumulative coal thickness in each of the major coal-bearing sequences on the UK continental shelf*

| Major coal-bearing sequences | Maximum cumulative coal thickness (m) (+ well identifier) | Number of coal penetrations (all wells) |
|---|---|---|
| Palaeogene and Neogene | | |
|   Westray Group | 24.1 (21/29-6) | 276 |
|   Stronsay Group | 94.0 (14/19-12) | 117 |
|   Moray Group | 44.2 (14/19-5) | 788 (655 in Beauly Mb) |
|   Montrose Group | 21.2 (15/19-3) | 181 (178 in Lista Fm) |
| Cretaceous | | |
|   Wealden facies | 10.6 (99/18-1) | 76 |
| Jurassic | | |
|   Piper Formation | 8.2 (15/17-12) | 18 |
|   Fulmar Formation | 11.4 (29/7-3) | 29 |
|   Stroma Mb, Pentland Fm | 22.6 (21/1-5) | 148 |
|   Pentland Formation | 46.1 (16/18-1) | 820 |
|   Hugin Formation | 14.4 (9/18-4) | 49 |
|   Ness Formation | 20.4 (3/9-4) | 583 |
|   Brora Coal Formation | 5.6 (12/21-3) | 34 |
|   West Sole Group | 9.2 (48/6-27) | 34 |
| Permian and Triassic | | |
|   Statfjord Formation | 9.1 (3/15-2) | 22 |
|   Permo-Trias, Western Approaches | 2.3 (73/8-1) | 20 |
| Carboniferous | | |
|   Westphalian | 74.4 (44/28-2) | 1537 |
|   Namurian | 29.6 (41/20-1) | 96 |
|   Yoredale Formation | 5.6 (42/10-1) | 22 |
|   Scremerston Formation | 21.3 (39/7-1) | 62 |
|   Firth Coal Formation | 98.6 (15/19-2) | 524 |

**Fig. 1.** Approximate distribution of main Carboniferous coal-bearing units on the UKCS.

is a complex function of the geographical spread and degree of stratigraphic penetration by oil/gas exploration wells, as well as the number of coal seams present in each coal-bearing unit.

The Westphalian coals extend from the southern North Sea, across the UK onshore and into the numerous partially preserved basins west of Britain (Fig. 1). In fact, there is tentative evidence that some of the thickest Westphalian coals on the UKCS have yet to be explored in such areas as the Eubonia Basin of the East Irish Sea, south of the Isle of Man. These widespread coal deposits are indicative of deltaic and fluvial sedimentation across a very extensive and broadly stable coastal plain.

In contrast, Mesozoic coals (particularly those of the Jurassic) are more restricted, having been deposited within a graben system of limited areal extent (Fig. 2). Somewhat enigmatically, despite the contrasting sedimentary regimes, some of the Jurassic coal-bearing units have very similar coal thickness distributions to those of the Carboniferous, whereas others appear significantly different.

The situation for the main Tertiary lignite-bearing units tends to be similar to those of the Jurassic coal-bearing units, in that they are also relatively restricted in their geographical distribution. However, it is recognized that as further drilling information is released, it may become apparent that they are well represented

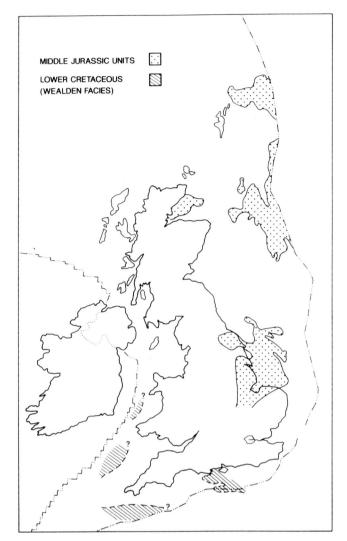

**Fig. 2.** Approximate distribution of main Mesozoic coal- and lignite-bearing units on the UKCS.

north and west of Britain (Fig. 3). There is also the problem that in many older wells lignites may be present, but have gone unrecorded because there was no interest in the shallow section so far above the exploration targets of the day.

## Coal thickness distributions

### General observations

The main reason for investigating coal thickness distributions within particular coal-bearing units on the UKCS is the need to estimate coal volumes (and hence gas-generating capacity) based primarily on wireline log data, in which the resolution of coals, especially thin coals, is less than ideal. This provided the impetus to investigate whether coalbed thicknesses followed any specific statistical distribution sufficiently well for deviations due to measurement or sampling bias to be apparent, with the ultimate aim being to compensate for any such deviations.

Distortions in coalbed thickness frequency distributions can arise from biases in thickness measurement in individual wells and from the spatial distribution of wells, both geographically and in stratigraphic penetration. The various aspects of these are summarized below.

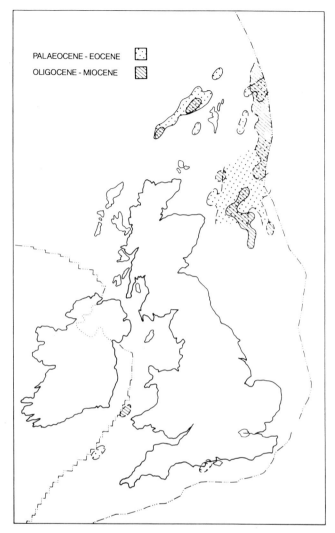

**Fig. 3.** Approximate distribution of main Tertiary lignite bearing units on the UKCS.

*Measurement bias.* The design and operation of the wireline logging tools used in oil and gas exploration wells is such that coal seams less than approximately 2 ft (0.6 m) thick are not fully resolved. This gives rise to three main problems. Firstly, data truncation due to thin coalbeds not being recorded. Coalbeds significantly thinner than the wireline tool resolution will tend not be recorded and this will lead to a complete truncation of very thin coalbeds plus a greatly reduced apparent frequency of occurrence of thin coalbeds just below the resolution limit; see Fig. 4. Secondly, measurement bias of poorly resolved coal near the tool resolution limit. Coalbeds of sufficient thickness to generate an incomplete wireline tool response appear as a relatively broad, low-amplitude response which may be identified as a slightly thicker coalbed (possibly interpreted to be of poor quality). This can result in a systematic positive bias in the thickness distribution close to the tool resolution limit. Finally, positive thickness bias due to unresolved dirt partings or other thin non-coal beds not fully resolved. Its cause is directly analogous to that for the first problem, but the effect is different. This will again result in a positive bias with a single thick coalbed recorded instead of a number of thinner coalbeds. The sampling strategy used in the present exercise was to split coalbeds whenever a clear dirt parting could be recognized.

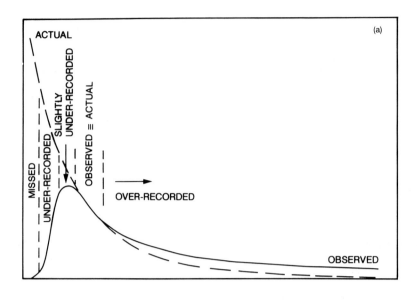

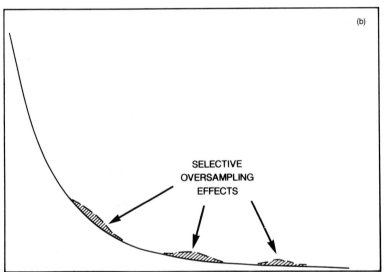

**Fig. 4.** (a) Measurement and (b) sampling distortions expected in coal thickness distributions estimated from offshore wireline log measurements.

It should be stressed that general (i.e. random) measurement inaccuracy is not a source of bias, even though it may be relatively large for thinner coalbeds. However, it is inevitably bound up in bias of the same order due to rounding errors. There may be a tendency when interpreting logs to round coalbed thicknesses to the nearest practical measurement unit, i.e. 1 ft or 0.5 m on metric logs. This is evident as a slight 'granular' effect visible on the cumulative frequency plots, but does not produce a significant distortion or deviation in any of the distributions.

*Bias introduced by poor spatial sampling (geographical or stratigraphic).* The oil/gas wells were almost exclusively drilled to investigate specific potential hydrocarbon accumulations and are therefore often spatially clustered, with the possibility of selective *oversampling* of specific coalbeds. The other problem is evident from the fact that even in the southern North Sea Basin, where it is routinely penetrated, relatively few wells extend more than 100 ft (33 m) into the Westphalian. Therefore the stratigraphic penetration is almost invariably incomplete, and normally only a small percentage of the coal-bearing section is sampled.

It should also be noted that we are considering composite distributions made up from numbers of individual coal seams, each with their own thickness variations, superimposed on the distribution of average thicknesses for the population of coal seams as a whole. In the situation that specific seams are penetrated only by wells at widely seperated locations, the distribution would probably be indistinguishable from that of a truly discrete population, but there may be considerable bias where multiple penetrations are close together.

Because of such possible distortions, even where sample sizes are large it cannot be determined categorically what the type of distribution is.

The next observation to make is that, as expected, a comparison of the offshore and onshore Westphalian data sets (which are samples from different parts of the same total population) confirms the under-recording of thin coalbeds in the offshore set (Fig. 5). Because of this truncation effect, there may be a superficial tendency to consider the offshore distributions to be log-normal, but this can be demonstrated not to be the case. After considering various other distributions belonging to the exponential family, in conjunction with a number of the data sets (not just those of the Westphalian), they are strongly suggestive of an exponential distribution approximating to the data.

*Sample quality and procedures*

Only those coal-bearing units from Table 1 which have the largest numbers of individual coal penetrations have been analysed. Table 2 presents basic information on the quality of the sample for these coal-bearing units, namely the total number of coal penetrations and average number of penetrations per well. Also shown (Fig. 6) is a comparison between the offshore and onshore Westphalian data sets in terms of how many wells or boreholes penetrate particular numbers of coals. The majority of offshore wells penetrating the Westphalian record six or fewer coals and very few wells more than 20 individual coals. In contrast, the onshore boreholes, although there are fewer in our sample, have a more even spread in terms of numbers of coals penetrated in any given hole. In general, the sample quality for other units tends to follow similar patterns to the offshore Westphalian, although some younger formations naturally have a higher proportion of complete penetrations.

The thickness data for each formation were first examined by plotting thickness against cumulative percentage frequency distribution, the latter plotted on a log scale, as this is a good test for the exponential distribution, with a straight line being obtained for distributions that conform well. In general, for most of the formations tested, acceptable approximations to exponential distributions are inferred. However, at small thickness values there is evidence of departure from a linear relationship which may be attributed to under-recording of thin coalbeds due to tool resolution. Some of the plots also tend to show other deflections and/or changes of slope, particularly at larger thickness values; the significance of such features proved difficult to assess from these initial plots.

Following this, and recognizing that the assessment of coal volumes is the primary

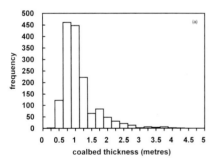

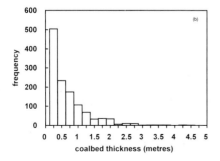

**Fig. 5.** Westphalian: (**a**) offshore and (**b**) onshore. Coal thickness frequency distributions as histograms.

**Table 2.** *Sample size and quality for coal-bearing units analysed*

| Formation | Number of wells recording coals | Total number of coal penetrations | Coal penetrations per well (average) |
|---|---|---|---|
| Westphalian onshore | 66 | 1227 | 18.6 |
| Westphalian offshore* | 145 | 1537 | 10.6 |
| Firth Coal Formation* | 49 | 524 | 10.7 |
| Ness Formation | 144 | 583 | 4.1 |
| Pentland Formation | 125 | 820 | 6.6 |
| Stroma Member (Pentland Formation) | 82 | 148 | 1.8 |
| Lista Formation | 49 | 178 | 3.6 |
| Beauly Member (Moray Group) | 133 | 655 | 4.9 |
| Westray Group (Olig-Neogene) | 63 | 276 | 4.4 |

objective, each distribution has been examined in terms of its *thickness-weighted* form, which is directly proportional to volume for any given area. These thickness-weighted distributions have been derived by summing coal thicknesses within a window about any given thickness value, giving greater weights to those values closest to the central value. In other words, this is a convolution process with a filtering (smoothing) effect which, if correctly designed, should compensate for the effects of measurement and rounding errors in individual observations.

A general error function approximating to a normal distribution was used, which allows both for the basic measurement error in the wireline tool and the possible accuracy of reading from the logs. For all the hard coals (and including the Lista Formation, in which the coals appear to be transitional from lignite into hard coal), where the majority of determinations are from sonic logs, a standard deviation of 0.5 ft (0.15 m) has been assumed, implying that nearly 70% of all measured values would be within this range from the true value, and about 95% would be within ±1 ft (0.3 m). This approach has the added advantage that it is possible to generate a valid semi-continuous representation of the distribution, without the disadvantages of arbitrary conventional histogram size classes.

For the Tertiary lignites, a very much larger proportion of the thickness values are taken from the lithology columns recorded on the composite logs (wireline logs are often not run through these intervals, and when they are, lignite responses may be difficult to distinguish from those of interbedded clastic sediments). As a result, the two main lignite-bearing units have 59% (Beauly Member) and 68% (Westray Group) of thicknesses from lithology rather than wireline logs. This poses some difficulty in that the errors from the lithology determinations will tend to be larger and more variable than the wireline log determinations with which they are mixed. The compromise adopted was to double the standard deviation of the error function used to generate the semi-continuous thickness-weighted distributions. At 1 ft (0.3 m), this implies nearly 70% of observations within ±1 ft and about 95% within ±2 ft (0.6 m).

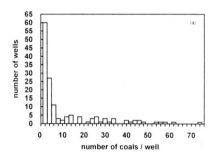

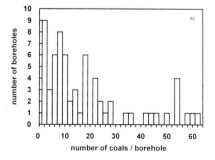

**Fig. 6.** Westphalian: (**a**) offshore and (**b**) onshore. Graphical representations of sample quality.

For comparison purposes, theoretical exponential distributions, also thickness-weighted, were generated and displayed in a comparable manner by cross-multiplying the frequency distribution with the thickness values. When this is done, the resulting distribution is equivalent to a particular form of the gamma (or Erlang) distribution.

In all instances except the onshore Westphalian, no reliable mean was available directly from the data itself because of the under-recording effects illustrated in Fig. 4. For this reason, means were estimated (assuming an exponential distribution) from that part of the frequency distribution curve that was considered to be sufficiently free of any of the biases shown in Fig. 4. The resultant theoretical exponential curve (or, in practice, its thickness-weighted equivalent) was then fitted and scaled to the observed distribution using the same part of the curve. This has proved a particularly useful technique for highlighting where the significant measurement problems and/or sampling bias occur. In particular, it can also give a tolerably good idea of the relative volumes of coal over- or under-estimated in various thickness ranges, and the net amount for the coal-bearing unit as a whole.

## Other observations

Of the various distributions belonging to the exponential family, it was noted that the generalized Pareto distribution can be used to give a generally better fit to the high-end tails (large values of coal thickness) of some of the observed data sets than the 'simple' exponential distribution, which is asymptotic to zero on both the $x$ and $y$ axes. However, although the generalized Pareto is really a more general form of the exponential distribution, it does not adequately predict some small, but nevertheless geologically realizable, thickness values between zero and the modal value. For this reason we have been reluctant to adopt it.

It is not clear why coal thickness distributions should approximate to exponential distributions. However, an analogy can be drawn from the realms of more conventional statistical applications which may have some relevance. This concerns the complementary relationship between the Poisson distribution as applied to randomly occurring discrete events and the observation that the lengths of the intervening time intervals, given a sufficient overall time interval, tend to fit an exponential distribution. The Poisson process is typically invoked for phenomena such as free traffic flow on roads past any given point. The vehicles passing are discrete events, which when classified by frequency within a given time interval conform to a Poisson distribution. The varying durations of the time intervals which separate the events of the Poisson process tend to be exponentially distributed.

Given the relatively long periods of quiescent and stable conditions that are required for coal formation, accumulation and preservation, compared with more 'discrete' (i.e. rapid and equivalent to the passing vehicle of the analogy) geological events liable to interrupt it, such as clastic influx or marine transgression, it may be that coal thickness distribution could, over time, develop in a broadly exponential manner.

The following section examines each of the data sets in turn and particularly in terms of the inferences that can be drawn regarding likely coal volumes.

## Coal thickness distributions by unit

### Westphalian: onshore

A sample of 1227 coalbed penetrations from 66 boreholes and shafts drilled in the Westphalian and topmost Namurian of the Durham, East Midlands, Lancashire, North Wales, Staffordshire and Warwickshire onshore coalfields was collated. The data were taken from the following British Geological Survey Memoirs (sheet numbers in parentheses): Durham and West Hartlepool (27), Barnard Castle (32), Sheffield (100), East Retford (101), Ollerton (113), Bradford and Skipton (69), Leeds (70), Preston (75), Rochdale (76), Liverpool (96), Stockport (98) and Lichfield (154). All the boreholes were fully cored throughout the coal-bearing sequence and coalbed thicknesses tended to be given to the nearest 1 in (0.025 m). Coals as thin as 1 in (0.025 m) were routinely recorded. The geographical and stratigraphic distribution of the coalbeds recorded are considered typical of the onshore Westphalian. The boreholes were deliberately selected to avoid particular spatial clustering as far as possible.

The Westphalian onshore coal boreholes show a distribution that conforms well to a theoretical exponential, both in the frequency and thickness-weighted displays (Fig. 7). The most notable feature is a distinct excess in thickness distribution between 1.5 and 2 m, with a maximum between 1.7 and 1.8 m. This is almost certainly due to a pronounced over-sampling effect due to the disproportionate

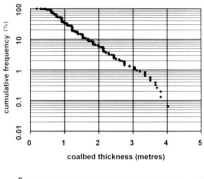

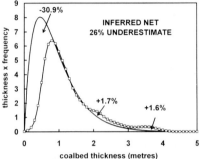

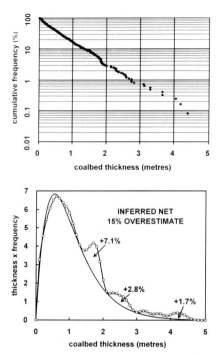

**Fig. 7.** Westphalian onshore. Frequency and thickness-weighted (approximately equivalent to coal volume) distributions.

**Fig. 8.** Westphalian offshore. Frequency and thickness-weighted (approximately equivalent to coal volume) distributions.

number of boreholes that have been targetted specifically to locate mineable coalbeds in that thickness range. It alone accounts for a probable overestimate of coal volume of about 7% and a smaller, similar, oversample peak of about 3% may be indicated between 2.2 and 2.9 m. There may be some similar oversample effects on thicker seams above 3 m, but above 4 m the decline of the fitted exponential distribution is probably departing from the natural distribution, and the overestimate is probably not as great as implied.

### Westphalian: offshore

The estimated theoretical distribution of the offshore coals is closely similar to that of the onshore coals, despite the obvious truncation of the lowest values in the observed data set in Fig. 8 compared with Fig. 7. The fit achieved between the observed and theoretical thickness-weighted distributions implies minor overestimation of thicknesses above 2 m, but most strikingly, an inferred 31% underestimate of coals less than 0.8 m thick. The net underestimate for the whole observed thickness range is inferred to be 26%. This is actually larger than

first-pass estimates of just under 20%, which were based on an approximate scaling between onshore and offshore frequency distributions as histograms. This possibly relates to some ambiguity in the latter method introduced by the 1.7–1.8 m oversample peak in the onshore distribution and the estimate based on the thickness-weighted displays is preferred.

### Firth Coal Formation

The Firth Coal Formation represents a smaller sample from a formation with a more limited areal extent than either of the subsets of the Westphalian. It also has some particular geographical clusters of drilling density, most notably on the southwestern flank of the Witch Ground Graben (Claymore area), and relatively few complete or major penetrations relative to its probable total coal-bearing thickness. Notwithstanding this, its thickness-weighted distribution (Fig. 9) fits that resulting from a theoretical exponential distribution very well between 0.7 and 2.2 m. From 2.2 to 3.9 m there is an overestimate in the observed distribution which may be attributable to some combination of oversampling (indicated peak at 2.4 m),

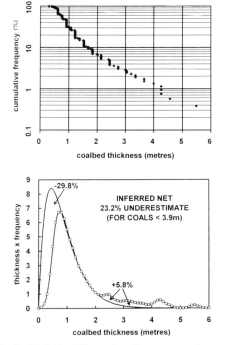

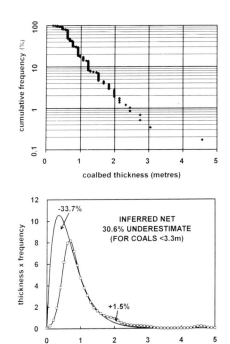

**Fig. 9.** Firth Coal Formation. Frequency and thickness-weighted (approximately equivalent to coal volume) distributions. Note: one additional coal at ≈9 m is not shown on the lower graph.

**Fig. 10.** Ness Formation. Frequency and thickness-weighted (approximately equivalent to coal volume) distributions.

possible overestimate of thicknesses by aggregating separate coals and decline of the theoretical exponential distribution tending towards zero (even when thickness-weighted). The inferred underestimate of the thin coals (below 0.7 m) is 30% and the implied net underestimate is 23%.

One feature of the observed distribution for the Firth Coal Formation is that a handful of coals is recorded above the effective limit of the theoretical distribution. These appear to account for about 6.5% of the observed distribution. In our view, these coals (if really of this thickness and not including unresolved dirt partings) should simply be regarded as an addition, occurring so sporadically as not to fit a distribution in the way the thinner coals can be regarded as doing. They are, however, fairly likely to be subject to oversampling or 'aggregate' measurement errors, and in the latter case should in reality transfer to the main part of the distribution. This extended but sporadic tail beyond the main distribution is a feature of a number of the distributions for other formations and similar comments apply in those cases.

## Ness Formation

The Ness Formation is another example of a well-behaved fit of observed to theoretical distribution (Fig. 10). Only 1% (in one coal!) of the observed distribution lies outside the effective distribution limit. The implied underestimate of thin coals is 34% and the net underestimate is nearly 31%. There may be a minor oversampling effect at about 2 m thickness.

## Pentland Formation

Analysis of the Pentland Formation sample has given probably the least satisfactory result of any of the units with hard coals (see Fig. 11). It has several coals beyond the effective distribution limit of 6.7 m, up to a maximum of 14.6 m, and accounting for 8% of the observed distribution. Several of these coals have closely similar thickness values, however, and it is possible that some spatial oversampling is involved. Many of the wells sampling the Pentland are in close geographical proximity, and the thicker coals tend to be a particularly

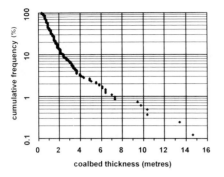

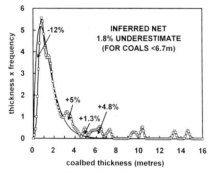

**Fig. 11.** Pentland Formation. Frequency and thickness-weighted (approximately equivalent to coal volume) distributions.

local development (notably, a small number of blocks in Quadrants 9 and 16).

It could also be inferred that there are some factors involved in the deposition of the thicker Pentland coals, such that their thickness distribution, cannot be approximated by the single distribution and that a combination of distributions is involved. The high end tails might then be approximated by a flatter, lower valued, adjunct to the main distribution.

The main part of the distribution is not a particularly good fit to the theoretical curve and most of the thickness range from 3 to 6.7 m appears to be overestimated. Again, oversampling and/or 'aggregate' measurement errors are probably involved. Underestimating of thin coals is implied to be less important than for the Carboniferous formations, at 12%, but the observed distribution is particularly uneven between 0.6 m and 1.8 m and this may have affected the fit adversely. Two prominent peaks in this range could represent oversampling (although this is usually interpreted to be a problem of thicker seams). Given all these factors, the inferred net underestimate of 1.8% is not considered to have much meaning beyond that of being a 'ranging shot'. On the available

data and the current analysis, thickness-weighted distribution (and hence volume) for the Pentland Formation is probably estimated approximately correctly by the observed distribution, but could well be within ±10% or possibly more.

## Stroma Member

With only 148 coalbed penetrations the Stroma Member had the smallest data set considered for analysis; see Table 2. The Stroma Member of the Pentland Formation is anomalous in that it contains a small number, typically six, often thick coalbeds, but very patchily distributed. Close spacing of exploration wells also leads to spatial clustering of sample points. For these reasons no attempt was made to fit a theoretical distribution to the Stroma coalbed penetrations.

## Lista Formation

As the oldest of the Tertiary formations considered, the Lista Formation is generally of higher rank than the lignites of the younger

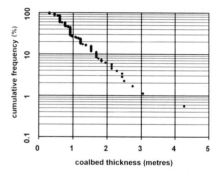

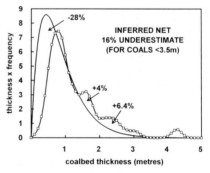

**Fig. 12.** Lista Formation. Frequency and thickness-weighted (approximately equivalent to coal volume) distributions (case 1, mean 0.43 for theoretical distribution).

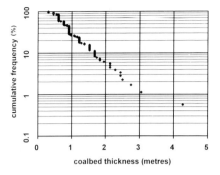

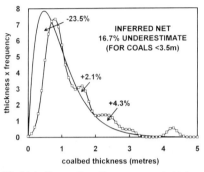

**Fig. 13.** Lista Formation. Frequency and thickness-weighted (approximately equivalent to coal volume) distributions (case 2, mean 0.47 for theoretical distribution).

Beauly Member (Moray Group) or Westray Group. Rank appears to grade from lignite into a sub-bituminous hard coal as a function of present day burial depth. Overall (Figs 12 and 13), the Lista coals appear to be similar to most of the hard coals which have been considered.

The thickness-weighted distribution has an uneven appearance and, at least initially, it appears difficult to derive and fit an equivalent theoretical distribution. However, despite a rather narrow 'well-behaved' portion of the distribution, reasonable results can be obtained. There are significant overestimates for the ranges 1.4–1.9 and 2.0–3.2 m and an underestimate for thin seams (<0.7 m) of either 23.5 or 28% (depending on the fit used).

In this case, to fit the observed data to the theoretical distribution, two variants of the latter were tried, with means calculated from slightly different overlapping portions of the frequency distribution curve. These means differed by almost 10% and the fits to the observed thickness-weighted distribution are clearly different. However, it is particularly noteworthy that there is a difference of only 0.7% in the inferred net underestimate of coal volume.

## Beauly Member

The low end of the Beauly Member distribution of lignite thicknesses (up to 5 m, say) may be a tolerable fit to an exponential distribution (Fig. 14). Beyond 10 m, there is no correspondence, as the theoretical distribution is asymptotic and negligible, and it appears that the approach is not applicable to this type of distribution, where numbers of particularly thick locally developed lignites occur. Probably, a large component of oversampling and 'aggregate' effects are involved, but more fundamentally it may be that the population is simply not extensive enough in geological time and repetition of coal mire conditions, or geographical extent, to be treated as having 'achieved' a distribution, except for the smaller thicknesses.

There is also evidence that some of the thick Beauly Member lignites split basinward, so that the concept of a distribution may be valid in certain areas where the beds are relatively thin and numerous, but not where there are only a few thick individual lignites.

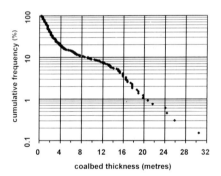

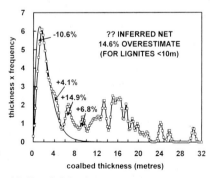

**Fig. 14.** Beauly Member. Frequency and thickness-weighted (approximately equivalent to coal volume) distributions.

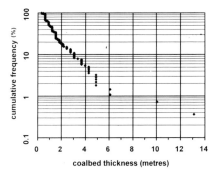

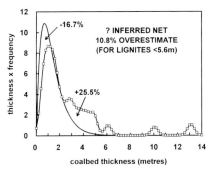

**Fig. 15.** Oligocene–Neogene (Westray Group). Frequency and thickness-weighted (approximately equivalent to coal volume) distributions.

## Westray Group

The Westray Group lignites give a less noisy and superficially 'better' distribution (Fig. 15) than those of the Beauly Member, but the same remarks may well apply. Again, a tolerable fit can be achieved for the thinner part only (up to 2.5 m).

## Conclusions

(1) Given adequate sample sizes in coal-bearing formations which represent reasonably long periods of geological time during which coal formation occurred periodically and a sufficient number of times, coal thickness distributions approximate to theoretical exponential distributions. In general terms, this is predictable according to the statistical analogy in which an exponential distribution describes the varying lengths of quiescent periods observed in a Poisson process.

(2) The main reason for underestimating coal volumes within any given coal-bearing unit offshore is the under-recording of thin coals beyond the resolution of the oil

and gas logging tools used. In general, this leads to a significant overestimate of the mean thickness of the coalbed distribution.

(3) On the other hand, the main reason for overestimating coal volumes is an oversampling effect, where the same seam may be penetrated several times in close geographical proximity, and record very similar thicknesses.

(4) A possible subsidiary cause of overestimating coal volumes may be a thickness measurement effect in which the thickness of coals at the very limit of logging tool resolution is overestimated by a small amount.

(5) From that part of the observed coal thickness distribution for a given unit which is reasonably well approximated by a theoretical exponential distribution, it is possible to estimate the mean of the distribution and by fitting this to the observed distribution to make a first-order prediction of the amount of the over- and underestimating of coal volume involved.

(6) The observed pattern of over- and underestimating particular values of coal thickness is much better understood in the context of thickness-weighted distribution (proportional to coal volume), rather than standard frequency distributions.

(7) The data have been analysed with respect to various distributions of the exponential family, and of the generalized Pareto, which may give a better fit to the high thickness tails.

(8) It may be that the distributions with pronounced high thickness tails are best approximated by the combination of two exponential distributions, with a secondary flatter and generally lower-valued distribution fitting the high end tail.

(9) A more pragmatic view is that most of the high thickness tails cannot and should not be fitted to any theoretical distributions, as they are intrinsically incomplete. Given either more thorough sampling and/or a greater span of geological time, these parts of the distributions would have the opportunity to 'fill in'.

(10) In practice, the extent to which these isolated very thick coals or lignites fit a distribution is of limited importance, as the distributions which they might fit are unlikely to be realized. They are important in volume terms, but their contribution will (usually) be just what it appears

**Table 3.** *Summary of results for coal-bearing units analysed*

| Formation | Observed coals *not* in range (%) | Largest deviation observed (and reason) (%) | Net over-/under-estimate (%) |
|---|---|---|---|
| Westphalian onshore | 0 | +7.1 oversampling | +15 |
| Westphalian offshore* | 0 | −30.9 sub-resolution | −26 |
| Firth Coal Formation* | 6.5 | −29.8 sub-resolution | −23.2 |
| Ness Formation | 1 | −33.7 sub-resolution | −30.6 |
| Pentland Formation | 8.3 | −12.0%(?) sub-rseolution | −1.8 (? ±10) |
| Stroma Member (Pentland Formation) | — | No results possible | — |
| Lista Formation | 2.6 | −23.5 sub-resolution | −16.7 |
| Beauly Member† (Moray Group) | 42.8(??) | +14.9 ?oversampling | +14.6(??) |
| Westray Group† (Olig-Neogene) | 8.6 | +25.5 ?oversampling | +10.8(?) |

* Very few wells penetrate the formation fully and most penetrate only a small percentage.
† Lignites.

to be; fitting to a theoretical distribution is hardly possible and seems unlikely to allow any valid adjustment to volume estimates. They are potentially subject to oversampling error, but on a scale that would require correlation of individual coals to resolve.

(11) The net effect of over- or under-estimating coal volume for each formation analysed is summarized in Table 3.

(12) It is particularly evident from Table 3 that the behaviour and predictability of the lignites is in marked contrast with most of the hard coals. However, it is possible to envisage that 'difficult' lignite distributions could be transformed to 'well-behaved' hard coal distributions purely because of the far greater compaction possible in the lignite/coal transition than for non-coal lithologies.

(13) Consideration of the spatial aspects of these distributions is clearly a next step and some of the geostatistical techniques established onshore in coal reserve estimation (Whateley 1992) may be applicable to parts of this data set; but the poorer resolution and sampling, and tendency to bias, of the offshore data sets must be taken into account.

We are particularly grateful to the Department of Trade and Industry for facilitating access to basic data, to Ashley Francis of IKODA Limited for pointers and comments on some basic statistical and geostatistical tenets, and to Claire Hayes and Eileen Young for assistance in preparation of the manuscript and illustrations.

## References

CAMERON, T. D. J., 1993a. *Lithostratigraphic Nomenclature of the UK North Sea—4. Triassic, Permian and Pre-Permian of the Central and Northern North Sea.* BGS and UKOOA.

——1993b. *Lithostratigraphic Nomenclature of the UK North Sea—5. Carboniferous and Devonian (Southern North Sea).* BGS and UKOOA.

DEPARTMENT OF TRADE AND INDUSTRY 1994. *Development of the Oil and Gas Resources of the United Kingdom (DTI 'Brown Book').* HMSO, London.

KNIGHT, J. L., DOLAN, P. & EDGAR, D. C., 1994. Coals on the United Kingdom Continental Shelf. *Geoscientist,* **4**(5), 13–16.

KNOX, R. W. O'B. & HOLLOWAY, S. 1992. *Lithostratigraphic Nomenclature of the UK North Sea—1. Palaeogene of the Central and Northern North Sea.* BGS and UKOOA.

RICHARDS, P. C., LOFT, G. K., JOHNSON, H., KNOX, R. W. O'B. & RIDING, J. B. 1993. *Lithostratigraphic Nomenclature of the North Sea—3. Jurassic of the Central and Northern North Sea.* BGS and UKOOA.

WHATELEY, M. K. G. 1992. The evaluation of coal borehole data for reserve estimation and mine design. *In*: ANNELS, A. E. (ed.) *Case Histories and Methods in Mineral Resource Evaluation.* Geological Society, London, Special Publication, **63**, 95–106.

# Assessment of West German hardcoal resources and its relation to coalbed methane

## DIERK JUCH

*Geological Survey of North Rhine Westfalia, POB 10 80, D-47710 Krefeld, Germany*

**Abstract:** To assess West German hardcoal deposits, a 2.5-dimensional computer model was developed. Based on a comprehensive geological evaluation it rendered the automatic construction of approximately 240 000 sectional coalbeds. A combination of this database with data describing coal properties, gas contents and other parameters relevant to the recovery of coalbed methane may provide reliable and detailed estimations of the gas in-place and recoverable gas volume. At the present state of knowledge only rather generalized estimates can be made. Thus a total of $454 \times 10^9 \, m^3$ of coal in the West German occurrences may be multiplied by various factors between 1 and $10 \, m^3/t$ to obtain a range of more or less realistic volumes of gas in-place.

As a consequence of the energy crisis in the 1970s, efforts have been made to assess coal resources on an international scale (e.g. Fettweis 1977; IEA 1983). One of the greatest problems to be solved was how to ensure that comparable resource figures were provided, as the limiting criteria of the various previous assessments differ from one deposit to another and change through time (Gregory 1977).

To find a solution to this problem, at least within a national frame, a project was started in 1978 at the Geological Survey of North Rhine Westfalia to assess the hardcoal deposits in West Germany based on a completely new method involving a computerized system. This system includes a mathematical–geometrical model of the coal deposits and a computer-based resource assessment system (Juch & Working Group 1984; 1989; Juch *et al.* 1994). The close connection between its development and its application guaranteed its practicability.

This combined system allows the assessment of each individual sectional coal bed of a deposit and enables the calculation of all classes of resources or reserves with a free choice of the cut-off values of the limiting parameters (e.g. coal thickness, depth). The additional effort required to digitize and process the data for this type of assessment proves to be rather small compared with the initial normal geological evaluation.

After completion of the data assessment in 1988 until very recently, no further development of the computer system has been possible. The system is currently being transformed to a modern computer platform using advanced graphical software (Thomsen, pers. comm.). The advantage of this system compared with other models or resource assessment systems is the simple transformation of stratified geological bodies, which are folded and faulted, into a relatively small database. The system has allowed the calculation of the geometry and volume of approximately 240 000 sectional coal beds. The results of these calculations have subsequently facilitated the evaluation of the West German coal deposits for various purposes, such as the estimation of the volume of coalbed methane (Juch *et al.* 1994).

## Main principles of the assessment system

Conventional assessment systems of coal resources are based either on the calculation of total coal-bearing sequences utilizing percentages of coal content or on individual seam calculations. The first method is mainly suited for a coal-in-place assessment, the second more for recoverable reserves calculations for mines. Further computerized evaluation systems, which rely on the input of point data, are mostly designed for calculations of individual seams. The assessment system discussed here attempts to combine both methods. The objective of the resource assessment system developed at the Geological Survey was to combine these methods such that both resources can be estimated and geological structure maps generated.

(1) Stratigraphic and tectonic subdivision of a coal deposit results in the definition of relatively homogeneous units of individual coal beds. Limiting elements of each sectional coal bed are preferably faults and planes of maximal curvature of the strata such as axial planes of folds.

(2) As the parallel folds in the Variscan foreland basins are mostly chevron-type or box-like (Brix *et al.* 1988), the sectional coal beds can be approximated by planar

*From* Gayer, R. & Harris, I. (eds) 1996, *Coalbed Methane and Coal Geology,*
Geological Society Special Publication No 109, pp 59–65.

tectonic surfaces. Thus a certain number of seams in a stratigraphic sequence which are limited by the same tectonic elements often form a relatively homogeneous structural unit with similar dips displayed by all sectional coal beds. Such a unit, denominated 'block,' is the basic unit of the entire system (see Fig. 1).

(3) The digitization of the top and bottom surfaces of a seam provides all elements which define the geometry of a block (e.g. Thomsen 1984): (a) three or more depth values for the top and the bottom of the seam are the basic data for the calculation of the plane and its dip; (b) the limiting elements of the sectional coal beds are defined by closed polygons with numbered angles of 'border points.' These frontier points must correspond to each other in the top and bottom seams of a block. The straight lines connecting these points describe the lateral boundaries of a block.

(4) The interior of a block, consisting of a number of intermediate sectional coal beds, is constructed and calculated in the following manner: (a) a representative stratigraphic column ('type section') containing the intermediate seams is established for the block; (b) the planes representing the top and bottom surfaces of the seam are calculated by projecting their depth points on a least-squares approximation plane; (c) the corresponding planes of the intermediate seams are assumed to vary proportionally with their distances from the top and bottom seams of a block; and (d) the intersections of the 'frontier points' connecting lines with these planes define the lateral limit of all intermediate sectional coal beds.

A summary of this assessment method , called the KVB model (KVB = Kohlenvorratsberechnung = coal resources calculation) is characterised as follows: the input data are based on geological interpretations; only a small portion of all sectional coal beds (1/20) need to be assessed digitally—the remainder are constructed automatically by the computer program; the respective sizes of the blocks depend

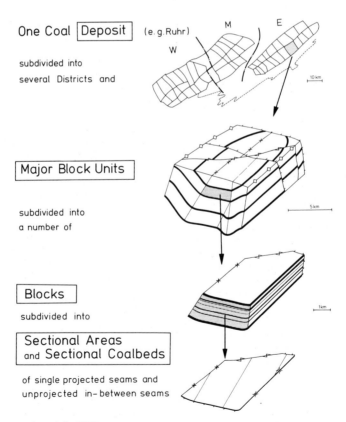

One Coal Deposit (e.g. Ruhr)

subdivided into
several Districts and

Major Block Units

subdivided into
a number of

Blocks

subdivided into

Sectional Areas
and Sectional Coalbeds

of single projected seams and
unprojected in-between seams

**Fig. 1.** Spatial organization of the KVB assessment system.

only on the tectonic structure of the deposit, the degree of geological confidence or other aspects and constraints; and the geometric simplification of the KVB model yields only small estimation errors of the coal volume.

The assessment system of the KVB model has been developed by A. Thomsen in close co-operation with project geologists.

## Basic geological informations

Most information about the various geological parameters to be assessed for the construction of the KVB-model has been gathered from mine descriptions and exploration results. With the exception of the seam property data, the immense quantity of information on the mining and exploration areas required a comprehensive geological evaluation before the digital assessment was undertaken An indication of the amount of data that exists for such an evaluation is exemplified in the paper by Freudenberg *et al.* (this volume).

### Stratigraphy and coal thickness

Stratigraphic sections are mostly found in the descriptions of underground boreholes, mine shafts and exploration holes. Lateral correlation of the seams and their integration into the local or regional stratigraphic system are indispensable requirements for any further evaluation. Depending on the data density, either one representative profile was constructed or selected from the various sections for a block, or the only existing section of a larger area was used for several blocks. The latter was the most common case.

The assessment of the stratigraphic profiles of the KVB model includes the distances between seams and their individual net coal thickness, excluding dirt. To identify the stratigraphic position of each seam, a codification is used based on a lithostratigraphic system. In North Rhine–Westfalia this is a generalized section for the entire coal-bearing sequence of the Upper Carboniferous (Hedemann *et al.* 1972; Freudenberg *et al.* this volume). The codification system of the 'stratigraphic height' (Leonhardt 1970) is derived from this section. It is based on the standardized distance of a respective seam from the base of the coal-bearing sequence (i.e. Namurian C). An additional code describes the split or combination situations of the respective coal beds and leaves.

In the Saar district a similar codification system is based on the stratigraphic position of a seam (Weingardt 1966). In both instances the code numbers have four digits and ascend from the base to the top. These are therefore well suited for computer manipulation.

The minimum net coal thickness of seams to be assessed by means of the KVB model is 20–30 cm in most mining areas. In coal exploration zones thinner seams are often integrated in case they correspond to seams with greater thickness in the southern mining zone. In the Münsterland zone, north of the Ruhr mining area and in the northern Saar area, with only limited or no basic data from boreholes, the coal seams have generally been assumed to be thinner in accordance with the known trends of northwards decreasing coal thickness (Drozdzewski 1993).

### Tectonics

In addition to information from mining and exploration records, tectonic evaluation is based on research into compressive tectonics that affect the Carboniferous coal basins (e.g. Drozdzewski & Wrede 1994), and which is documented by numerous cross-sections. However, to fulfill the requirements of the KVB model, all structural observations have to be transformed into seam projections. In the well documented mining areas detailed mine plans of the lowermost mined seams have been used. The projection scale was generally 1:10 000, whereas in the adjacent areas with only occasional boreholes and seismic sections the projections were commonly performed at a scale of 1:50 000.

In accordance with the scale differences and because of the varying realibility of geological knowledge, the block boundaries may be defined with variable degrees of precision. Mostly two, three or four seams have been projected over large areas. Projections of other seams in smaller areas have been necessary to fill in the irregular shapes of the deposit blocks that result from tectonics and to fill the angular gap between the dipping Carboniferous strata and the overlying unconformity. After digitizing the projected top and bottom sectional coal beds, some of the unprojected intermediate seams have been constructed automatically by the KVB model program and plotted to verify the geological and geometric plausibility of the respective blocks and block boundaries.

### Seam property data

Seam and coal thickness and quality data from mine and borehole records have been assessed as basic point data. With the exception of stratigraphic correlation and codification, the data can be digitized without further geological evaluation or transformation as is necessary with the tectonic and stratigraphic parameters. In some instances the local evaluation of coal thickness data using computer programs helped to improve the composition of the stratigraphic profile of the KVB model.

A comprehensive evaluation of coalification data (Juch 1991) has led to a completely new picture of coalification in the Ruhr Carboniferous, documented by maps of 16 seams comprising a stratigraphic section of approximately 2500 m thickness. Ribbert & Wefels (in press) developed a method for the semi-automatic evaluation of the abundant coalification data of the Ruhr district.

In addition to the established increase of coalification towards older strata and certain lateral variations due to 'syn-orogenic' influences (e.g. Teichmüller, 1987), the coalification data can be interpreted to show (see Fig. 2): (1) a clear correlation between compressive tectonic structures and coalification in the middle and eastern part of the Ruhr district indicating a syn-kinematic component of coalification; (2) the pre-kinematic coalification in the western part shows no correlation with tectonic structures; (3) a slight lateral decrease in coalification towards the northwest or north on both local and regional scales. Two alternative explanations are possible, a cooler crust at the external margin of the Variscan foreland basin or a tilting of the basin towards the southeast at an early stage of compressive tectonism, which moved diachronously towards the northwest; and (4) post-orogenic or post-Variscan coalification of more than local significance as postulated by Lommerzheim (1991; 1994) cannot be supported by the coalification analysis in the Ruhr Carboniferous.

## Gas content data

Gas content data for coal and other physical parameters required for the calculation of coalbed methane volumes can easily be combined with the assessment system of the KVB model by use of spatial coordinates and the stratigraphic codification system (pers. comm. Schütz & Lou).

In the Ruhr district von Treskow (1986) and Freudenberg et al. (this volume) have documented a very irregular distribution both laterally and vertically of gas contents. Its relation to other parameters such as coalification, stratigraphy, depth or distance from the overburden remains uncertain. Thus the gas content

distribution reflects a complex geological history of degasification superimposed on early thermogenic gas generation and followed by biogenic gas generation (Freudenberg et al. this volume). In general, a mean value of gas content can be derived around $5 \, \text{m}^3/\text{t}$.

In the Saar coal basin district Schloenbach (1994) reported a gas content value of $7 \, \text{m}^3/\text{t}$ for the Sulzbach strata. Demuth & Müller (1983) estimated a volume of $600 \times 10^9 \, \text{m}^3$ of gas in all coal beds down to a maximum depth of 1800 m. As this corresponds roughly with the assessed coal volume of $38–50 \times 10^9 \, \text{m}^3$ or (Juch et al. 1994) a gas content of $11.5 \, \text{m}^3/\text{t}$ might be calculated. This value indicates an overestimation compared with the more recent gas content data of Schloenbach (1994). Indeed, the estimation of the gas volume is based on a higher proportion of coal in the Westphalian strata of the Saar basin than in the basic data of the author's assessment. The principal reasons for the difference are that in the coal assessment made by Juch et al. (1994): seams with less than 20 cm of coal were neglected; coal content in dirt was omitted; and a significant decrease of coal thicknesses towards the northwest into the unknown parts of the Saar coal basin was assumed.

In contrast Demuth & Müller (1983) included all coal substance in their basic stratigraphic sections and probably extrapolated it laterally towards the northwest without coal thickness change.

## Conclusions: application and results of the 'KVB' assessment system

The assessment concentrated on well-known hard coal occurrences in the Ruhr district including the adjoining Münsterland, the Ibbenbüren Carboniferous horst, the district of Aachen–Erkelenz and the Saar district. In total, approximately 8000 blocks were assessed and 237 000 individual sectional coal beds were calculated. They contain an estimated total coal volume of $405 \times 10^9 \, \text{m}^3$.

Excluded from the assessement by the KVB model were mostly very thin seams (less than 20–30 cm thick) and marginal occurrences near the base of the coal-bearing sequence and in abandoned and worked out areas. These marginal volumes can be estimated to approximately $60 \times 10^9 \, \text{m}^3$. Thus a total coal volume of $465 \times 10^9 \, \text{m}^3$ was determined for the hard coal occurrences in Northrhine–Westphalia and the Saarland, from which $11 \times 10^9 \, \text{m}^3$ exploited coal must be subtracted (Juch et al. 1994).

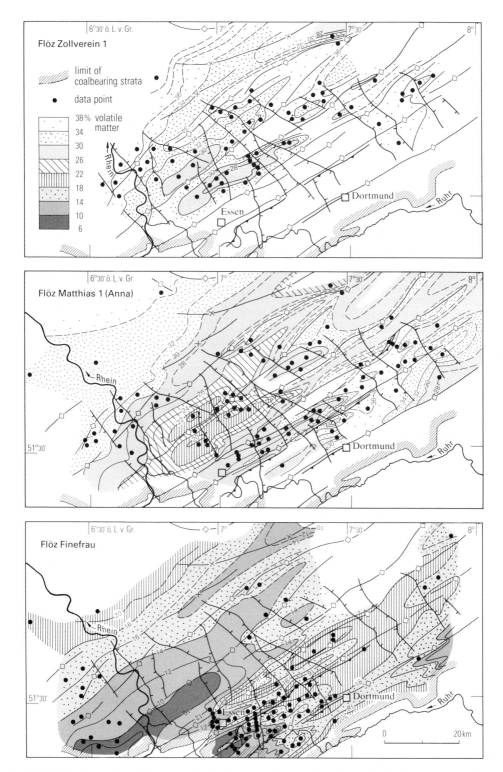

**Fig. 2.** Coalification maps of three seams in the Ruhr coal basin, with iso-lines of volatile matter; vertical distances between seams are 350 and 720 m.

A combination of the sectional coal beds with the coal seam property data has been tested successfully in a few instances. However, due to technical constraints a comprehensive automatic evaluation of all data has not been made to date. The relationship between total seam thicknesses and net pure coal thicknesses was recently evaluated by Daul (1995), who considered the coal resources of the Ruhr district in relation to mining purposes.

A transformation of the coal resources data into volumes of gas in-place or even recoverable gas appears possible in a similar manner by linking the respective individual coal beds and appropriate gas content data.

Owing to technical constraints and the limited availibility of data, a calculation of the coalbed methane potential is currently restricted to a generalized multiplication of a more or less differentiated coal volume with an average value for gas contents. For example, assuming a minimum net coal thickness of seams of 60 cm, a maximum target depth of 1500 m for the enlarged Ruhr area and a mean gas content of $5 \, m^3/t$, a coal volume of $62 \times 10^9 \, m^3$ results, which might contain a gas in-place volume of $400 \times 10^9 \, m^3$. The corresponding volumes for the Saar coal basin are $16.6 \times 10^9 \, m^3$ of coal and $151 \times 10^9 \, m^3$ of gas in-place, based on the above-mentioned gas content of $7 \, m^3/t$. The difference between this figure and the gas volume of $600 \times 10^9 \, m^3$ estimated by Demuth & Mueller (1983) can be explained by different cut-off values of both assessments and different extrapolations of coal seam thicknesses to the unexplored areas northwest of the mining zone of the Saar district.

The gas volume figures serve only as a realistic indication of the magnitude of the coalbed methane potential in the hardcoal deposits, as the geological conditions that control the gas content and its recoverability have not yet been investigated in detail and proved by tests.

With increasing knowledge related to coalbed methane in the west German Carboniferous measures, the presented coal resources assessment system (KVB) may provide a reliable and effective tool for the calculation of the coalbed methane potential of the area.

## References

BRIX, M., DROZDZEWSKI, G., GREILING, R. O., WOLF, R. & WREDE, V. 1988. The N Variscan margin of the Ruhr coal district (Western Germany): structural style of a buried thrust front? *Geologisches Rundschau*, **77**, 115–126.

DAUL, J. 1995. Untersuchung über Verteilung und Veränderung von Steinkohlenvorräten im Ruhrgebiet und über deren Ausnutzung, *Diss. Montanuniv. Leiben*.

DEMUTH, H., & MÜLLER, R. 1983. *Gasgewinnung aus dem Steinkohlengebirge mittels Bohrungen von ber Tag*. BMFT-FB-T, **83-055**.

DROZDZEWSKI, G. 1993. The Ruhr coal basin (Germany). Structural evolution of an autochthonous foreland basin. *International Journal of Coal Geology*, **23**, 231–250.

—— & WREDE, V. 1994. Faltung und Bruchtektonik—Analyse der Tektonik im Subvariscikum. *Fortschritte in der Geologie von Rheinland und Westfalen*, **38**, 7–187.

FETTWEIS, G. 1979. *World Coal Resources—Methods of Assessment and Results*. Developments in Economic Geology, **10**, Elsevier, Amsterdam.

GREGORY, K. 1977. *The Approach of the IEA Coal Research to World Coal Resources and Reserves*. IIASA Proceedings Series, **6**, Pergamon Press, Oxford, 114–124.

HEDEMANN, H.-A., FABIAN, H.-J., FIEBIG, H. & RABITZ, A. 1972. Das Karbon in marin-paralischer Entwicklung. In: *Congrès International de Stratigraphie et Géologie du Carbonifère, 7, Krefeld*, **1**, 29–47.

IEA COAL RESEARCH 1983. *Concise Guide to World Coalfields*. IEA Coal Research, London.

JUCH, D. 1991. Das Inkohlungsbild des Ruhrkarbons—Ergebnisse einer Übersichtsauswertung. *Glückauf-Forsch.-H.*, **52**, 37–47.

——, ROOS, W.-F. & WOLFF, M. 1994. Kohleninhaltserfassung in den westdeutschen Steinkohlenlagerstätten. *Fortschritte in der Geologie von Rheinland und Westfalen*, **38**, 189–307.

—— & WORKING GROUP 1984. New methods of coal resources calculation. In: *Congrès International de Stratigraphie et Géologie du Carbonifère, 10, Madrid*, **4**, 117–124.

—— & WORKING GROUP 1989. Development and application of a new computer based assessment system of the hard coal resources in the Federal Republic of Germany. In: *Congrès International de Stratigraphie et Géologie du Carbonifère, 11, Beijing*, **5**, 320–331.

LEONHARDT, J. 1970. Möglichkeiten der Erfassung, Verschlüsselung und Auswertung geologischer Daten im Steinkohlenbergbau. *Glückauf-Forsch.-H.*, **30**, 309–319.

LOMMERZHEIM, A. 1991. Die geothermische Entwicklung des Münsterländer Beckens und ihre Bedeutung für die Kohlenwasserstoff-Genese in diesem Raum. *DGMK-Berichte*, **468**, 319–372.

——1994. Die Genese und Migration der Erdgase im Münsterländer Becken. *Fortschritte in der Geologie von Rheinland und Westfalen*, **38**, 309–348.

RIBBERT, K.-H. & WEFELS, H. G. EDV-gestützte Auswertung und Darstellung von Inkohlungsdaten aus dem Ruhrkarbon. *Geologisches Jahrbuch*, **A144**, in press.

SCHLOENBACH, M. 1994. Coal bed methane resources in Germany's Saar Basin and current activities.

*In*: *CBM-Extraction, Conference Documentation*, IBC Technical Services Ltd, Royal School of Mines.

TEICHMÜLLER, M. 1987a. Recent advances in coalification studies and their application to geology. *In*: SCOTT, A. C. (ed.) *Coal and Coal-bearing Sequences: Recent Advances*. Geological Society, London, Special Publication, **32**, 127–169.

THOMSEN, A. 1984. Digital representation of geological information and geostatistics in coal resourcescalculation in the F. R. G. *Sciences de la Terre, Sér. Inf.*, **21**, 79–105.

TRESKOW, A. VON 1985. Die Zusammenhänge zwischen dem Gasinhalt und der Geologie im Ruhrevier. *Glückauf*, **121**, 1744–1755.

WEINGARDT, H. W. 1966. Probleme und Methoden der Flözgleichstellung im Saarkarbon. *Zeitschrift der Deutschen Geologischen Gesellschaft*, **117/1**, 136–146.

# Main factors controlling coalbed methane distribution in the Ruhr District, Germany

U. FREUDENBERG[1], S. LOU[2], R. SCHLÜTER[1],
K. SCHÜTZ[2] & K. THOMAS[2]

[1] *Montan-Consulting GmbH, Baumstrasse 31, 47198 Duisburg, Germany*
[2] *Conoco Mineraloel GmbH, Ringstrasse 51, 45219 Essen, Germany*

**Abstract:** The Carboniferous coal deposits in the Ruhr District of Germany have been evaluated for their coalbed methane (CBM) potential utilizing data from more than 600 wells collected during the last 20 years of coal exploration. It can be shown that the distribution of methane (and other gases) and therefore the commercial value of the deposit are controlled by several factors that are a result of a combination of processes related to burial during the Carboniferous, subsequent phases of erosion/exposure and, finally, re-burial. Therefore, the distribution and accumulation are not only a function of coal petrology and geochemistry, but also of basin configuration (e.g. tectonics, hydrology, stratigraphy). Thus the assessment of the CBM potential of the basin requires a multidisciplinary approach.

Coals of Carboniferous age have been mined for centuries in western and eastern Europe. Although the mining activity has frequently been associated with disastrous accidents caused by underground gas explosions, and European Carboniferous coals are known to have provided the hydrocarbon source for a large number of gas fields (e.g. the giant Groningen Field, the fields in the southern UK gas basin), only limited interest has in the past been given to coal bed methane (CBM) as a potential energy resource. However, the success of CBM production in the USA has resulted in a significant interest in European coals as a potential future source of natural gas and several coal basins have been studied extensively, including the Ruhr Basin in Germany (see Fig. 1), to establish the CBM potential within these basins.

Although the mining industry in the Ruhr District has collected a considerable volume of gas content data, little (Teichmüller *et al.* 1970; Hinderfeld *et al.* 1989) has been done to determine regional variations in the gas content and the reason for such variations. The objective of this paper is to characterize the gas distribution and present a model that accounts for some of the vertical and lateral variations in gas content within the Ruhr Basin.

## Geological evaluation

Results obtained from more than 600 core wells from Ruhrkohle AG were used to conduct a study of the gas distribution within an area in the Ruhr District that is currently planned for CBM production by a partnership consisting of Conoco Mineraloel GmbH, Ruhrkohle AG and Ruhrgas AG. Figure 1 highlights the locations of these wells. Although the wells appear to be restricted to a narrow east–west band across the Ruhr District, they provide data covering several structural elements (synclines/anticlines) as the structural trend is approximately SW to NE (see Fig. 2). As a result, the wells can be considered as providing a 'structural two-dimensional coverage'. The analytical data from the coal seams penetrated in these wells were obtained from many sources including Ruhrkohle AG, the Geologisches Landesamt Nordrhein-Westphalia (GLA), Deutsche Montan Technologie (DMT), Ruhranalytik and Conoco Mineraloel GmbH.

### Study area

The area studied is referred to as the Lower Rhenish–Westphalian coal region and is located north of the River Ruhr and between the German–Dutch border in the west and the city of Hamm in the east (see Fig. 1). The folded and thrusted coal-bearing Carboniferous strata crop out in a belt up to 15 km wide along the River Ruhr. With a modest slope, the erosional surface of the Carboniferous is tilted towards the Münsterland Basin in the north and mining currently occurs down to a depth of approximately 1500 m. North of a line from Duisburg to Dortmund this surface is unconformably overlain by unfolded transgressive strata of the Upper Cretaceous, which increase in thickness towards the north.

### Structural setting

The Münsterland Basin and the Ruhr District are the northern extension of the Rhenish–Westphalian massif, which is an approximately

*From* Gayer, R. & Harris, I. (eds) 1996, *Coalbed Methane and Coal Geology*,
Geological Society Special Publication No 109, pp 67–88.

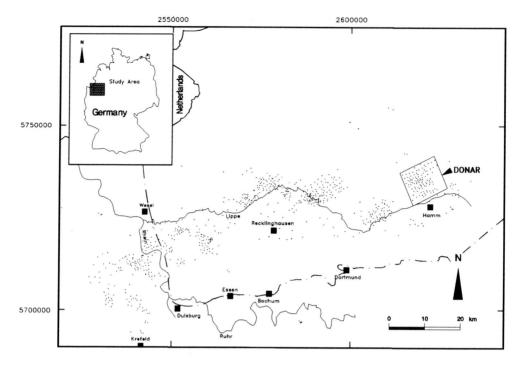

**Fig. 1.** Location map, including coal exploration well locations and Donar mining area.

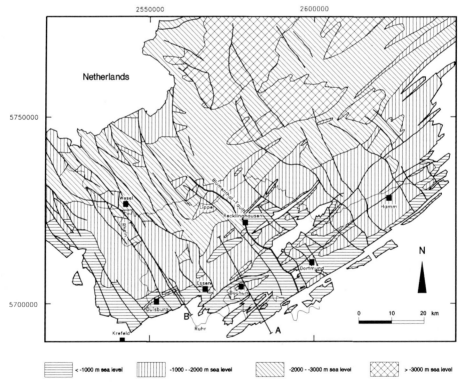

**Fig. 2.** Top structure map of coal seam 'Sonnenschein' (stratigraphic height 1211). The Carboniferous folding trending WSW–ENE and the NW–SE trending cross-faults are easily identified. Based on Juch (1994).

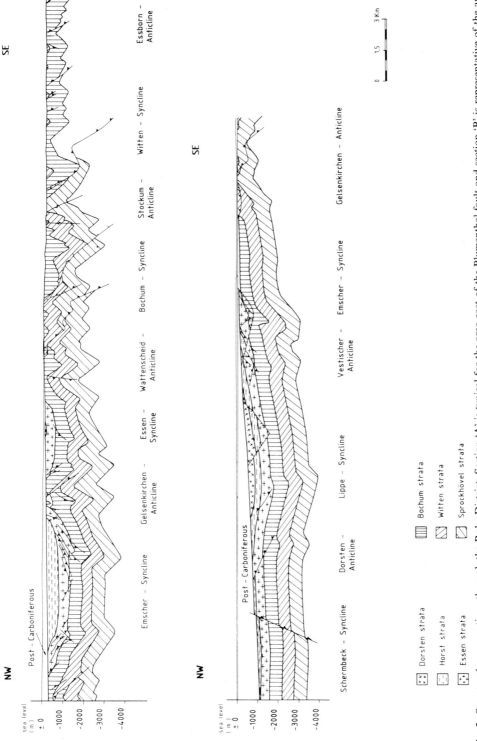

**Fig. 3.** Structural cross-sections through the Ruhr District. Section 'A' is typical for the area east of the Blumenthal fault and section 'B' is representative of the area west of the fault. The section highlights the differences in structural complexity on the western and eastern side of the fault. Location of sections are shown on Fig. 2. (source Geologisches Landesamt Nordrhein–Westfalen 1982).

200 km by 100 km large plate fragment slightly dipping to the north, covered by up to 6000 m thick Devonian, Carboniferous and younger sediments. To the far south, erosion has resulted in exposure of Devonian strata followed northwards by the outcrop of the Carboniferous strata. To the north these strata are covered by Upper Cretaceous sediments in the east and by Permian to Tertiary sediments in the west. The northeastern boundary of the plate is a major fault delineating the Teutoburger Wald. This fault, interpreted to be of strike-slip nature, displaces the Carboniferous more than 1000 m near Ibbenbüren. This fault also marks the northern limit of the Münsterland Basin.

Interpretation of seismic, well data and information obtained from mine and surface exposures indicates that the area was influenced by two major tectonic trends; one trending ENE–WSW and a second trending NW–SE.

Within the Carboniferous section the ENE–WSW direction is associated with broad synclines separated by long narrow anticlinal structures including thrusts. The top structure map (Fig. 2) of the coal seam 'Sonnenschein' (modified after Juch 1994) highlights the internal structural configuration of the Carboniferous. In general, it is widely accepted that the anticlines

and the overthrusts are closely associated and are better expressed at deeper depth levels (Drozdzewski *et al.* 1980).

The synclines and anticlines are further compartmented into larger blocks by the NW–SE trending cross-faults.

*Structural evolution*

After deposition of the Carboniferous sediments a compressional regime resulted in folding and thrusting with an ENE–WSW strike direction (Figs 2 and 3). The main movements occurred during late Carboniferous times (Drozdzewski *et al.* 1980; Drozdzewski 1994) and the resulting structures may partly be related to strike-slip/transfer movements.

The NW–SE trending normal faults appear to be primarily a result of extensional movements although some lateral movements have occurred along these faults. These faults have affected the Carboniferous section slightly later and to a significantly lesser degree than the ENE–WSW trending fold/thrusts. However, the 'Blumenthal fault' (see Figs 2 and 4) appears to be associated with a significant displacement and subdivides the Ruhr area into a western area where the

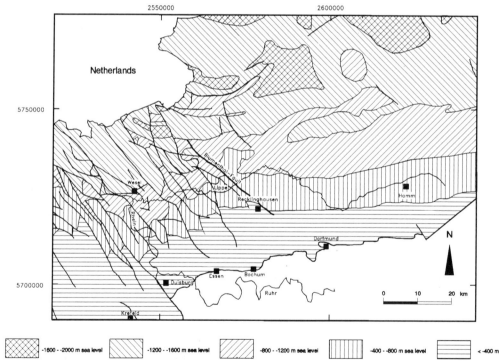

**Fig. 4.** Structure map at level of the top of the Carboniferous. The gently northward sloping surface extends to a depth of approxiametely 2500 m. Largely based on Juch (1994).

Carboniferous section (with the exception of an area in the southeastern corner) is mostly flat-lying (see section 'B' on Fig. 3) and covered by sediments of Zechstein and Triassic age, and an eastern area where the Carboniferous section is structurally very complex and heavily folded (similar to the appearance seen on section 'A' on Fig. 3). Zechstein and Triassic sediments are absent in this area and the Carboniferous is directly overlain by Upper Cretaceous sediments. In addition to the lateral variation in structural style, different vertical structural levels ('stockwerks') exhibiting variation in structural style can also be observed in the Ruhr District (see e.g. Drozdzewski et al. 1980). The lower level (stockwerk) dominates the area to the south, followed in a northward direction by areas dominated by the intermediate (central Ruhr area) and upper levels showing less structural complexity. The difference seen in tectonic style across the Blumenthal Fault may in part be a result of a change in the dominating 'stockwerk' across this fault.

It seems that these NW–SE extensional elements were already active during the late Carboniferous, but were probably rejuvenated during the Jurassic and Cretaceous and, to a lesser degree, in the Tertiary (Drozdzewski 1994). Through the peneplanation starting in Permian times, which levelled most displacements, the evolution of these faults is difficult to reconstruct.

In conclusion, in a simplified description the Münsterland area consists of broad, long, north dipping synclines separated by narrow fold/thrust zones compartmented into larger blocks by NW–SE trending cross-faults. The surface of the Carboniferous unconformity (Fig. 4) appears as a peneplain formed by extensive erosion during the Late Carboniferous, Permian, Triassic, Jurassic and earliest Cretaceous. This erosion has resulted in a significant reduction in the preserved Carboniferous section and today approximately 3500 m can be studied in the Ruhr District/Münsterland Basin (Figs 3 and 5).

## Stratigraphy and depositional setting

### Late Carboniferous

During the Late Carboniferous approximately 5000–6000 m of strata were deposited north of the Rhenish Massif in a molasse basin by a broad delta system. The sediments were derived from a southern source area dominated by clastic material, deposited initially in a marine environment and subsequently in an environment of more lacustrine and fluvial character (Strack & Freudenberg 1984). The marine influence decreased continuously during the Late Carboniferous; sedimentation became more terrestrial, interrupted by marine (or

| System | Series | North American Stratigraphy | Stage | Substage | Zone | Stratigraphic Boundary (Seams) | Stratigraphic Height ('m') | Total Thickness (m) | Range of $R_o$ (Volatile Matter) (%) |
|---|---|---|---|---|---|---|---|---|---|
| Carboniferous | Upper Carboniferous (Silesian) | Pennsylvanian | Westphalian | C | Lembeck strata | Dickenberg | 3475 | > 500 | 0.6 - 1.0 (43 - 34) |
| | | | | | | Odin4 | 3105 | | |
| | | | | | Dorsten strata | Nibelung | 3077 | 330 - 450 | 0.6 - 1.2 (43 - 28) |
| | | | | | | Baldur | 2682 | | |
| | | | | B | Horst strata | Ägir | 2637 | 300 - 375 | 0.7 - 1.3 (42 - 25) |
| | | | | | | M | 2299 | | |
| | | | | | Essen strata | L | 2288 | 450 - 600 | 0.8 - 1.6 (40 - 19) |
| | | | | | | Viktioria 4 | 1824 | | |
| | | | | A | Bochum strata | Katharina | 1781 | 550 - 750 | 0.9 - 2.3 (37 - 9) |
| | | | | | | Schöttelchen 2 | 1139 | | |
| | | | | | Witten strata | Plaßhofsbank | 1105 | 400 - 600 | 1.5 - 2.8 (21 - 7) |
| | | | | | | Fink | 742 | | |
| | | | Namurian | C | Sprockhövel strata | Sarnsbank | 654 | 670 | > 2.3 (< 9) |
| | | | | | | (Grenzsandstein) | 0 | | |

Fig. 5. Generalized stratigraphy for the Lower Rhine–Westfalian coal mining region (modified after Hahne & Schmidt 1982).

sometimes only brackish) ingressions (e.g. Katharina, 'L' and Aegir horizons; see Fig. 5).

A consistent scheme has been developed for correlation within the Ruhr District. This scheme (adopted in this study) utilizes seam reference numbers ('Flözkennzahlen') assigned to the individual coal seams. This reference number or 'stratigraphic height' should be envisaged as a measure of the distance from the seam to the 'Grenzsandstein' (marking the base of the coal-bearing Carboniferous/Namurian 'C') based on a complete Carboniferous (type) section for the area. As some thickness variations exist within individual sequences across the Ruhr District, the distance from the Grenzsandstein to a specific and correlatable (identical stratigraphic height) coal seam varies without affecting the 'stratigraphic height'. As stratigraphic height is measured from the base of the coal-bearing sequence, the number increases as the coal seams become stratigraphically younger (see Fig. 5).

## Post-Carboniferous

There is an erosional gap (or hiatus) above the Carboniferous Westphalian sediments and in the western area, west of the Blumenthal fault, the Carboniferous is covered by Zechstein, Triassic (mainly Buntsandstein), and locally by Jurassic, Cretaceous and Tertiary deposits. The Zechstein and Triassic sediments are preserved furthest to the south within graben (bounded by NW–SE faults) where erosion has been less extensive. East of the Blumenthal fault only Upper Cretaceous sediments are present and they increase in thickness up to 2500 m towards the north (Fig. 4). The southern erosional boundary of the Upper Cretaceous is located south of the line Duisburg–Essen–Bochum–Dortmund–Soest.

## Coal characteristics

From a CBM perspective, the coals serve both as source and reservoir for methane gas. The gas is generated from coals as the temperature increases during the burial of the plants/coals and as the maturity or rank of the coals increases. During this process, the adsorption capacity for methane (and other gases) on the internal coal surface increases and results in a large gas storage capacity. The following sections describe some of the source rock and maturity characteristics of the Carboniferous coals in the Ruhr District.

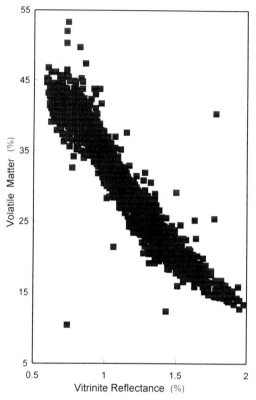

**Fig. 6.** Volatile matter versus vitrinite reflectance for the central Ruhr District. The graph highlights the excellent correlation between the two maturity parameters.

## Coal composition

The major constituents (in weight per cent) of the Carboniferous coals are carbon, volatile matter, ash, minerals, sulphur, phosphorous, and moisture. The carbon content shows a general increase from 45 to 90% as the volatile matter decreases (from approximately 45 to 5%) and as the maturity (e.g. expressed as stratigraphic height) of the coals increases. The concentrations of the remaining compounds are generally low, with a sulphur content of approximately 1% (showing a slight decrease with increasing maturity), phosphorous content generally less than 1% and ash and minerals generally of the order of 2–7% (showing no systematic variation with, e.g. maturity, depth). The ash and mineral matter contents relate to geological factors such as depositional environment and the content of these constituents can only be predicted on a local scale. The limited data that are available related to the moisture

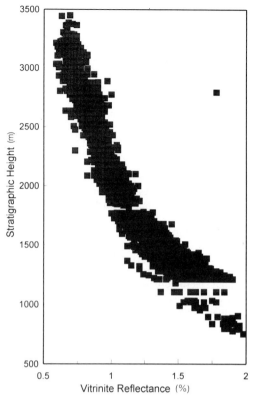

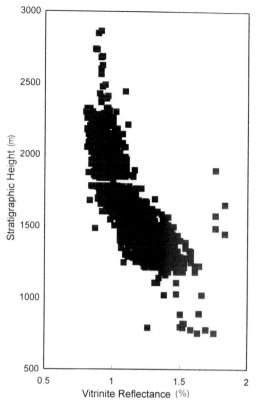

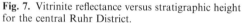

**Fig. 7.** Vitrinite reflectance versus stratigraphic height for the central Ruhr District.

**Fig. 8.** Vitrinite reflectance versus stratigraphic height for the eastern Ruhr District. The graph shows the lower maturity level in the east relative to the central area to the west.

content of the coals suggest values in the range 1–2% (Opahle pers. comm. 1993). These variations in coal composition may influence the gas adsorption capacity and gas generation potential of the coals. Additionally, the composition of the macerals may also significantly influence the adsorption capacity of the coals (see e.g. Faiz *et al.* 1992; Lamberson & Bustin 1993).

## Maturity

The maturity of the Carboniferous coals can primarily be derived from two types of measurements; vitrinite reflectance ($R_o$%) and volatile matter content (or fixed Carbon). These measurements have been made on a large number of samples from all stratigraphic levels and an excellent correlation exists, showing a decrease in volatile matter percentage as the vitrinite reflectance increases (see Fig. 6). Therefore the

volatile matter content may be used as a reliable measurement for maturity.

In addition to the above-mentioned relationship, similar correlations exist between vitrinite reflectance and stratigraphic height (see Figs 7 and 8). These figures also clearly indicate that the present day measured vitrinite reflectance (maturity) was reached during the maximum burial in Carboniferous times before folding/inversion of the Carboniferous section occurred. In the event that any significant later re-burial (after the Carboniferous) to greater depth/temperatures had taken place (after folding/inversion), then vitrinite reflectance values would show: (a) a significantly larger spread and (b) no correlation with the stratigraphic height (approximate age).

The vitrinite reflectance measured on samples from Westphalian 'A' to 'C' indicates values in the range 0.6–2.8% and the same samples show a decrease in volatile matter from 45 to 8% (see Fig. 6). The approximate values of vitrinite reflectance, volatile matter and stratigraphic height for the individual Westphalian chronostratigraphic units are summarized in Fig. 5.

In addition, the graphs of vitrinite reflectance versus stratigraphic height for the central and eastern parts of the Ruhr District (Figs 7 and 8, respectively) clearly indicate a slightly higher maturity level in the central area relative to the adjacent eastern area. This is important to note as it has been postulated, to explain the higher gas content seen in this area, that a post-Carboniferous re-heating of the eastern area has occurred. However, Figs 7 and 8 support our interpretation that none of the areas was heated subsequent to Carboniferous times and that the central area actually was heated to a higher temperature (during the Carboniferous) than the eastern area.

### Maturity/basin modelling

Modelling of the maturity trends versus stratigraphic height (related to past maximum burial) suggests that the base of the Carboniferous section (stratigraphic height = 0) was buried to a depth of approximately 5200 m (and possibly deeper in certain areas) before uplift and erosion and that a temperature of 210°C was reached at this level. This temperature would be equivalent to a geothermal gradient during the Carboniferous of approximately 40°C/km.

The temperature gradient derived from this type of modelling has been used to determine the adsorbed methane at the time of maximum burial (for further detail, see the sections on adsorption capacity/gas saturation).

### Hydrocarbon generation and migration

Since the Carboniferous coals reached their present day maturity state during or at the end of the Carboniferous, before the major phase(s) of folding and uplift occurred (see previous section and Juch 1991), it is possible that large amounts of gas were present within the sediment package (particularly the sandstone) during this time as the shales may have provided an effective vertical seal. This may have resulted in widespread areas of overpressured sediments and the accumulations of gas in traditional hydrocarbon traps.

However, gas generated during the phases of burial and trapped in overpressured sandstones, shales and coals was probably released during the tectonic disturbance, uplift and exposure at the end of the Carboniferous and during later periods of non-deposition and erosion. Therefore, degasification of a significant portion of the Carboniferous intervals is likely to have taken place, possibly to the depth of the groundwater table during these periods of exposure and erosion.

As no subsequent (post-Carboniferous) burial to greater depth (within the study area) than originally reached during the Carboniferous occurred, a second phase of gas generation from the Carboniferous coals never took place. This conclusion contradicts an earlier conclusion reached by Lommerzhein (1991; 1994). It should be emphasized that certain areas in the northernmost part of the area, where more than 2000 m of Cretaceous sediments and Westphalian 'C' are present, may have entered a second phase of gas generation.

Based on the burial and gas generation model outlined here, it may be expected that the coals in the Ruhr District (and other areas with a similar tectonic/basin evolution) may contain less gas than found in coal within basins with a 'simpler' geological history where, for example, overpressuring may significantly increase the gas potential of the coals (e.g. the San Juan Basin in the USA, Scott *et al.* 1994).

Additionally, the gas content in the coals in the Ruhr District may be expected to show significant lateral and vertical variation related to the position during uplift and erosion (i.e. groundwater level), coal maturity and later migration of gas stored in coal below the degasification zone (providing a means for secondary sourcing). However, the net result of these and other processes is to reduce the overall gas content in the basin.

### Reservoir evaluation

The gas distribution and content in coals may not only depend on past geological history (e.g. burial and temperature) and coal composition, but may also be a result of present day physical conditions such as temperature and pressure.

### Reservoir temperature

Temperature data from a number of exploration wells from the eastern part of the Ruhr District were compiled utilizing corrected bottom hole data from wireline logs (Fig. 9). The present day temperature increases rapidly with depth in the post-Carboniferous (Upper Cretaceous) section of the eastern Ruhr District, resulting from the lower conductivity of this section, and the

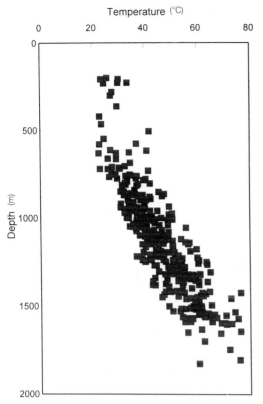

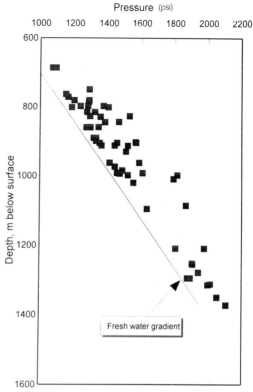

**Fig. 9.** Extrapolated bottom hole temperatures versus depth in Ruhr District coal exploration wells. The data indicate a geothermal gradient of approximately 32°C/ 1000 m.

**Fig. 10.** Estimated Carboniferous formation pressure versus depth. Data calculated from available mud weights in coal exploration wells. A pressure gradient of approximately 0.97 bar/m (0.43 psi/ft) may be derived from the data.

geothermal gradient is therefore high, in excess of 50°C/km. In the Carboniferous sediments, a significantly lower gradient from 20 to 40°C/km is seen due to the higher conductivity. In general, assuming a geothermal gradient of approximately 32°C/km to any given depth in the Carboniferous (in the eastern Ruhr District) may result in a maximum absolute temperature error of approximately 10°C. The correlation of temperature with depth indicates that variation in gas content as a result of subsurface temperature variation should equally well be seen as a variation versus depth.

## Reservoir pressure

Absolute pressure data from the Ruhr District are not available and pressure can therefore only be derived indirectly from mud weight information from coal exploration drilling. The data (Fig. 10) were obtained during drilling operations when a mud loss or a gas kick (small)

occurred during the drilling in the Carboniferous (mud column weight calculated as psi). The data suggest that, with the exception of some high calculated pressures resulting from being significantly overbalanced, the pressure gradient in the Carboniferous is hydrostatic with a gradient of approximately 1 bar/10 m (0.43 psi/ft). Therefore, as in the case of the temperature data, depth can provide a reasonably accurate estimate for pressure.

## Gas content

Gas in coal may be present in the cleat porosity (if not water-filled) and as gas adsorbed on the internal coal surface. Measurement of desorbed gas is usually subdivided into lost gas (gas lost during the recovery and handling of the coal sample/core), desorbable gas that is released by the coal after it has reached the surface and the sample has been sealed in a container, and residual gas that is trapped in the coal and not released during the time the coal is allowed to desorb in the container. The residual gas is determined at the end of the desorption time by crushing the coal sample and measuring the amount of gas released.

To establish the maximum gas that can be retained within a coal sample, adsorption analyses are frequently performed to determine an adsorption isotherm describing the amount of desorbable gas as a function of pressure. The analysis is performed by exposing the coal to increasing methane pressure and measuring the amount of gas adsorbed by the coal at each pressure level.

Measurement of desorbed gas reflects the actual gas content in the coal (assuming the measurements are performed correctly), whereas the adsorption isotherm provides a measure for the maximum amount of gas that may be present in the coal at a given pressure or depth and/or temperature.

In the following sections, gas content refers to the sum of lost, desorbed and residual gas and is reported on a dry, ash-free basis.

### Desorbed gas

Gas content data from more than 500 wells were available from the Ruhr District. Figure 11 shows the data versus present day depth from wells in the eastern Ruhr District and it is evident that no visible correlation exists between gas content and present day depth. This suggests that the vertical gas distribution is not controlled by present day temperature and/or pressure. Additionally, it is noted that the gas content generally is less than $12 \, m^3/t$ and mostly less than $10 \, m^3/t$.

Displaying the data versus depth below the top Carboniferous unconformity (Fig. 12) provides a similar picture to Fig. 11, although it

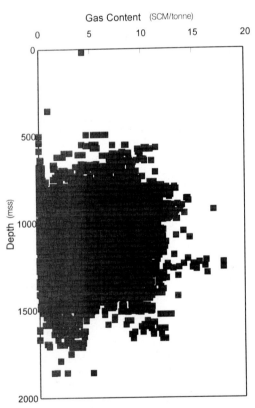

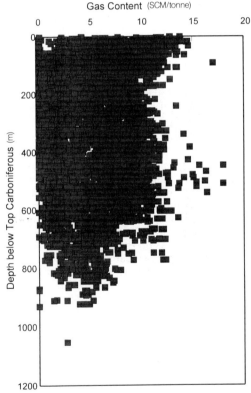

**Fig. 11.** Gas content versus present day depth (below sea level) in coal exploration wells in the eastern Ruhr District. No obvious correlation between gas content and present day depth (and therefore temperature and pressure) appear to exist.

**Fig. 12.** Gas content versus depth below the top of the Carboniferous in coal exploration wells in the eastern Ruhr District. The data seem to indicate a slightly higher gas content in the interval immediately below the unconformity at the top of the Carboniferous.

appears that slightly more data points with higher gas content occur below the unconformity, followed by a cluster of lower gas content in the interval from approximately 100–350 m below the unconformity.

Investigation of gas content as a function of maturity (expressed as stratigraphic height, Fig. 13) reveals that gas content is low (or zero) in low maturity coal with stratigraphic height greater than approximately 2500 m, with some lateral variation attributed to a slight variation in maturity at a certain stratigraphic height across the area. Relating the increase in gas content to vitrinite reflectance suggests that the onset of gas generation from Ruhr District Carboniferous coals generally occurred at an $R_o\%$ of approximately 1.0, equivalent to a volatile matter content in the range 30–38% (see also Jüntgen & Klein 1975 and Figs 6–8).

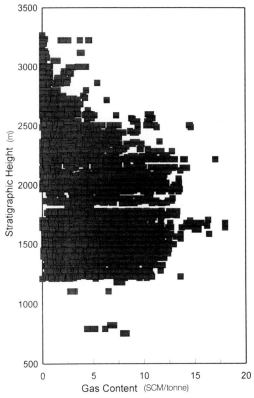

**Fig. 13.** Gas content versus stratigraphic height in coal exploration wells from the eastern Ruhr District. The data suggest an increase in gas content at a stratigraphic height of approximately 2500 (equivalent to a $R_o\%$ 0.9).

Below this maturity level, gas generation is insufficient to provide enough gas to saturate the coals. Although it may be concluded that the maturity of the coals have a major influence on the gas content, it is shown in the following sections that other factors also has a significant impact on the gas content of the coals in the Ruhr District.

To facilitate the investigation of both vertical as well as horizontal variations in gas content, data from individual wells east of the Blumenthal Fault are presented versus depth below the top of the Carboniferous in Fig. 14. An interval of approximately 200 m is recognized with a high gas content, followed by a section with considerably lower gas content that subsequently continues into a trend of increasing gas content with depth. Wells 5 and 6 in Fig. 14 are adjacent wells located within the Donar exploration area. Well 5 is from an area where little tectonic disturbance has occurred and the strata show very little dip. Well 6 is, on the contrary, from a tectonically complex zone (anticline) with beds dipping 45 degrees. The relationship between gas content and structural setting, showing a higher gas content in 'anticlines', is further highlighted in the section across the Donar mining area (Fig. 15, for location of Donar Mining Area, see Fig. 1). A similar pattern as described above for individual wells is also recognized when larger data populations for selected mining areas are considered. This higher gas content is interpreted to be a result of more mature coals (with higher gas content) being uplifted to a similar depth as less mature coals (with lower gas content) in the adjacent synclines.

The gas content data appear to show some correlation with maturity (expressed as stratigraphic height); however, it is also evident that some low maturity coals have rather high gas contents when within the first few hundred metres of the Carboniferous unconformity. Excluding gas measurements within the first 200 m below the unconformity, a trend of increasing gas content with increasing maturity emerges (Fig. 16A). It should also be noted that the same data indicate no relationship with depth (Fig. 16B). On the other hand, gas measurements within the uppermost 200 m also show a trend of decreasing gas content versus depth below the unconformity (see Fig. 17B) independently of the maturity level (stratigraphic height) of the coal (Fig. 17A).

In addition to the vertical variations in the gas distribution, a trend of increasing gas content from the west to the east is also observed on an individual seam basis (see Fig. 18).

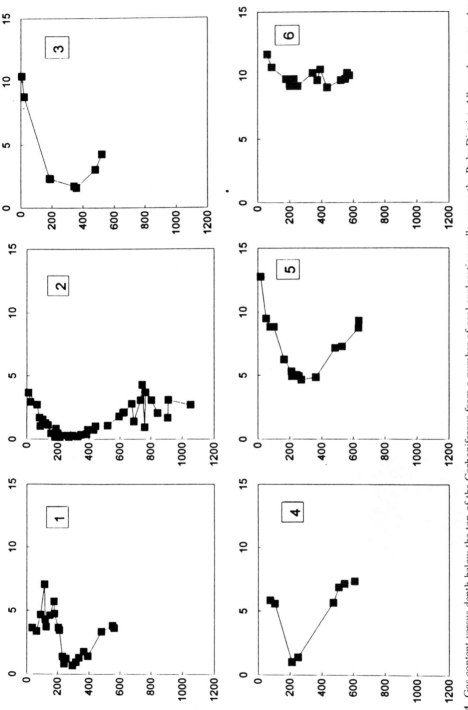

**Fig. 14.** Gas content versus depth below the top of the Carboniferous from a number of coal exploration wells across the Ruhr District. All graphs are standard cubic metres per metric tonne versus depth below Top Carboniferous. The gas content data clearly indicate a higher gas content immediately below the unconformity in the more eastern wells and a general higher gas content in the wells towards the east.

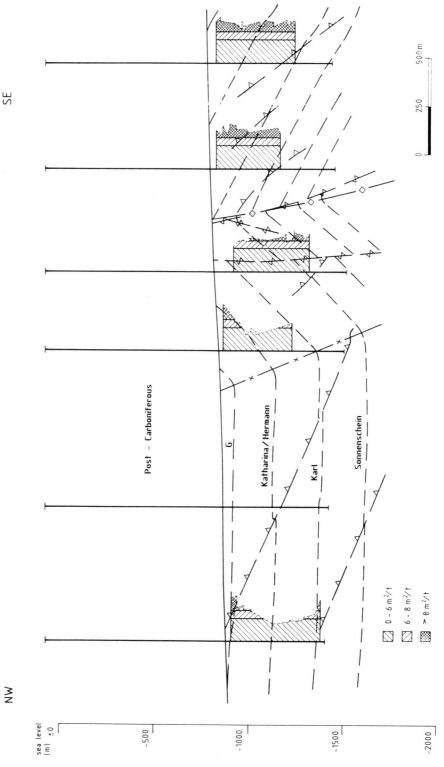

**Fig. 15.** Structural cross-section across the Donar Mining area where higher gas content is observed in the anticlimal strutures.

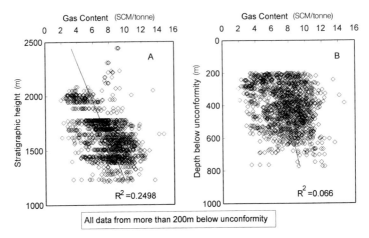

**Fig. 16.** Donar area gas content versus stratigraphic height (A) and depth (B) from in the interval in excess of 200 m below the top of the Carboniferous. An increase in gas content with increasing maturity (stratigraphic height) is observed.

## Adsorption capacity

The adsorption isotherms measured in a laboratory on Carboniferous coals from the Ruhr District indicate an increase in adsorption capacity with increasing maturity level (see Fig. 19 and Konrad 1981) and at reservoir conditions, methane adsorption capacity is estimated to be in the range 10 to 20 $m^3$/t.

The variation in the adsorption capacity (measured at a given temperature and pressure) is believed to be mostly a function of coal maturity. Adsorption capacity has also been documented (see e.g. Faiz *et al.* 1992; Lamberson & Bustin 1993) to vary as a function of maceral composition. However, the Carboniferous coals in the Ruhr area show rather limited variation in maceral composition, with the vitrinite content generally in the range of 60–85%. Moisture content may also affect the adsorption capacity (Joubert *et al.* 1973; Yee & Hanson 1993). The available data (Opahle, pers. comm. 1993) indicate limited variation in the moisture content and therefore the impact on the adsorption capacity is interpreted to be negligible.

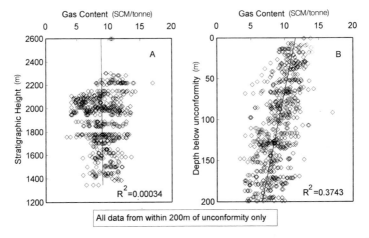

**Fig. 17.** Donar area gas content data versus stratigraphic height (A) and depth below the top of the Carboniferous (B) within the first 200 m below the top of the Carboniferous. The data indicate that the decreasing gas content with depth below the unconformity occur independently of the stratigraphic height (coal maturity).

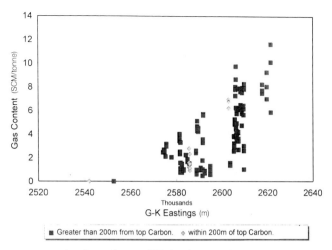

**Fig. 18.** Gas content versus easting for coal seam 1278 (coal seam 'Dickebank') highlighting the increase in gas content from west to east.

## Gas saturation

Measured gas contents in coal samples and the available methane adsorption isotherms indicate that the coals in the Ruhr District at present day depth (pressure and temperature) are undersaturated relative to the adsorption capacity.

Several factors may provide an explanation for the undersaturated nature of these coals, including degasification of the coals during periods of uplift/erosion, limited gas generation related to a low maturity level, reduced storage capacity in the past at the time the coals were at maximum burial depth (temperature) and remigration due to pressure differences, or a combination of the above.

In particular, the reduction in storage capacity with increase in temperature is highlighted in Fig. 20. The calculated isotherms indicate that during maximum burial to depths of more than 3000 m and temperatures in excess of 120°C, the adsorption isotherm was significantly lower (lower curve on figures) than present day laboratory measurements (at approximately 45°C). The figure also provides an indication of the shape of the isotherm (middle curve)

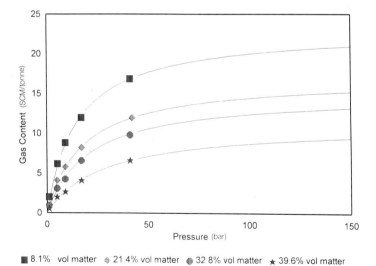

**Fig. 19.** Gas adsorption capacity versus pressure for a range of coal ranks (percentage volatile matter). The adsorption capacity increases with increasing coal rank (and pressure).

during the subsequent Cretaceous burial (in the eastern Ruhr District). The coal seams appear to have had a storage capacity of 8–10 m$^3$/t during the periods of maximum burial. Consequently, as no later generation within the area covered by coal mining exploration data has occurred, and overpressuring during late Carboniferous time is unlikely to have existed (due to the tectonic activity), the coals are unlikely to contain thermogenic gas in excess of approximately 10 m$^3$/t.

*Gas composition*

Gas compositional data have traditionally not been obtained by the mining companies and the following observations are based on data collected in a recent well in the eastern part of the Ruhr District. The data from the well indicate that the gas in the Carboniferous coals is dominated by methane with minor amounts of associated $CO_2$, nitrogen and hydrogen (see Fig. 21). Additionally, traces of CO and heavier hydrocarbons are also present. Generally, compositional variations indicate an increase in ethane, propane (and other heavy components) and $CO_2$ with depth. Gas dryness decreases with depth and the nitrogen and methane content may also be interpreted to decrease with depth. Additionally, a restricted interval immediately below the unconformity (from 750 to 800 m) is characterized by high methane, ethane and hydrogen contents similar to gas samples from significantly greater depth. In this interval, gas

dryness and nitrogen content are characteristically lower than in samples from slightly greater depth. In the following section from 800 to 1000 m, gas dryness and nitrogen content increase sharply and simultaneously the hydrogen, total gas, methane, ethane, propane (and possibly $CO_2$) decrease.

With the exception of the samples immediately below the Carboniferous unconformity, the above-mentioned trends in compositional data indicate that the gas to a depth of approximately 1100 m (e.g. where the gas dryness decreases) is of biogenic origin, possibly mixed with an increasingly thermogenic gas component with depth. With the exception of the topmost gas samples, the biogenic process seems to involve reduction of $CO_2$ by hydrogen. This process is a common source for methane (see e.g. Rice 1993; Whiticar 1992; Wiese & Kvenvolden 1993) and may partly account for the higher gas content below the top Carboniferous unconformity. However, if the higher ethane content measured in the samples directly below the unconformity was originally characteristic of the subsequent interval down to approximately 1100 m, then biogenic gas may also have originated from the destruction of heavier hydrocarbon components.

*Carbon isotopes*

Carbon isotopic composition of desorbed gas from the Ruhr coals indicate an increase in the $\delta^{13}C$ with depth below the Carboniferous unconformity (see Fig. 22, based on Teichmuller *et al.*

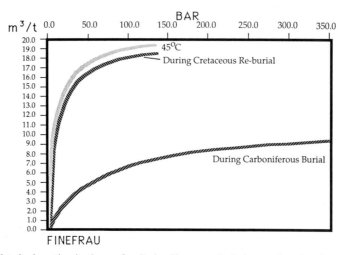

**Fig. 20.** Calculated adsorption isotherms for Carboniferous coals. It is seen that the adsorption capacity was significantly reduced during maximum burial at the end of the Carboniferous. The adsorption capacity during the subsequent Cretaceous re-burial seems to be only slightly lower that data measured in the laboratory at 45°C.

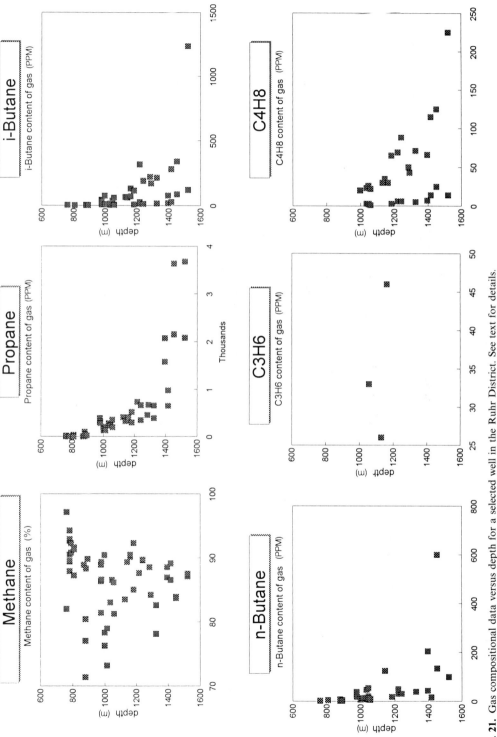

**Fig. 21.** Gas compositional data versus depth for a selected well in the Ruhr District. See text for details.

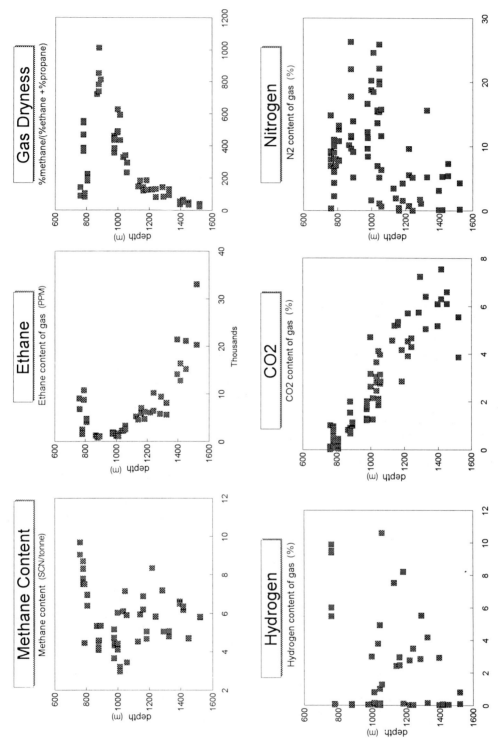

**Fig. 21.** Continued.

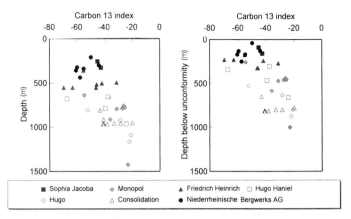

**Fig. 22.** $\delta^{13}C$ ratio versus present day depth (A) and depth below the unconformity (B) from a number mine locations in the Ruhr District. The $\delta^{13}C$ values indicate a biogenic (or partly) origin of the gases to a depth of approximately 650 m below the top of the Carboniferous. Data from Colombo *et al.* (1970) and Teichmüller *et al.* (1970).

1970; Colombo *et al.* 1970). Work by Faber (1987) indicates that gases with $\delta^{13}C$ ratio heavier than −30‰ (increasing with maturity) are likely to be of thermogenic origin. The isotopic composition of the methane in the Ruhr District suggests that the gases in an interval of approximately 600–700 m below the top of the Carboniferous unconformity may be of biogenic origin or, alternatively, a mixture of biogenically and thermogenically derived gas. At depths in excess of 700 m below the unconformity, the isotopic composition suggests a thermogenic origin of the methane. This interpretation is further supported by the increase in $\delta^{13}C$ values with depth below the unconformity in coals of similar maturity level.

*Basin aquifers/hydrology*

An active aquifer system may significantly influence the distribution and composition of coalbed gas. Carbon dioxide may be dissolved in water and thereby removed from the coals. Simultaneously, meteoric water may carry microbes that may subsequently interact with heavier hydrocarbon components in the coals or the adsorbed gas phase and produce methane and other gases. In the Ruhr District, only limited water is produced during the mining of the coals. However, mining is intentionally avoided within the uppermost approximately 70 m of the top of the Carboniferous to prevent water (Cretaceous?) entering the mines. Although the generally limited water influx in the mines may indicate a restricted aquifer system in the Carboniferous, an active system

may exist at the base of the post-Carboniferous section or within the uppermost Carboniferous. Michel & Rüller (1964) showed that formation water salinity increases with depth. Additionally, to a depth of approximately 600 m below sea level, near the Carboniferous outcrop at the southern edge of the basin, salinity is also correlated with the stratigraphic level, with the older strata containing more saline formation water than younger strata at similar depth. At depths in excess of 600 m, data (Puchel, 1964; Michel & Rüller 1964; Michel *et al.* 1974) indicate that there is no correlation with stratigraphy, although the increase in salinity continues to greater depth. This may indicate that complete mixing of saline Carboniferous formation water with fresh surface water occurs close to the Carboniferous outcrop and that mixing at greater depth (at further distance from the outcrop) is more restricted.

**Discussion and conclusions**

The data presented in the preceding sections indicate a significant variation in gas content vertically and laterally within the Ruhr District/ Münsterland Basin. Simultaneously, most geological parameters also vary significantly and no one single parameter appears to correlate directly with the variation in gas content.

However, several variables can be shown to have little or no influence on the gas distribution. The present day physical conditions (temperature and pressure) show little lateral variation and a systematic and predictable variation with depth (Fig. 9 and 10). The gas content shows no indication of a correlation with present day depth and may increase or decrease at a certain depth although other variables (e.g. coal maturity) stay constant. The absence of a correlation with both temperature and pressure indicates that the coals for the most part are undersaturated, i.e. contain less gas than predicted from the available adsorption isotherm data. Therefore the coals may to some degree have been degasified during periods of uplift/erosion and/or the gas content reflects the storage capacity during the period of maximum burial/temperature.

Elements such as phosphorous and sulphur can also be shown to have little influence on the gas content and the relative invariability of the maceral composition cannot explain the variation in the gas content across the area.

The comparison of the maturity information and the gas content distribution may, on a regional scale, be interpreted to indicate that the gas content varies independently of the maturity level (stratigraphic height) of the coals. However, east of the Blumenthal fault and below approximately 200 m below the top of the Carboniferous, the gas content increases with increase in maturity (Fig. 16A). West of the Blumenthal fault, a similar trend is also seen including the interval immediately below the unconformity. Interestingly, the increase with increasing maturity may in certain areas (e.g. Donar) reach the storage capacity equivalent to the physical conditions during maximum burial (Fig. 20). It should also be noted that although the gas content data from individual regions indicate an increase with depth (excluding the interval immediately below the top of the Carboniferous), significant variations exist in the absolute values between regions, i.e. at a specific maturity level of the gas content may vary from near 0 to $15 \text{ m}^3/\text{t}$. However, the gas content generally increases in an eastward direction in the area east of the Blumenthal fault (Fig. 18). In summary, the gas content appears on a local scale to be largely a function of coal maturity; however, on a regional scale, the gas content is controlled by other factors.

East of the Blumenthal fault, the uppermost approximately 200 m of the Carboniferous show a decrease in gas content as a function of depth below the unconformity independently of coal maturity (Fig. 17B). Additionally, the general level of the gas content within this interval increases from west to east. This interval is also characterized by a lighter carbon isotopic composition, a depletion in heavier hydrocarbons, hydrogen and possibly $CO_2$ and a higher nitrogen content (Figs 21 and 22). Regional formation water chemistry indicates that a gradual mixing of Carboniferous formation water and fresh surface water with depth occurs. Currently, we see two possible explanations for this observation. Microbes in surface water entering the Carboniferous along the unconformity may result in reduction of $CO_2$ by hydrogen producing dry methane gas. Simultaneously, heavier hydrocarbons may also have provided a source for biogenic methane. The vertical distribution of the carbon isotopes showing a very light isotopic composition towards the top of the Carboniferous (characteristic of biogenic methane) and becoming increasingly heavier with depth, supports the interpretation of a biogenic source for the gas in the first few hundred metres below the unconformity. Alternatively, the higher gas content may also be a result of a second generation stage within an area to the north where younger Carboniferous coals (Westphalian 'C/D') have recently entered the gas generation stage, with the subsequent expulsion and migration of generated methane along the Carboniferous unconformity in a southerly direction. The Cretaceous sediments are in this instance interpreted to provide an effective seal for the upward migration of hydrocarbons. The light isotopic composition may be a result of the maturity level in the northern source area and fractionation during subsequent migration. The variation in gas composition may be a result of the long distance migration with methane being more mobile than some of the heavier hydrocarbon components. The patterns in total gas content, gas composition, carbon isotopic composition and formation water chemistry may also be a result of a combination of the above two processes.

The absence of a 'secondary' gas peak below the unconformity in the area west of the Blumenthal fault may be attributed to either a different aquifer system active above the Carboniferous unconformity in, e.g. the Triassic (and therefore never carrying microbes into contact with Carboniferous coals), or to gas migrating from a northern source area taking a similar path through Triassic sandstones and therefore never charging the Carboniferous coals with methane gas.

The alternative models outlined here or a combination of these processes may equally well explain the vertical distribution in gas content within certain areas. However, in the event that long distance migration from a northern source area has occurred, then the level of lateral difference in the absolute gas content is difficult to explain in the event that the migration route was along the unconformity. Alternatively, if these lateral variations are a result of surface water flowing into the basin, then local pressure differences, permeability variations, etc. may all have influenced the flow of water and therefore the microbial activity.

Finally, the area west of the Blumenthal fault is generally structurally less complex and has been less affected by compressional forces with the exception of the southeastern area (see section 'A', Fig. 3). In this area, gas appears in certain areas to increase abruptly within a short stratigraphic interval. This may be related to downwards degasification to a level where an effective seal has resulted in gas being trapped at greater depths.

The above provides an adequate model for the distribution of coal gas within specific and limited areas in the Ruhr Basin. However, the variations seen between areas are currently difficult to explain and in the event that these differences relate to the past basin configuration (i.e. aquifer system), it may be difficult to determine the main geological factors that control this variation due to the missing geological record from the Carboniferous to early Cretaceous.

The authors thank J. Cole and D. Juch for critical reviews of the manuscript. We are also thankful to Ruhrkohle AG, Ruhrgas AG and Conoco Mineraloel GmbH for granting permission to publish this paper.

# References

COLOMBO, U., GAZZARINI, F., GONFIANTINI, R., KNEUPER, G., TEICHMÜLLER, M. & TEICH-MÜLLER, R. 1970. Carbon isotope study on methane from German coal deposits. *In*: HOBSON, G. D. & SPEERS, G. C. (eds) *Advances in Organic Geochemistry*. Oxford, Pergamon Press, 1–26.

DRODZEWSKI, G. 1994. The Ruhr coal basin (Germany): structural evolution of an autochthonous foreland basin. *International Journal of Coal Geology*, **23**, 231–250.

——, BORNEMANN, O., KUNZ, E. & WREDE, V. 1980. *Beiträge zur Tiefentektonik des Ruhrkarbons*. GLA, Krefeld.

FABER, E. VON 1987. Zur Isotopengeochemie gasformiger Kohlenwasserstoffe. *Erdol, Edgas, Kohle*, **103**, 210–218.

FAIZ, M. M., AZIZ, N. I., HUTTON, A. C. & JONES, B. G. 1992. Porosity and gas sorption capacity of some eastern Australian coals in relation to coal rank and composition. *In*: *Proceedings, Symposium on Coalbed Methane and Development in Australia, November 1992 Townsville, Australia*, **4**, 9–20.

GEOLOGISCHES LANDESAMT NORDRHEIN–WEST-FALEN 1982. *Geologische Karte des Ruhrkarbons, 1 : 100 000, Krefeld.*

HAHNE, C. & SCHMIDT, R. 1982. *Die Geologie des Niederrheinisch–Westfälischen Steinkohlegebiets*. Verlag Julius Springer, Berlin.

HINDERFELD, G., KUNZ, E., OPHALE, M. & STENGEL, H. 1989. Die Gasführung des Ruhrkarbons in ihrer räumlichen Verteilung und Entstehung zur Prognose der Ausgasung. *Berichte aus Forschung und Entwicklung*, **14**, 1–85.

JOUBERT, J. L., GREIN, C. T. & BIENSTOCK, D. 1973. Effect of moisture on methane capacity of American coals. *Fuel*, **53**, 186–191.

JUCH, D. 1991. Das inkohlingsbild des Ruhrkarbons—Ergebnisse einer Ubersichtsauswertung. *Gluckauf Forschungshefte*, **52**, 37–47.

——1994. Kohleninhalterfassung in den westdeutschen steinkohlenlagerstatten. *Fortschritte in der Geologie von Rheinland und Westfalen*, **38**, 189–307

JUNTGEN, H. & KLEIN, J. 1975. Entstehung von Erdgas aus Kohligen Sedimenten. *Erdol und Kohle, Erdgas, Petrochemie*, **28**, 65–73.

KONRAD, P. VON 1981. Weiterentwicklung von Verfahren zur Prognose und Verhütung von Gasausbrüchen. *Glückauf*, **117**, 5–9.

LAMBERSON, M. N. & BUSTIN, R. M. 1993. Coalbed methane characteristics of Gates Formation coals, northeastern British Columbia: effect of maceral composition. *American Association of Petroleum Geologists*, **77**, 2062–2076.

LOMMERZHEIM, A. 1991. Die geothermische entwicklung des Munsterlander Beckens (NW-Deutschland) und ihre bedeutung fur die Kohlenwasserstoffgenese in diesem raum. *Deutsche Wissenschaftliche Gesellschaft für Erdol, Erdgas und Kohle e. V.*, **468**, 319–372.

——1994. Die genese und migration der erdgase in Munsterlander Bekken. *Fortschritte in der Geologie von Rheinland und Westfalen*, **38**, 309–348.

MICHEL. G. & RÜLLER, K. H. 1964. Hydrochemische Untersuchungen des Grubenwassers der Zechen der Hüttenwerk Oberhausen AG. *Bergbau-Archiv, Jahrgang*, **25**, 21–27.

——, RABITZ, A. & WERNER, H. 1974. Betrachtungen über die Tiefenwasser im Ruhrgebiet. *Fortschritte in der Geologie von Rheinland uund Westfalen*, **20**, 215–236.

PUCHEL, H. 1964. Zur Geochemie des Grubenwassers im Ruhrgebiet. *Zeitschrift der Deutschen Geologischen Gesellschaft*, **116**, 167–203.

RICE, D. D. 1993. Biogenic gas: controls, habitats, and resource potential. *In*: HOWELL, D. G. (ed.) *The Future of Energy Gases*. US Geological Survey Professional Paper, **1570**, 583–606.

SCOTT, A. R., KAISER, W. R. & AYERS, W. B. 1994. Thermogenic and secondary biogenic gases, San Juan Basin, Colorado and New Mexico—Implications for coalbed gas producibility. *American Association of Petroleum Geologists*, **78**, 1186–1209.

STRACK, Ä. & FREUDENBERG, U. 1984. Schichtenmächtigkeiten und Kohleinhalte im Westfal des Niederrheinisch–Westfälischen Steinkohlenreviers. *Fortschritte in der Geologie von Rheinland und Westfalen*, **32**, 243–256.

TEICHMÜLLER, R., TEICHMÜLLER, M., COLOMBO, U., GAZZARRINI, F., GONFIANTINI, R. & KNEUPER, G. 1970. Das Kohlenstoff-Isotopen-Verhältnis im Methan von Grubengas und Flözgas und seine Abhängigkeit von den Geologischen Verhaltnissen. *Geologische Mitteilungen*, **9**, 181–206.

WIESE, K. & KVENVOLDEN, K. A. 1993. Introduction to microbial and thermal methane. *In*: HOWELL, D. G. (ed.) *The Future of Energy Gases*. US Geological Survey Professional Paper, **1570**, 13–20.

WHITICAR, M. J. 1992. Stable isotope geochemistry of coals, humic kerogens, and related natural gases. *In*: *Proceedings, The Canadian Coal and Coalbed Methane Geoscience Forum, February 1992, Parkville, Canada*, 149–172.

YEE, J. P. S. & HANSON, W. B. 1993. Gas sorption on coal and measurement of gas content. *In*: LAW, B. E. & RICE, D. D. (eds) *Hydrocarbon from Coals*. American Association of Petroleum Geologists, Studies in Geology **38**, 203–218.

# Opportunities for the development and utilization of coalbed methane in three coal basins in Russia and Ukraine

JAMES S. MARSHALL, RAYMOND C. PILCHER & CAROL J. BIBLER

*Raven Ridge Resources, Inc., 584 25 Road, Grand Junction, CO 81505, USA*

**Abstract:** The coal mining regions of Russia and Ukraine liberate almost $4.5 \times 10^9$ m$^3$ of methane annually, of which only 3.9% is being utilized. Preliminary estimates suggest that the methane resources contained in coal mines of the Donetsk, Kuznetsk and L'vov-Volyn basins range from 0.6 to $1.1 \times 10^{12}$ m$^3$. Various opportunities exist for increasing the quantity and quality of methane recovered from coal mines in these basins. The most attractive options for utilization of this methane include generating heat and power for use at mine facilities and nearby industries or residences.

The Commonwealth of Independent States (CIS) is the third largest producer of coal in the world behind China and the USA. In 1990, coal mining operations in the CIS emitted an estimated $7.2-8.9 \times 10^9$ m$^3$ (4.8–6.0 Tg) of methane to the atmosphere, which represented about 20% of world coal mine methane emissions (US EPA, 1993). Between 80 and 90% of these emissions were liberated by underground mining operations, which are primarily located in the republics of Ukraine, Russia and, to a lesser extent, Kazakhstan.

The energy sectors of Russia and Ukraine are at a turning point. The coal mining industry, in particular, is undergoing restructuring, a process which includes decreasing or eliminating subsidies. Mines are thus being compelled to become more efficient in order to increase profitability. In addition, the countries are seeking to mitigate environmental problems resulting from mining and use of coal. Consequently, there is a desire to decrease dependency on low grade coal and utilize more natural gas. Increased recovery and utilization of coalbed methane in Russia and Ukraine is a potential means of improving mine safety and profitability while meeting the republics energy and environmental goals.

This paper is a synopsis of a report written for the US EPA entitled 'Reducing methane emissions from coal mines in Russia and Ukraine: the potential for coalbed methane development', which is part of an ongoing US EPA programme being conducted in Russia and Ukraine to reduce emissions of methane (a potent greenhouse gas) while increasing its utilization (Marshall *et al.* 1994).*

* The EPA report may be obtained by contacting Karl Schultz at EPA's Global Change Division, telephone (202) 233 9468, fax (202) 233 9550, E-mail schultz.karl@epamail.epa.gov. Specify 'Reducing methane emissions from coal mines in Russia and Ukraine: the potential for coalbed methane development', No. EPA-430K-94-003.

## Energy consumption and production

In Russia, natural gas provided 46% of the total energy consumption in 1992, whereas oil accounted for 27% (US DOE, EIA 1994). Coal comprised only 18% of the fuel mix; hard coal accounted for about 83% of the total energy derived from coal and the remainder was from lignite. In Ukraine, natural gas likewise dominated the fuel mix in 1992, accounting for 42% of total energy consumption. The proportions of oil and coal consumed in Ukraine were nearly reverse that of Russia; coal accounted for 29% of Ukraine's energy mix, whereas oil accounted for 19%. Hard coal accounted for more than 98% of all the coal consumed in Ukraine (PlanEcon 1993).

## Natural gas: the dominant fuel

The CIS is the largest producer, transporter, consumer and exporter of natural gas in the world, producing almost 40% more gas than the next largest producer, the USA. In 1993, the CIS produced nearly $761 \times 10^9$ m$^3$ of natural gas, which represented a decline of 3% from the previous year. This is relatively stable compared with the large decline in oil production.

The sale of natural gas currently accounts for about 40% of hard currency earnings in the CIS, with $101 \times 10^9$ m$^3$ exported in 1993 (PlanEcon 1994*a*). CIS exports account for about 35% of world natural gas exports and represent approximately 25% of Europe's natural gas consumption.

Although natural gas dominates the fuel mix of both Russia and Ukraine, the two republics differ greatly in natural gas production. Russia produced $640 \times 10^9$ m$^3$ in 1992, more than 30% of which was exported to other republics and European countries. Ukraine, in contrast, produced only $21 \times 10^9$ m$^3$ of natural gas in

*From* Gayer, R. & Harris, I. (eds) 1996, *Coalbed Methane and Coal Geology*, Geological Society Special Publication No 109, pp 89–101.

1991, which satisfied less than 19% of its natural gas demand, so it was forced to import an additional $90 \times 10^9 \text{ m}^3$ of natural gas to meet its energy needs.

## Hard coal

Hard coal production in the CIS is split between the European part of the country in the west (Ukraine), with its underground bituminous coal mines, and the Asian part in the east (Russia and Kazakhstan). In 1992, the republics of Russia and Ukraine accounted for 73% of total hard coal production in the CIS.

In Russia, hard coal production fell from $274 \times 10^6$ tonnes in 1988 to $198 \times 10^6$ tonnes in 1993, a decrease of 28% (PlanEcon 1994*a*). The decline in Russian coal production has occurred for several reasons, including increasingly difficult mining conditions, equipment shortages caused by the break-up of the former Soviet Union, increasing labour costs and organizational and managerial problems stemming from the changing political system. The coal production decline has lessened in recent years (output in 1993 was only 8% less than in 1992), but it is likely that production will continue to decrease.

Russia's coal industry has been heavily subsidized for many years. The Russian government is attempting to restructure its coal industry and has started to reduce subsidies in real terms. As a result, the most uneconomic mines will be closed, while other mines must seek ways to become more efficient.

In Ukraine, coal production continues to decline rapidly. Preliminary figures indicate that Ukrainian coal production was only about $93 \times 10^6$ tonnes in 1994, about 20% less than what was achieved in 1993 (PlanEcon 1994*b*). The problem stems in part from a severe decline in technical inputs to the sector as a result of severe financial constraints plaguing the industry. Declining production is also a result of labour unrest, which is considerably stronger in Ukraine than Russia; the key issues are low pay, non-payment of wages and job guarantees.

Ukraine is not moving as quickly as Russia to restructure its coal industry. Although Ukraine did raise coal prices in 1994, thus reducing the degree to which mines must be subsidized, it still subsidizes 20% of production costs. Ukraine's Coal Minister says that the industry will continue to receive subsidies, despite the enormous burden they place on the republic's economy.

## Role of coalbed methane

Given the current condition of the energy sectors in Russia and Ukraine, there are likely to be many opportunities for coalbed methane to contribute to energy needs. The potential contributions of coalbed methane include: improved mine safety and profitability; improved local environmental quality; and an additional natural gas resource.

Opportunities exist for increased recovery and utilization of methane in conjunction with coal mining in Russia and Ukraine, as well as development of the coalbed methane resource independent of mining. As stated earlier, coal production is declining in the CIS. However, the recoverability of coalbed methane should not be adversely affected by decreasing coal production because, in most cases, methane continues to flow into abandoned mine workings for several years after a mine is closed. The technologies for producing coalbed methane, whether it is recovered in conjunction with coal mines or produced independently, are well developed, although they remain to be demonstrated in the CIS.

## Coalbed methane resources of Russia and Ukraine

Methane is explosive in concentrations of 5–15% in air and, in most mines, the concentrations are maintained below 1% to ensure safety. Many of the coal seams that are mined in Russia and Ukraine have high methane contents and methane drainage and ventilation have been used for many years to ensure safety.

Coalbed methane can be a significant gas resource in its own right. Although various countries have long used this resource on a small scale, primarily for on-site needs at mines, it has been only within the past decade that coalbed methane has gained widespread recognition as a viable energy source. Mining industry and government officials in Russia and Ukraine have stated their desire to utilize more coalbed methane, but funding and technology transfer are needed for these countries to initiate new coalbed methane projects.

There are four coal basins in Russia and Ukraine where hard coal is mined and which have the potential for coalbed methane development (Fig. 1). They are: the Donetsk Basin (Donbass), located in southeastern Ukraine and western Russia; the Kuznetsk Basin (Kuzbass), located in western Siberia; the L'vov-Volyn Basin, located in western Ukraine, which is the southeastward extension of the Lublin Basin in

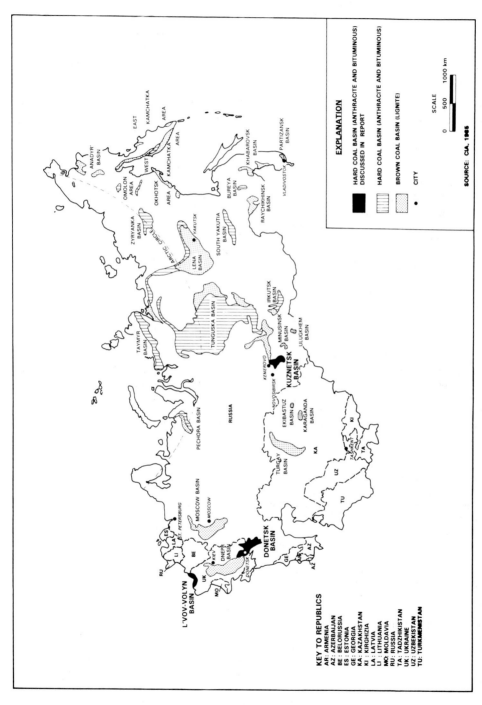

**Fig. 1.** Major coal basins of the former Soviet Union. Source: CIA (1985).

Poland; and the Pechora Basin, located in northern Russia, almost entirely above the Arctic Circle. This paper focuses on the Donetsk, Kuznetsk and L'vov-Volyn basins (the Pechora basin has not been evaluated).

A stratigraphic correlation chart of coal-bearing formations in the three basins studied is shown in Fig. 2. The L'vov-Volyn and Donetsk Coal basins produce only from Carboniferous formations, whereas the Kuznetsk Basin produces coal from both Permian and Carboniferous strata.

Of the three basins, the Donetsk and Kuznetsk appear to have the largest near-term potential for coalbed methane development. Both of these regions are heavily industrialized and should offer many markets for coalbed methane. The L'vov-Volyn region is predominantly agricultural and it is likely that coalbed methane use would primarily be limited to the mines. Characteristics of each basin are summarized in Table 1 and a more detailed description of the relevant features of each basin is provided in the following.

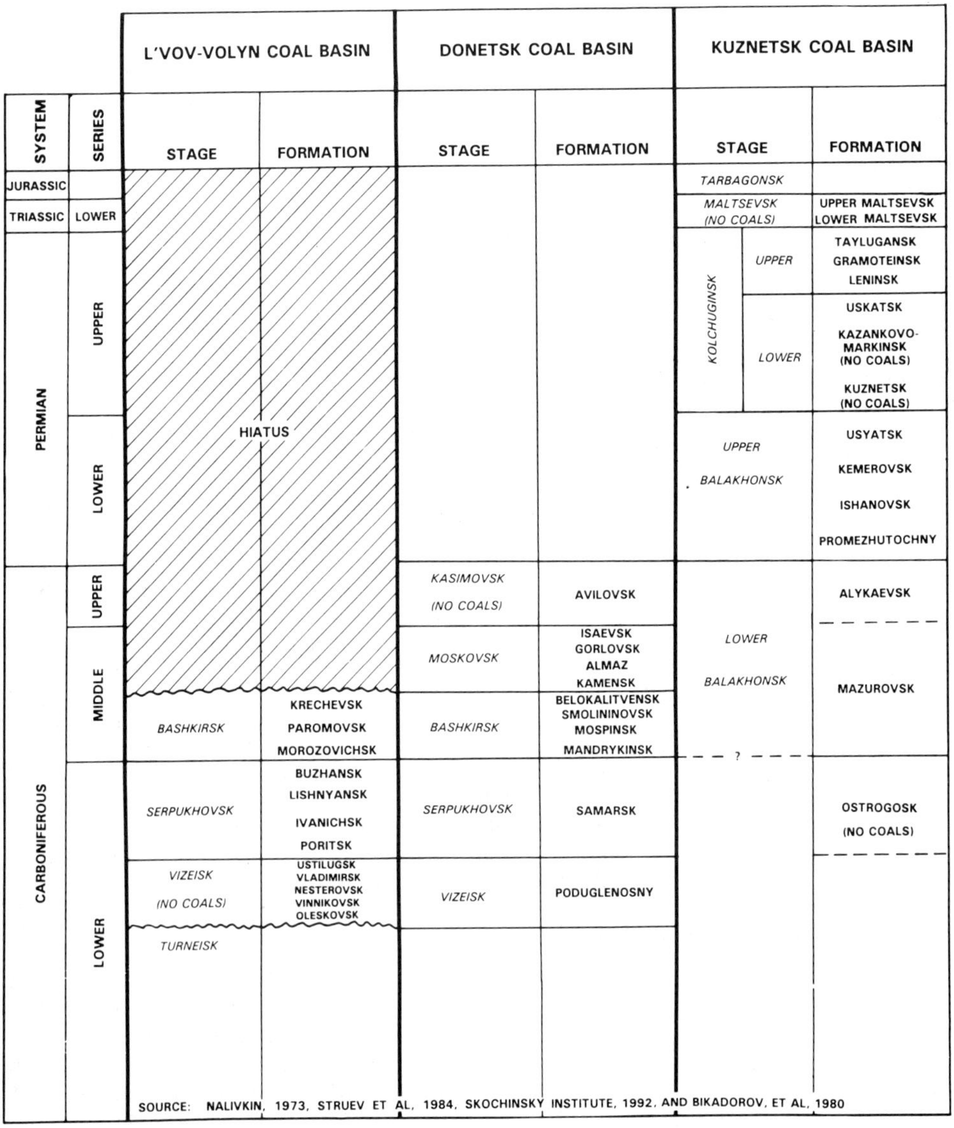

**Fig. 2.** Stratigraphic correlation of coal-bearing formations in the L'vov-Volyn, Donetsk and Kuznetsk coal basins, CIS. Source: Nalivkin (1973), Struev *et al.* (1984), Skochinsky Institute (1992) and Bikadorov *et al.* (1980).

**Table 1.** *Summary of selected coal basin characteristics (1991 data except where otherwise noted)*

| Characteristic | Donetsk coal basin | L'Vov-Volyn coal basin | Kuznetsk coal basin | Total |
|---|---|---|---|---|
| General characteristics | | | | |
| Basin asrea (1000 km$^2$) | 60 | 3.2 | 26.7 | 89.9 |
| Total documented coal reserves (billion tonnes)* | 140.8 | 2.1 | 637.0 | 779.9 |
| Coal production (million tonnes) | 150.4 | 9.5 | 58.9 | 218.8 |
| Average methane content (m$^3$/tonne) | 14.7 | 4.9 | 11.9 | |
| Mine data | | | | |
| Number of mines | 308 | 17 | 71 | 396 |
| Number of mines that emit methane | 211* | 13 | 71 | 295 |
| Number of mines that drain methane | 100 | 4 | 32 | 136 |
| Number of mines that utilize methane | 17 | 0 | 0 | 17 |
| Methane liberation data | | | | |
| Total methane liberated (Mm$^3$) | 3390.4 | 153.0 | 942.3 | 4485.7 |
| Methane liberated from mines with drainage systems (Mm$^3$) | 2452.4 | 63.8 | 992.6 | 3508.8 |
| Methane captured by drainage systems (Mm$^3$) | 538.9 | 6.3 | 243.5 | 788.7 |
| Methane utilized (Mm$^3$) | 170.2 | 0 | 0 | 170.2 |
| Specific emissions (m$^3$/tonne)† | 22.5 | 12.8 | 21.0 | |

Sources: Zabourdyaev (pers. comm.), Kozlovsky (1986; pers. comm.) and Skochinsky Mining Institute pers. comm.
* 1990 data.
† 'Specific emissions' are the volume of methane liberated per unit weight of coal mined.

Some data relevant to coalbed methane development were unavailable (e.g. details on formation water, water disposal, gas composition and coal permeability) These types of data will be obtained in the course of future EPA-sponsored studies.

### Donetsk Basin

The northwestern portion of the Donetsk Coal Basin is in Ukraine, whereas the southeastern portion is in Russia (Fig. 3). Coal mining began in the basin in 1723, and in 1991 there were 24 coal production associations operating 308 underground mines (Skochinsky Mining Institute, pers. comm. 1993). The Carboniferous deposits contain over 300 seams, of which 100 are considered workable. Most of the workable coal seams are found in sediments of the Moskovsk and Bashkirsk stages of the Middle Carboniferous, but potentially workable coals can also be found in the Serpukhovsk Series of the Lower Carboniferous. Seams vary in thickness from 0.45 to 2.5 m, averaging 0.9 m. Structural attitude of the coal beds ranges from horizontal to dips greater than 35°.

On average, Donetsk Basin coal contains 19.2% ash, 6.5% moisture, up to 4% sulphur, and has a heating value of 25.4 MJ/kg. The coal ranges from sub-bituminous to anthracite, but most of the coal mined is high-volatile bituminous A to anthracite. Coal mines in the Donetsk are extremely deep; over 40% of the mines have workings deeper than 700 m, and one-third have workings deeper than 1000 m (FBIS 1992).

*Methane liberation.* Coal mines in the Donetsk Basin are among the gassiest in the world. Of the basin's 24 coal production associations, 19 contain mines that liberated methane in 1991. Nearly $3.4 \times 10^9$ m of methane were liberated, with 16%, or $539 \times 10^6$ m$^3$, captured by methane drainage systems in 100 mines. Only $170 \times 10^6$ m$^3$ of this methane were used (exclusively in boilers at the mines), thus $3.2 \times 10^9$ m$^3$ were vented to the atmosphere. Methane used by the mines represents 32% of the total drained methane and only 5% of total methane liberated. These data suggest that there are good opportunities for increased methane drainage and utilization at Donetsk Basin mines.

### L'vov-Volyn Basin

The L'vov-Volyn Coal Basin, located in western Ukraine (Fig. 4), is the southeastern extension of the Lublin Coal Basin in Poland. Coal mining began in 1954 and there is currently one coal production association, Ukrzapadugol, operating 17 mines. The main coal-bearing horizons are found in Lower Carboniferous sediments of the

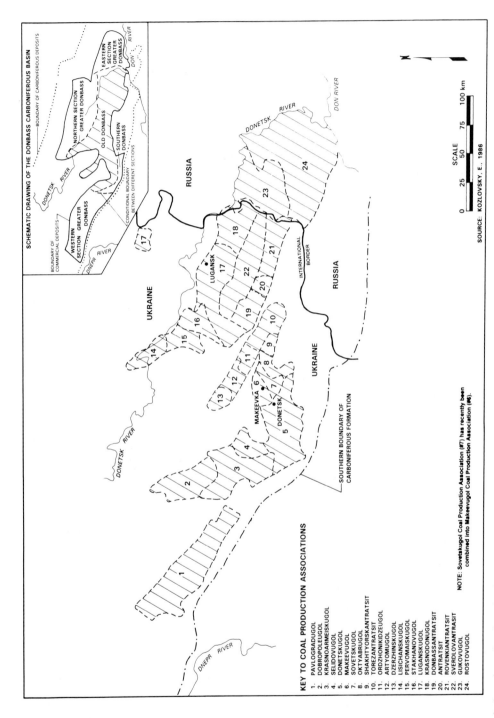

**Fig. 3.** Location of coal production associations in the Donetsk coal basin, Ukraine and Russia.

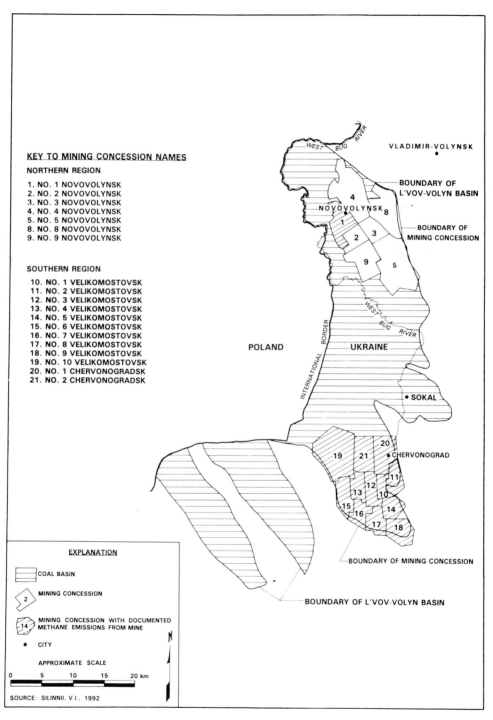

**Fig. 4.** L'Vov-Volyn coal basin, Ukraine. Source: Silinni (pers. comm. 1992).

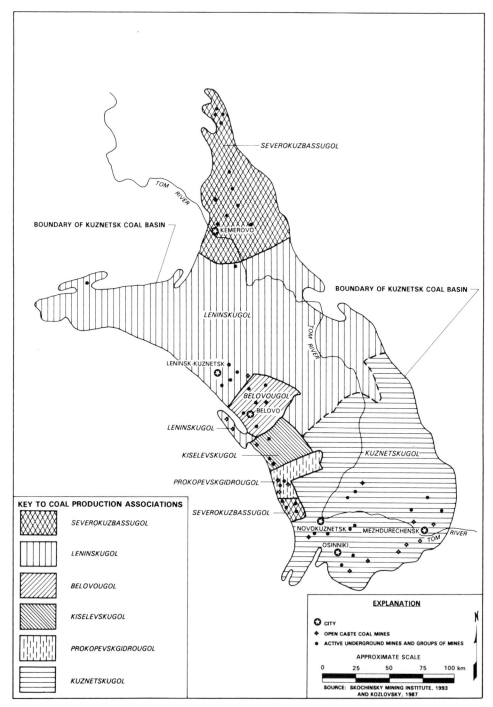

**Fig. 5.** Location of coal production associations in the Kuznetsk coal basin, Ukraine. Source: Kozlovsky (1987) and Skochinsky Mining Institute (pers. comm. 1993).

Serpukhovsk stage and Middle Carboniferous sediments of the Bashkirsk stage. Seam thickness averages 0.95 m and mining depth ranges from 300 to 600 m (Kozlovsky 1987).

On average, coal of the L'vov-Volyn basin contains 5–10% moisture, 23–42% ash, 36–39% volatile matter, 3–5% sulphur and has a heating value of 32–35 MJ/kg. The coal is predominantly high-volatile bituminous B and C.

*Methane liberation.* The 12 mines in the southern part of the basin, together with one in the northern part, emitted $153 \times 10^6 \, m^3$ of methane in 1991. This amount represents only 5% of total coal mining methane emissions from Ukraine. Although four of the mines have methane drainage systems, only 6% of all methane liberated is captured by those drainage systems and none is utilized.

Potential project opportunities exist at several mines. No. 10 Velikomostovsk, for example, produced more than $1 \times 10^6$ tonnes of coal in 1991, and emitted $28 \times 10^6 \, m^3$ of methane to the atmosphere. Of this, nearly $3 \times 10^6 \, m^3$, or about 10%, were medium-quality gas (typically 40–63% methane) recovered by mine drainage systems, and the remaining $25 \times 10^6 \, m^3$ were released in mine ventilation air.

## Kuznetsk Basin

The Kuznetsk Coal Basin is the second largest coal basin in the CIS and is located in central Russia (Figs. 1 and 5). Kemerovo, located in the northern end of the basin, and Novokuznetsk in the south are important industrial centres. In 1991, the basin contained 71 mines operated by six coal production associations (Fig. 5). Total coal thickness ranges from 57 to 423 m. The main coal-bearing horizons occur in Permian–Carboniferous sediments deposited during Bala-

khonsk and Kolchuginsk time, which crop out around the perimeter of the basin. Coal seams also occur in Jurassic sediments of the Tarbagansk Series.

More than 100 seams, with an average thickness of 2.5 m, are being mined. The depth of the underground mines ranges from 300 to 800 m (Airuni 1991). On average, Kuznetsk Basin hard coal contains 10% moisture, 19% ash, less than 0.5% sulphur and has a heating value of 23.2 MJ/kg. Coals closer to the surface are lower in sulphur and higher in moisture. Predominantly high-volatile bituminous B coal is mined; some high-volatile bituminous A and medium-volatile bituminous coal is also produced.

A typical Kuznetsk Basin mine produces between 400 and 1700 $m^3$ of water per hour, with a mineral content of about 2.3 g/l. Detailed data on saline water management were unavailable, but all Kuznetsk Basin mines reportedly have various facilities for mine water purification, such as sedimentation tanks and filtering stations.

*Methane liberation.* In 1951, the Kuznetsk Coal Basin became the first area to utilize methane drainage systems in underground mining in the USSR. All of the active underground mines are gassy. They liberated more than $1.2 \times 10^9 \, m^3$ in 1991. In spite of highly gassy conditions, only $211.9 \times 10^6 \, m^3$ (17%) of this gas were removed by drainage systems in 32 mines. All of this gas was then emitted to the atmosphere.

## Coalbed methane resource estimates

Coalbed methane resource estimates were based on data provided by the Skochinsky Mining Institute in Lyubertsy, Russia and by mining enterprises and research institutes that have collected data for purposes of producing coal

**Table 2.** *Summary of estimated methane resources associated with balance reserves of coal in the Donetsk, L'Vov-Volyn and Kuznetsk coal basins*

| Coal basin | Estimated methane resources ($10^9 \, m^3$) associated with balance reserves of coal, calculated according to: | | |
| --- | --- | --- | --- |
| | All mines | Active mines | Industrial reserves |
| Donetsk | 433–788 | 144–279 | 109–209 |
| L'Vov-Volyn | Not available | 1–3 | Not available |
| Kuznetsk | 194–342 | 199–204 | 69–120 |
| Total | 627–1131 | 264–486 | 178–329 |

Balance coal reserves are roughly akin to 'proved' and 'probable' reserves; they meet specific criteria related to quantity, quality, mining technology, geological conditions and mining conditions. 'Industrial' coal reserves comprise that portion of the balance reserves that is designated for extraction according to mine plans.

and maintaining mine safety. These estimates should be considered preliminary, but mining experience in Russia and Ukraine and available data indicate that the coalbed methane resource is large. It is clear that more detailed data collection is warranted to better assess the resource and identify the prospective areas.

Preliminary estimates suggest that the coalbed methane resources contained in mines of the Donetsk, Kuznetsk and L'vov-Volyn Coal Basins are substantial, ranging from 0.6 to $1.1 \times 10^{12}$ m$^3$ (Table 2). Additional methane resources are present in areas beyond the boundaries of the coal production associations and their associated mines. According to data from the former USSR Academy of Sciences (1990) and the Eastern Mine Safety Research Institute (pers. comm. 1992), the total estimated coalbed methane resource contained in just the coal seams of the three basins is greater than $7.8 \times 10^{12}$ m$^3$. Additional methane resources are contained in the partings and strata surrounding the coal seams. The above-mentioned institutions estimate that the total methane resource contained in the basin is more than $52.1 \times 10^{12}$ m$^3$.

To evaluate fully the coalbed methane development potential of each coal basin, it will be necessary to estimate coalbed methane resources and assess what percentage is recoverable using available technologies. This effort will require detailed information on the coal reserves, including geological and reservoir characteristics, which would be generated by an exploration programme.

Such information is currently unavailable for the coal reserves of the Donetsk, Kuznetsk and L'vov-Volyn basins. Thus in this paper the coalbed methane resources associated with the balance coal reserves of these basins were estimated based on available data using two approaches. To reflect the uncertainties associated with preparing such estimates where data are limited, the coalbed methane resource estimates are presented as ranges. For the low end of the range, estimates were based on the measured methane contents of coal reserves in each basin. The high-end estimates were developed using specific emissions (i.e. the amount of methane liberated per tonne of coal mined).

## Coalbed methane resources of the Donetsk Basin

The coalbed methane resources associated with the balance coal reserves of all mines in the Donetsk coal basin are estimated to range from 430 to 430 to $790 \times 10^9$ m$^3$. Of these resources,

an estimated 140 to $280 \times 10^9$ m$^3$ of methane are associated with the coal contained in active mining areas and 109 to $210 \times 10^9$ m$^3$ are associated with the basin's industrial coal reserves (i.e. those scheduled for mining).

Additional methane resources are associated with non-balance reserves of coal, as well as with coal resources located beyond the boundaries of the coal production associations. According to the former USSR Academy of Sciences (1990) the total estimated methane resources in the coal seams in the Donetsk basin are $1.2 \times 10^{12}$ m$^3$ and the total estimated methane resources contained in the basin (including the coal seams, partings and surrounding strata) are $25.4 \times 10^{12}$ m$^3$. If these estimates are accurate, the methane resources contained in the coals account for less than 5% of the total methane resources, indicating that there are potentially numerous other types of prospective gas reservoirs within the basin, including conventional stratigraphic and structural traps.

## Coalbed methane resources of the L'vov-Volyn Basin

The coalbed methane resources associated with the balance coal reserves of all mines in the L'vov-Volyn coal basin are estimated to range from 1.0 to $3.3 \times 10^9$ m$^3$. As coal resource data for all mines and for industrial reserves were not available, methane resource estimates could not be calculated for these categories.

Additional methane resources are associated with non-balance reserves of coal, as well as with coal resources located beyond the boundaries of the mines ('reserve regions'). Based on average gas contents calculated from data supplied by the Skochinsky Mining Institute (1991; pers. comm. 1993) and coal reserve data contained in Struev *et al.* (1984), we estimate that the volume of these additional methane resources may be $1.9 \times 10^9$ m$^3$.

## Coalbed methane resources of the Kuznetsk Basin

The coalbed methane resources associated with balance coal reserves of all mines in the Kuznetsk coal basin are estimated to range from 194 to $342 \times 10^9$ m$^3$. Of these resources, an estimated 119 to $204 \times 10^9$ m$^3$ of methane are associated with the coal contained in active mining areas and 69 to $120 \times 10^9$ m$^3$ are associated with the basin's industrial coal reserves (i.e. those scheduled for mining).

Additional methane resources are in non-balance reserves of coal, as well as in coal resources located beyond the boundaries of the coal production associations. According to the Eastern Mine Safety Research Institute, the total estimated methane resources contained in coal seams of the Kuznetsk Basin are $6.6 \times 10^{12}\,\mathrm{m}^3$, and the total estimated methane resources contained in the basin (including the coal seams, partings and surrounding strata) are $26.7 \times 10^{12}\,\mathrm{m}^3$. If these estimates are accurate, the coalbed methane resources contained in the coals of the basin account for less than 25% of the basin's total methane resources.

## Coalbed methane recovery

Many opportunities for increased recovery of coalbed methane exist in Russia and Ukraine. Nearly $4.5 \times 10^9\,\mathrm{m}^3$ of methane were liberated from coal mining activities in the Donetsk, Kuznetsk and L'vov-Volyn coal basins in 1991 (Table 1). Of the 396 mines operating in these three basins, 136 mines, or 34%, had methane drainage systems in 1991. Drainage systems at these mines recovered $740.7 \times 10^9\,\mathrm{m}^3$, or 17%, of the methane liberated by mining, but only $170 \times 10^6\,\mathrm{m}^3$ (3.8% of total liberated) were utilized, resulting in methane emissions of over $4.3 \times 10^9\,\mathrm{m}^3$.

Methane that is being drained and vented to the atmosphere could be used rather than wasted and significantly more gas could be available for utilization with an integrated approach to methane recovery in conjunction with mining operations. Experience elsewhere has shown that expanded methane drainage can be a profitable means of reducing the methane concentration in ventilation air, in that ventilation requirements are reduced, coal can be more rapidly extracted and gas recovered by drainage is often of commercial quality.

### Methane drainage methods

Detailed descriptions of the principal methane recovery techniques used in each coal basin are now presented.

*The Kuznetsk Basin.* Of the 71 mines in the Kuznetsk Coal Basin, 32 have methane drainage systems. The principal systems of drainage include the use of small diameter boreholes drilled from the surface into the coal seam targeted for mining, and large diameter boreholes drilled from the surface into the roadway located behind the retreating longwall. The small diameter boreholes drilled into the coal in advance of mining provide dual service by pre-draining methane and then being converted into gob wells after mining has passed. Portable pumping stations on the surface are connected to these wells to ensure the evacuation of the gas from the gob, and can be moved to other wells as mining progresses. The large diameter boreholes are fitted with a manifold and fan system to remove the gas that is being produced in the gob areas. This methane is vented to the surface, which keeps it from migrating back into the active working face area.

In many active mines of the Kuznetsk Basin, underground drainage methods are insufficient to drain the quantities of methane needed to meet safe mining standards. This is due to a pressure differential between the boreholes and the mines, resulting from excessive suction by conventional ventilation systems and booster auxiliary fans. In many mines, these systems prevent the efficient flow of methane into in-mine drainage systems, which cannot create a sufficient vacuum.

*Donetsk and L'vov-Volyn Basins.* Of the 269 mines in the Donetsk Coal Basin, 100 have methane drainage systems; four of the 17 mines in the L'vov-Volyn Coal Basin drain methane. Because of the depth of mining, underground drainage is the preferred methane recovery technique used in both basins. The two most widely used methods are cross-measure boreholes drilled from the roadways into surrounding strata and boreholes drilled from the roadways into gob areas. Underground booster pumping stations are often used to drain the methane from the boreholes. The methane is either discharged into the ventilation system away from the active mining areas or pumped to a central vertical well and piped to the surface via surface vacuum pumps.

### Options for increased recovery

A variety of methane recovery methods are used in Russia and Ukraine, but many mines confront technical difficulties in their application. In addition, in many cases, the necessary investment capital is not available to support the development of a fully integrated system of methane recovery. The quality and quantity of gas recovered in Russia and Ukraine could be improved through a variety of measures. The optimum methods to use depend on site-specific conditions, but general types of measures are presented here.

Projects to improve the quality of gas recovered could include: extending the length of the standpipe beyond the yield zone in pillars and ribs to prevent entry of air into the borehole through this zone; repair of leaks in the in-mine and surface gas gathering systems, including sealing the standpipes using proper grouting techniques and materials; gas quality monitoring to prevent large fluctuations in methane concentration; reducing suction to a level that eliminates or minimizes the entrainment of air through cracks and cleats in the face; monitoring the oxygen or nitrogen content in the gas stream produced from gob wells and adjusting the flow-rate to reduce or eliminate the level of contamination; and positioning boreholes to maximize gas quality (both vertical and lateral distance to the open mine area are important).

Projects to improve the quantity of gas recovered may include: intensifying pre-drainage efforts, with multiple seam completions and stimulation techniques; expanding in-seam drainage systems, especially in longwall panels; optimizing spacing of gob wells; employing borehole completion technology that allows mining-through of pre-drainage and gob wells; and maximizing drainage time for vertical wells and in-mine drainage systems by implementing drainage programs well ahead of mining.

## Coalbed methane utilization

The best utilization options for methane from coal mines will vary from region to region, depending on the quality and quantity of the gas and local energy markets. In the Donetsk and the L'vov-Volyn basins, coalbed methane could be used as an alternative to imported natural gas. Ukraine is a net importer of natural gas, so any domestic natural gas would reduce its dependence on imports. In the Kuznetsk Basin, high-quality coal is exported out of the region whereas low-quality coal is consumed locally. Thus in this region coalbed methane use could replace low-quality coal consumption, which would improve the local air quality. The principal utilization options are discussed in more detail in the following.

### *Direct utilization options*

The Donetsk and Kuznetsk basins are heavily industrialized regions and coal is used extensively for steam and electrical generation. The largest consumers of energy in these regions are machine factories, petrochemical plants, metallurgical factories and, of course, the coal industry. The mining, power and industrial complexes which dominate both regions were originally developed with an emphasis on large-scale production, often at the expense of efficiency, profitability and the environment.

Coalbed methane utilization would clearly benefit these regions by helping them meet their energy needs with a clean-burning, local fuel source. Some fuels that coalbed methane could replace are brown coal, low-quality hard coal and coke oven gas. Specific uses will depend on the conditions in the vicinity of the mines, but could include on-site coal drying, heating of mine facilities and heating or refrigeration of ventilation air. Methane could also be used by nearby industries to displace brown coal and low-quality hard coal, coke oven gas and (especially in Ukraine) imported natural gas.

### *Power generation options*

Currently, there are only 17 mines in the Donetsk Basin that utilize coalbed methane and all of them use it in boilers at the mine site. No mines in the L'vov-Volyn or Kuznetsk regions utilize coalbed methane. Opportunities exist for the generation of electricity and steam at mine power plants; electrical power is used at all coal mines and thermal heat is supplied to the surrounding communities for district heating. Currently, these mines generate most of their electricity and steam from coal. Coalbed methane could displace the burning of coal which pollutes, and in regions such as the Donetsk Basin is in increasingly short supply. Several power generation options are: converting boilers to intermittent use of gas; gas turbines; internal combustion engines; and co-firing with natural gas.

## Summary and recommendations

Russia and Ukraine face many problems in their energy sectors which are generally related to difficulties in making the transition from a centrally planned economy to a market economy. These problems include energy shortages and a struggling coal industry. The long-standing practice of subsidizing coal mines provided no incentives for mines to operate efficiently. The most uneconomic mines are now being forced to close and others are seeking ways to improve their profitability. For some of these mines, increasing recovery and utilization of coalbed methane may be a means of increasing productivity and profitability.

The present study suggests that coalbed methane will be most valuable when used locally by the mine from which it is recovered, or in nearby industries. The value of methane as a substitute for other fuels increases when enrichment, drying and compression of the gas are not required. Of course, where large amounts of conventional natural gas are imported, upgrading coalbed methane for injection into pipelines may be a viable economic option.

Economic feasibility studies are needed to identify the mines that would probably benefit most from coalbed methane recovery and utilization. The costs and benefits of coalbed methane projects at these mines must then be carefully evaluated. Economic studies should also consider regional benefits, such as jobs and an increased domestic energy base, that will occur as coalbed methane recovery and use increases.

Several Russian and Ukrainian government agencies, as well as several of the coal production associations, are actively investigating opportunities to expand methane recovery and use. Such agencies should develop appropriate policies and incentives that will help make economic development and utilization of coalbed methane a part of the energy sector restructuring programmes of the two republics.

Foreign governments and international agencies, as well as foreign companies, can assist Russia and Ukraine with this process by providing financial and technical assistance for coalbed methane projects. Efforts should be made to inform, educate and train government and industry personnel to raise awareness of the coalbed methane resource and the available technologies for its recovery and utilization.

## References

AIRUNI, A. T. 1991. *Sposoby bor'by s vydeleniem metana na ugol'nykh shakhtakh.* TsNIEIugol, Moscow [in Russian].

BIKADOROV, B. C., NIKITINA, Y. B., LOMTEVA, L. M., SVEREHKOVA, A. I., STADNITSKAYA, M. C. & OZFEROVA, C. O. 1980. *General stratigraphic section of the coal bearing sequence of the Kuznetsk Coal Basin, Russia.* Ministry of the Coal Industry of the USSR.

CENTRAL INTELLIGENCE AGENCY (CIA) 1985. *USSR Energy Atlas.* CIA, Langley.

FOREIGN BROADCAST INFORMATION SERVICE (FBIS) 1992. *Data on Safety Conditions in Mining Sector/Ukraine.* FBIS-USR-92–085, 8 July 1992, 94.

KOZLOVSKY, E. (ed.) 1986. *Gornaya Encyclopedia.* Vol. 2. Sovietskaya Encyclopedia, Moscow.

——1987. *Gornaya Encyclopedia.* Vol. 3. Sovietskaya Encyclopedia, Moscow.

MARSHALL, J. S., PILCHER, R. C., BIBLER, C. J. & KRUGER, D. W. 1994. *Reducing methane emissions from coal mines in Russian and Ukraine – the potential for coalbed methane development.* USEPA, Washington, No. EPA 430-K-94-003

NALIVKIN, D. V. 1973. *Geology of the USSR.* University of Toronto Press.

PLANECON 1993. *PlanEcon Energy Outlook for the Former Soviet Republics.* PlanEcon, Washington.

——1994a. *Energy Overview, Former Soviet Republics and Eastern Europe Through the Final Quarter of 1993.* PlanEcon, Washington.

——1994b. *Energy Overview, Former Soviet Republics and Eastern Europe Through the Third Quarter of 1994.* PlanEcon, Washington.

SKOCHINSKY MINING INSTITUTE 1991. *Prognoznyi katalog shakhtoplastov Podmoskovnogo, Lvovsko-Volynskogo, i Dneprskogo Ugolnykh Basseinov s kharakteristikoi gorno-geologicheskikh i gorno-tekhnicheskikh faktorov na 1995 i 2000.* The Skochinsky Mining Institute, Moscow [in Russian].

STRUEV, M. I., ISAKOV, V. I., SHPAKOVA, V. B., KARAVAEV, V. Y., SELINNI, V. I. & PAPEL, B. S. et al. 1984. *L'vovsko-Volynskii kamennouogol'nyi bassein, Geologo-promyshlennyi ocherk.* Naukova Dumka, Kiev, Ukraine [in Russian].

US DOE, EIA 1994. *International Energy Annual 1992.* USDOE, Washington.

US EPA. 1993. *Options for reducing methane emissions internationally,* Vol. 2. USEPA, Washington, No. EPA 430-R-93-006B.

USSR ACADEMY OF SCIENCES 1990. *Gazobil'nost' kamenougol 'nykh shakht SSSR: Kompleksnoe osvoenie gazonosnykh ugol'nykh mestorozhdenii.* Nauka, Moscow [in Russian].

# Coal clasts in the upper Westphalian sequence of the South Wales coal basin: implications for the timing of maturation and fracture permeability

R. A. GAYER[1], J. PEŠEK[2], I. SÝKOROVÁ[3] & P. VALTEROVÁ[4]

[1] Department of Earth Sciences, University of Wales Cardiff, PO Box 914,
Cardiff CF1 3YE, UK

[2] Faculty of Science, Charles University, Albertov 6, 12843 Prague 2, Czech Republic

[3] Institute of Rock Structure and Mechanics, Academy of Sciences, V Holešovičkách 41,
18209 Prague 8, Czech Republic

[4] Geofond, Kostelní 26, Prague 7, Czech Republic

**Abstract.** Coal clasts, common within channel lag deposits within the upper Westphalian C and Westphalian D Upper (Pennant) Coal Measures of the South Wales coal basin, represent reworked previously deposited coal-forming material. Analysis of clast shape, coal petrology and palynology at 16 localities within the eastern part of the South Wales coalfield suggests that three types of clast are present: (1) large elongate rafts of coal, with axial ratios up to 90:1, with irregular, often 'fish-tail' terminations and showing post-depositional compaction relative to the enclosing sandstone; (2) rafts of similar dimensions to (1), but showing no evidence for differential compaction and terminated by cleat fractures; (3) small near-equidimensional pebbles that either show no evidence for differential compaction or in which the surrounding sandstone shows greater compaction than the coal pebble. Both types 1 and 2 rafts have miospore assemblage ages that are indistinguishable from coal seams in adjacent sediments and show a similar range of vitrinite-dominated, maceral group composition. The pebbles have miospore assemblages suggesting derivation from coal seams ranging from a similar age to those of the rafts to Westphalian A (Lower Coal Measures) and with a wider range of maceral group composition than the rafts. Type 1 rafts were derived by erosion of partially consolidated and lithified peat deposits from the contemporary alluvial plain, and cleat fractures were developed *in situ* during compaction. They formed as extensional cracks parallel to the regional NNW–SSE Variscan compressive stress. In type 2 rafts and pebbles cleat partly controlled the shape of the clasts and was developed before erosion of the clasts from their source coal seams. Vitrinite reflectance ($R_m\%$) indicates that maturation of the rafts to bituminous coal rank developed *in situ* within the Pennant Measures, but in the pebbles maturation to bituminous rank may have been developed in the source coal seam before erosion. It is suggested that this maturation, and the concomitant compaction and cleat development, occurred within $\approx 1$ Ma of deposition after rapid burial to $\approx 1$ km in a high heat flow regime, asssociated with hot thrust-guided fluids in a foreland basin setting. Rapid maturation of type III kerogen into the oil and gas generation windows, together with the development of cleat, forming effective migration pathways, has important implications for coalbed methane exploration.

In most coalbed methane (CBM) prospective sites it is assumed that the CBM is retained in the coal seam after generation and that variations in the gas content of coals are either a consequence of differing levels of gas generation, reflecting coal type, or due to gas lost during a later (often recent) uplift event. The possibility that variations in the CBM content of coals may reflect loss of gas by migration out of the coal during gas generation appears to be considered only rarely. However, the fact that many traditional gas fields (e.g. those of the southern North Sea) contain gas sourced from underlying coals, suggests that CBM migration is a common feature.

The retention of CBM within a coal seam is critically dependent on the timing of gas generation within the coal seam relative to burial and to the development of fractures that allow its migration. If both the gas and migration pathways are formed before the coal seam is sealed by the development of impermeable roof rocks, or a hydrostatic head is able to prevent migration, much of the CBM will be lost. Studies of the maturation processes of coal, from peat through lignite and bituminous coals to anthracite, suggest that the alkane gases are generated by progressive hydrocarbon cracking during the maturation stages from medium-volatile bituminous coal to anthracite (Levine 1993). The

*From* Gayer, R. & Harris, I. (eds) 1996, *Coalbed Methane and Coal Geology*,
Geological Society Special Publication No 109, pp 103–120.

principal fracture permeability formed in coal is a closely spaced bed-normal joint system, termed cleat, that is thought to form and intensify during the bituminous coal stages of maturation (Levine 1993) in response to stresses acting across a coal basin (Gayer & Pesek 1992). Thus CBM generation and the development of cleat that will allow the gas to migrate occur simultaneously. Provided the coal seam is sealed by an impervious roof rock at this stage, most of the gas will be retained. If, however, a porous roof rock is present (e.g. sandstone or overpressured clays), some or all of the gas generated will be lost from the coal seam.

This paper reports an investigation to determine the relative timing of coal maturation and the development of cleat. It describes a study of coal clasts present at several levels in the Westphalian C and D Pennant Coal Measures of the South Wales coal basin (Fig. 1). The study shows that the coal seams from which some of the clasts were derived had been sufficiently matured and compacted to allow cleat to develop before erosion and clast formation. In some instances clasts show weathering and mineralization, which is thought to have

occurred within the seam before erosion. Using palynology to date both the seam from which the clasts were derived and the level at which the clasts occur, a maximum time for maturation and cleat formation can be estimated.

## South Wales coal basin

The South Wales coalfield forms an elliptical outcrop within a major E–W trending, Late Palaeozoic Variscan synform (Fig. 1) and represents the structurally controlled erosional remnant of the original coal basin. The coal-bearing sequence ranges from Westphalian A through to Westphalian D and lies above a non-coal-bearing clastic Namurian sequence that in turn lies unconformably above Lower Carboniferous platform carbonates.

The main coal-bearing interval comprises the Lower and Middle Coal Measures, which formed during the Westphalian A and Westphalian B to Lower C, respectively, in a largely fluvial coastal plain environment dominated by overbank mudrocks that represent deposition in extensive interdistributary lakes. Relatively rare,

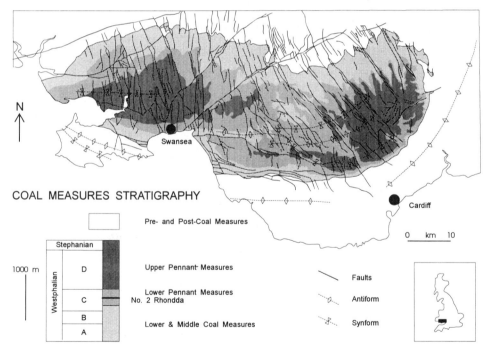

**Fig. 1.** Geological map of the South Wales coalfield showing the outcrop of the Westphalian Coal Measures (including the Pennant Sandstone) from which the coal clasts were collected (for sample site locations see Fig. 2), the principal Variscan structures and a diagramatic stratigraphic column of the Westphalian Coal Measures.

mature, quartz-dominated sandstones are either channelized, representing fluvial meander belts, or laterally extensive sheet sands, representing crevasse splay flood events. Marine bands and freshwater bivalve bands above many of the coal seams represent frequent transgressive events (Hartley 1993).

The Upper (or Pennant) Coal Measures contain few coals and consist of thick, cross-bedded lithic and subarkosic sandstones, commonly with lag conglomerates above an erosional base. These have been interpreted as low-sinuosity channel systems developed in a braid plain (Jones 1989).

Palaeocurrent data from cross-bedding in the sandstones suggest that those of the Lower and Middle Coal Measures were sourced from a mature region to the north of the coal basin, possibly second cycle Old Red Sandstones from the Welsh Massif (Kelling 1974; Jones 1989). Data from the Upper Coal Measures, Pennant Sandstones suggest that immature sediment was derived from the south and east, possibly from the rising Variscan orogenic hinterland (Kelling 1988; Gayer & Jones 1989; Jones unpublished data). The switch from a dominantly north-derived mature sandstone during the Westphalian A through early Westphalian C to immature south- and east-sourced Pennant Sandstones from late Westphalian C though Westphalian D together with high subsidence rates and a northward migration of the basin depocentre,

implies a foreland basin setting for the coal basin (Kelling 1988; Gayer & Jones 1989).

The coal basin has been strongly affected by post-depositional Variscan fold and thrust deformation (Woodland & Evans 1964; Owen 1974; Frodsham et al. 1993; Gayer & Nemcok 1994). The age of the deformation is uncertain, but must pre-date the late Triassic unconformity. The locus of deformation is thought to have migrated into the coal basin as deformation propagated from the orogenic hinterland towards the foreland. The fold and thrust deformation imposed a complex set of tectonic fractures on the coals that superimposed and partially obliterated a cleat fracture system (Hathaway & Gayer this volume).

Coal rank varies markedly across the coalfield (Fig. 2), with medium-volatile bituminous coals occurring in the east and south of the coalfield, but with rank rising progressively to the north and west, so that an area of anthracite lies in the northwest of the coalfield (White 1991). The rank was developed before the main Variscan deformation (White 1991). The origin of the coal rank and particularly of the anthracite is controversial. Recent theories have included variable burial depths within a constant heat flow (White unpublished data) and variable heat flow caused by hot geothermal fluids flowing into the basin along thrusts (Gayer et al. 1991). A study of the temperature and pressure of fluid inclusions in quartz crystals within an ironstone

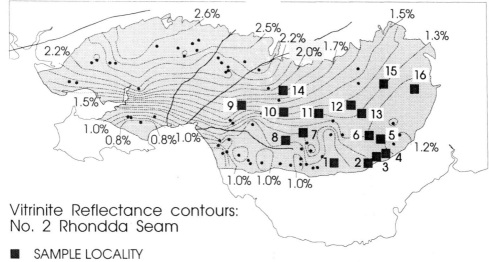

Vitrinite Reflectance contours:
No. 2 Rhondda Seam

■ SAMPLE LOCALITY

▢ OUTCROP OF COAL MEASURES

**Fig. 2.** Outline map of the South Wales coalfield, showing contours of vitrinite reflectance ($R_m$) for the No. 2 Rhondda Seam (base Rhondda Beds, Upper Westphalian C). Modified from White (1991).

nodule in the Middle Coal Measures in the south of the coalfield suggests an abnormally high geothermal gradient of $\approx 60°C\,km^{-1}$ (Alderton & Bevins in press).

## Coal clasts within the Pennant Sandstone

Coal clasts were collected from a range of stratigraphic levels, from upper Westphalian C to Westphalian D at 16 localities in the eastern half of the coalfield (Fig. 2). The sample sites were selected to allow a large number of clasts to be studied over as wide a range of the Pennant Measures stratigraphy as possible in areas of the coalfield where the rank of the coal was sufficiently low for miospores to be preserved. Table 1 lists the 16 localities with the stratigraphic level and coal rank (vitrinite reflectance) of each. Most of the horizons containing coal clasts occur as channel lag deposits associated with cross-bedded sandstones, thought to represent low-sinuosity, braided channel-fills. Occasionally clasts occur in cross-bedded conglomeratic units

interpreted as crevasse splay deposits. At all the localities sampled, palaeocurrents, determined from cross-bedding orientations in associated longitudinal sand bars, suggest water flow from the south and east (Jones 1989).

At most of the localities sampled, the conglomerates contain a range of clast types that include mature well-rounded pebbles of vein quartz, probably recycled from Namurian basal conglomerates or Upper Old Red Sandstone quartz conglomerates (Jones 1989), and immature siltstones, mudstones, ironstone nodules and tree trunks, locally derived from the adjacent floodplain. In addition, most localities contain a large number of coal clasts; in some instances the conglomerates display clast-supported fabrics in the channel lag, but in others the clasts are isolated within a coarse-grained sandstone matrix overlying the lag.

## Clast shape

The long axes of coal clasts from the total sample set range from less than 1 cm to over 4 m.

**Table 1.** *Details of sample localities, clast type and vitrinite reflectance* $(R_m)$ *of material described in this study. The* $R_m$ *means have been calculated by averaging the individual means obtained from either 50 or 100 measurements for each clast, and the standard deviations represent deviations from the quoted means*

| Location | | | | | Vitrinite reflectance $R_m$ (%) | | | |
|---|---|---|---|---|---|---|---|---|
| No. | Name | Grid reference | Stratigraphic horizon | Sample type | Mean | Standard deviation | Sample size | Closest seam |
| 1 | Llantrisant | ST 050 836 | Brithdir Beds | Raft | 0.84 | | 1 | 0.95 |
| | | | | Pebble | 0.735 | | 1 | |
| 2 | Nant Garw | ST 123 847 | Rhondda Beds | Raft | 0.865 | 0.0535 | 11 | 1.15 |
| | | | | Pebble | 0.821 | 0.0908 | 7 | |
| 3 | Caerphilly Common | ST 158 855 | Rhondda Beds | Raft | 0.83 | 0.0778 | 2 | 1.23 |
| 4 | Rudry Common | ST 181 875 | Hughes Beds | Pebble | 0.76 | | 1 | 1.07 |
| 5 | Trehir Quarry | ST 156 896 | Hughes Beds | Coal | 0.77 | 0.0462 | 3 | 1.07 |
| | | | | Raft | 0.77 | 0.0141 | 2 | |
| | | | | Pebble | 0.89 | 0.0299 | 4 | |
| 6 | Abertridwr | ST 131 890 | Hughes Beds | Raft | 0.74 | | 1 | 1.09 |
| 7 | Gilfach Goch | SS 978 902 | Rhondda Beds | Raft | 0.925 | 0.0493 | 4 | 1.22 |
| | | | | Pebble | 0.975 | 0.0414 | 6 | |
| 8 | Lewistown | SS 934 887 | Rhondda Beds | Pebble | 0.882 | 0.037 | 5 | 1.15 |
| 9 | Abercregn | SS 848 971 | Brithdir Beds | Coal | 1.2 | | 1 | 1.45 |
| 10 | Craig Ogwr | SS 937 946 | Rhondda Beds | Raft | 0.97 | 0.0408 | 7 | 1.45 |
| | | | | Pebble | 1.07 | | | |
| 11 | Tylorstown | ST 012 955 | Rhondda Beds | Raft | 1.10 | 0.099 | 6 | 1.45 |
| | | | | Pebble | 1.30 | 0.127 | 7 | |
| 12 | A470 | ST 081 965 | Hughes Beds | Pebble | 1.1 | | 1 | 1.20 |
| 13 | Abercynon | ST 095 955 | Hughes Beds | Pebble | 1.058 | 0.1228 | 5 | 1.13 |
| 14 | Treherbert | SS 933 995 | Rhondda Beds | Pebble | 1.28 | | 1 | 1.85 |
| 15 | Markham | SO 168 022 | Hughes Beds | Pebble | 1.11 | | 1 | 1.15 |
| 16 | Cefn Crib | SO 237 014 | Grovesend Beds | Coal | 0.83 | 0.0141 | 2 | 0.89 |
| | | | | Pebble | 0.86 | | 1 | |

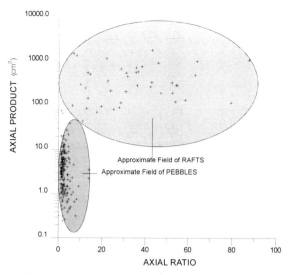

**Fig. 3.** Plot of clast axial ratio (clast elongation) versus clast axial product (for rectangular clast = clast area) for coal clasts in the Upper Coal Measures (Pennant Sandstone) of the South Wales coalfield. The shaded areas represent the fields of coal pebbles and coal rafts with low axial ratio and product and high axial ratio and product, respectively.

The clasts are clearly divided by shape into two populations: (a) relatively small clasts which are approximately equidimensional (pebbles); and (b) relatively large clasts which are typically highly elongated (rafts) (Fig. 3). The pebbles have cross-sectional areas of between 0.25 and 40 cm², and the rafts have a broad range of axial ratios from about 10:1 to 90:1 with cross-sectional areas from 50 to 2000 cm². In the case of the longer rafts, and where the exposure is limited, it is sometimes difficult to distinguish the rafts from coal seams. However, the rafts are always within a conglomeratic or coarse sandstone lithology and are never underlain by a rooted seat earth, which is usually found beneath a true autochthonous coal seam.

*Pebbles.* The coal pebbles vary in shape from subrounded to subangular—in the latter the ends of the pebbles are controlled by cleat joints normal to coal banding in the pebbles. In many instances coal pebbles have been broken along a cleat so that half of the pebble is subrounded, whereas the half with the cleat breakage is subangular (Fig. 4a). The long axis of the pebbles are commonly, but not always, oriented parallel to the bedding of the sandstone bed, but they also occur showing imbrication (Fig. 4b), with their long axis lying at an angle to bedding, sometimes as high as 60°. The coal pebbles show little evidence of differential compaction relative to the surrounding sandstone and where present

the enclosing sandstone is compacted around the coal clast (Fig. 4c).

*Rafts.* The coal rafts almost always lie with their long axes parallel to regional bedding of the enclosing sandstone. In some instances the rafts and the bedding in the sandstone are gently folded and this is attributed to differential compaction of underlying coal rafts. In many instances the terminations of the rafts show a complex interfingering of coal and adjoining sandstone, with the common development of a 'fish-tail' form (Fig. 5a). The fish-tail shape is further emphasized by thickening of the raft towards the terminations with the sandstone (Fig. 5b). Bedding in the sandstone immediately above the thinned central portion of the raft is commonly down-bowed and this is considered to be due to differential compaction of the coal raft relative to the sandstone matrix. Gayer & Pesek (1992) suggested that the complex terminations of the rafts is a consequence of differential compaction between the raft and the sandstone, estimated to be between 3:1 and 5:1. However, at least part of the interfingering of coal into sandstone is likely to be due to a complex original shape of the raft termination so that the true differential compaction is probably nearer to 2:1. This suggests that the material originally forming the clasts was likely to have been partly compacted and matured peat or lignite.

a)

b)

c)

Some of the rafts show no evidence of differential compaction and these usually have abrupt terminations defined by cleat surfaces (Fig. 5c). It is likely that these clasts were derived from well compacted and cleated coal that had achieved a significant level of maturation.

## Clast petrology

A total of 32 rafts and 43 pebbles from the 16 localities indicated in Fig. 2 were analysed for maceral group composition and vitrinite reflectance.

*Maceral group composition.* Observations under reflected light microscopy showed that both pebbles and rafts have lithologies typical of banded coal (Fig. 6). Figure 7 shows that the Upper Coal Measures coal seams and the coal rafts have a similar vitrinite dominated composition, with greater than 60% of the macerals belonging to the vitrinite group and with a slightly greater content of inertinite group than liptinite group of macerals. The coal pebbles have a greater spread in maceral group composition, with some pebbles having less than 25% vitrinite group of macerals. Several rafts were found to be composed of 100% vitrinite and these were interpreted to be tree debris.

In all clasts the inertinite group of macerals are dominated by fusinite, which is a characteristic feature of coals of the South Wales coalfield (Scott 1989). The liptinite group of macerals consists of sporinite, cutinite and resinite, with no clear trends in any group of clasts.

The similar composition of coal rafts and coal seams would be consistent with the rafts being derived from coals formed in similar environments to those of the Pennant Measures of South Wales. The broader range in compositions of the pebbles may imply derivation from coals formed in different environments from those of the Pennant Measures sequence.

*Coal rank.* The maturation state of the clasts was determined by random mean vitrinite reflectance ($R_m$) determinations and the results are presented by locality in Table 1 and compared with the $R_m$ values of the stratigraphically closest coal seam in Fig. 8. The $R_m$ values for both rafts and pebbles are in almost all instances lower than those of the stratigraphically adjacent coal seams. Such a relationship has been noted for disseminated vitrinite in clastic sediments associated with coals (Stach *et al.* 1982) and is thought to be the result of a chemical environmental control on vitrinite reflectance. In this instance the matrix of the clasts is porous sandstone, which would have allowed fluid permeability producing a distinct fluid chemistry to that of the coals. The majority of rafts show $R_m$ values consistently 0.3% lower than for the *in situ* coals and thus a pattern of $R_m$ values that matches the $R_m$ contours for the coalfield (Fig. 2). The pebbles show a wider scatter, with some pebbles having $R_m$ values slightly higher than those of coals lying just beneath the stratigraphic level of the pebbles. However, in general the $R_m$ of the pebbles follows the trend shown by the rafts. In most instances the average $R_m$ value of pebbles is higher than that of rafts from the same locality.

These results are interpreted to imply that maturation of the coal clasts occurred in common with that of the coalfield as a whole and thus that the clasts received their maximum

---

**Fig. 4.** Photographs of coal pebbles from the Upper Coal Measures (Pennant Sandstone) of the South Wales coalfield. (**a**) Subrounded coal pebble, broken along cleat fracture at high angle to bedding in enclosing sandstone (locality 8 Fig. 2). Coin is 2.1 cm diameter. (**b**) Near-equidimensional coal pebbles showing imbrication in bedding of enclosing sandstone. Cleat fractures are developed normal to primary coal banding, irrespective of present attitude of pebble. Rhondda Beds at Tylorstown (locality 11, Fig. 2) Coin is 2.2 cm in diameter. (**c**) Equidimensional coal pebble in sandstone of the Rhondda Beds at Gilfach Goch (locality 7, Fig. 2). Note well-developed cleat fractures normal to coal banding and the slight compaction of the enclosing sandstone relative to the coal clast. Coin is 2.1 cm diameter.

**Fig. 5. (Overleaf)** Photographs of coal rafts from the Upper Coal Measures (Pennant Sandstone) of the South Wales coalfield. (**a**) Coal raft with 'fish-tail' terminations within sandstones of the Rhondda Beds, Tylorstown (locality 11, Fig. 2), showing sag synform in overlying sandstones due to compaction in coal raft. Coin above the raft is 2.2 cm in diameter. (**b**) Detail of 'fish-tail' termination of a coal raft in sandstones of the Rhondda Beds, Tylorstown (locality 11, Fig. 2), showing increasing thickness towards termination of raft, and parallel deflection of bedding in overlying sandstone. Note bedding-normal cleat in raft, better developed in thinner portion and becoming weakly developed towards the termination. Pencil is 14.5 cm long. (**c**) Coal raft with well developed cleat and with sharp termination defined by cleat fracture. Note the lack of bedding deflection in sandstone above raft, implying minimal post-burial differential compaction. Rhondda Beds, Craig yr Allt, Nant Garw (locality 2, Fig. 2). The left-hand end of the raft is 2.6 cm thick.

a)

b)

c)

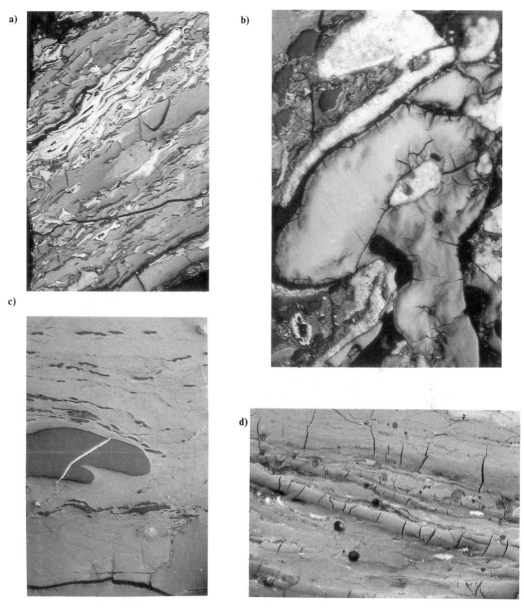

**Fig. 6.** Photomicrographs to show coal petrology of coal clasts from the Upper Coal Measures (Pennant Sandstone) of the South Wales coalfield. (**a**) Coal pebble showing trimacerite consisting of cracked vitrinite (desiccation cracks), dark liptinite (sporinite and fragments of cutinite) and pale inertinite (fusinite). Sample C63/93 from the Rhondda Beds, Craig yr Allt, Nant Garw (locality 2, Fig. 2). (**b**) Coal pebble showing heavily weathered and mineralized (siderite/hematite) coal. Sample C59/93 from the Rhondda Beds, Craig yr Allt, Nant Garw (locality 2, Fig. 2). (**c**) Coal raft showing a megaspore and other liptinitic sporinite and cutinite in vitrinite. Fracture through megaspore and vitrinite filled with siderite. Sample R51/93 from the Rhondda Beds, Craig yr Allt, Nant Garw (locality 2, Fig. 2). (**d**) Coal raft showing banded vitrinite with development of open cleat fractures. Black holes are sites of former pyrite framboids. Sample R21/92 from the Hughes Beds, Abertridwr (locality 6, Fig. 2).

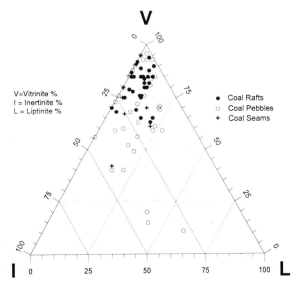

**Fig. 7.** Triangular plot of maceral groups within coal clasts and selected Pennant Measures coal seams of the study. Note the broadly similar range of coal composition of rafts and seams, but the greater spread of pebble compositions, with increased content of inertinite and liptinite groups. $N = 32$ rafts, 43 pebbles, nine seams.

$R_m$ values *in situ*, after burial in the Pennant Sandstone. It is possible that some or all of the clasts may have developed an earlier $R_m$ value before erosion, transport and burial, but, in the case of the rafts, this would have had to be at a lower rank ($R_m$ values at least 0.3% lower) than that acquired after burial in the Pennant Sandstone. However, the wider scatter in $R_m$ values of the pebbles from any one locality, some of the values being significantly higher than the $R_m$ value of the subjacent coal seam, e.g. Trehir Quarry and Tylorstown (localities 5 and 11 in Table 1 and Fig. 2), would appear to indicate that some of the pebbles may have been derived from coal seams that had $R_m$ values higher than the value developed by the pebbles as a result of burial.

### Cleat

Most coal clasts show well-developed cleat spaced between 2 and 5 mm, which is preferentially developed in bright, vitrinite-rich coal bands. In most clasts two sets of mutually perpendicular cleat are developed, comparable with face and butt cleat in coal seams. Cleats are sometimes mineralized with carbonates, clay minerals and sulphides; most commonly in pebbles, but also in rafts. In some instances an iron oxide mineralization is present in the cleat and along weathering fractures in pebbles and this is thought to be due to weathering of an original iron sulphide cleat mineralogy.

*Cleat orientation.* The cleat is always developed perpendicular to the coal banding in the clast, irrespective of the orientation of the latter. The shape of many clasts is controlled by a combination of coal banding and cleat, resulting in rectangular outlines in cross-section.

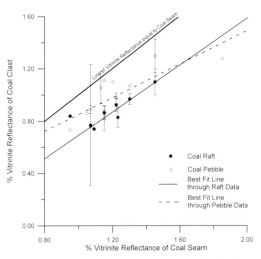

**Fig. 8.** Plot of mean vitrinite reflectance ($R_m$) of coal clasts, by locality, against $R_m$ of stratigraphically closest coal seam. All the raft data and most of the pebble data fall below line representing $R_m$ equal to the stratigraphically closest coal seam. The plot also shows best-fit linear regressions through the raft and pebble data.

The cleat dip in pebbles shows a wide variation: some with shallow dip, where the coal banding is steep, but most near-vertical, where the coal banding is parallel to bedding in the enclosing sandstone. The strike of the cleat is also variable, but with near-vertical cleat tending to form two clusters with poles trending WSW and NNW (Fig. 9a). Because most of the pebbles are small (long axis less than 5 cm), only a few cleats can be measured in one pebble and it has not been possible to compare cleat orientation between pebbles. However, the wide scatter of cleat orientation shown in Fig. 9a suggests that there is no consistent orientation.

The cleat in rafts shows a similarly random orientation of strike (Fig. 9b). Because the rafts tend to rest with coal banding parallel to the sandstone bedding, which generally has a low dip, most of the raft cleat is subvertical. Unlike the pebbles, some of the rafts are sufficiently large to obtain a reasonable cleat orientation data set from individual rafts. Two sites where rafts were particularly well developed have been chosen to illustrate the characteristics of raft cleat orientation: Tylorstown and Craig Ogwr (localities 11 and 10, respectively, Fig. 2). At Tylorstown the raft cleat strike orientation is randomly distributed, with poles forming individual clusters around the perimeter of the stereo plot (Fig. 10a). Each cluster is produced by poles to cleats in individual rafts, as shown by the plots for three rafts (Fig. 10b–10d). The raft whose cleat poles are shown in Fig. 10b is the type 1 raft illustrated in Fig. 5a, and which has

clearly been differentially compacted after deposition. The other two rafts belong to type 2 (Fig. 10c and 10d), showing no evidence for post-reworking differential compaction and one terminated by cleat fractures. The rafts at Craig Ogwr have a better defined cleat orientation (Fig. 11a), although there is a spread of data with poles ranging in azimuth from ENE to SE (cleat strike NNW to NE). Most of the rafts at this location show some evidence for differential compaction after reworking and the cleat orientation data from two of the rafts show a subtle change in orientation that appears to be related to syn-sedimentary deformation. The two rafts, illustrated in Fig. 12, occur so that raft A (Fig. 11b) lies directly above raft B (Fig. 11c). The lower raft B is tilted to the east, whereas the upper raft A is horizontal, so that it appears that raft B was either deposited on a slope (?a foreset) before raft A was deposited horizontally on a filling of sand, or that raft B originally deposited horizontally, was tilted by underlying differential compaction before raft A was later deposited horizontally. The cleat orientation appears to support the latter explanation, as cleat is formed normal to the coal banding in both clasts, suggesting that raft B acquired its cleat before it became tilted (Fig. 11b and 11c).

## Age of clasts

An attempt to date the coals from which the clasts were derived was made using micropalynology. Initially, material for palynological study of

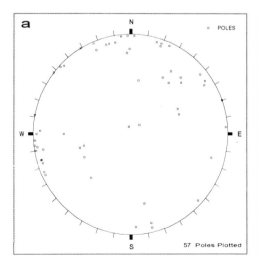

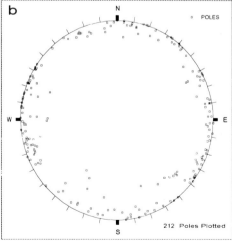

**Fig. 9.** Lower hemisphere stereographic plot of poles to cleats in coal clasts from all localities studied in the eastern half of the South Wales coalfield. (**a**) Poles to cleat in coal pebbles, showing wide scatter of cleat orientations, including some with shallow dip; $n = 57$. (**b**) Poles to cleat in coal rafts, showing most with near-vertical dip and with a wide range in strike; $n = 212$.

pebbles was collected at each locality by aggregating samples from several clasts, as insufficient coal was collected from each small clast and it was considered likely that all the clasts at one locality had a common source. However, preliminary results indicated a mixed age for these aggregates (Gayer & Pesek 1992), so that later separate samples were collected from each pebble. Miospores were found in most of the clasts. In some instances they were only able to indicate an Upper Carboniferous age, but in others the miospore assemblage was able to constrain the age to a particular Westphalian stage or range of stages. The detailed micropalynological analysis will be reported elsewhere. Here we give a summary of the findings and Table 2 lists specific miospores

found in some of the clasts that highlight a particular stage or range of stages. It is emphasized that a full miospore assemblage is required to assign an age to a sample and that the information in Table 2 serves merely as a summary.

In all instances where a unique stage was demonstrated, the age of the rafts was the same stage as that of the stratigraphic unit containing the rafts. For example: sample 1 represents a raft from the Rhondda Beds at Nant Garw (locality 2, Fig. 2), for which the miospore assemblage indicates an upper Westphalian C age (highlighted in Table 2 by the miospore *Vestispora laevigata*); sample 19, a raft from the Hughes Beds at Trehir Quarry (locality 5, Fig. 2), gives

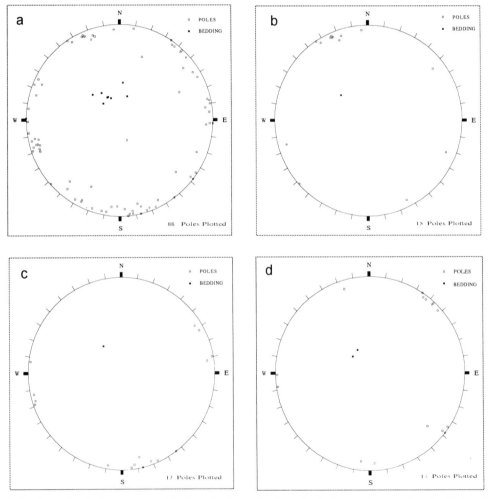

**Fig. 10.** Lower hemisphere stereographic plots of poles to cleat in coal rafts from Tylorstown (locality 11, Fig. 2). (**a**) Poles to cleat from all measured rafts; $n = 88$. (**b**) Poles to cleat from raft A illustrated in Fig. 5a; $n = 15$. (**c**) Poles to cleat from raft B; $n = 17$. (**d**) Poles to cleat from raft D; $n = 17$. See text for explanation.

an assemblage indicative of Westphalian D (highlighted in Table 2 by the presence of the miospore *Triqitrites spinosus*); and sample 30, a raft from the Brithdir Beds at Llantrisant (locality 1, Fig. 2) contains an upper Westphalian C or Westphalian D miospore assemblage (highlighted in Table 2 by the miospore *Torispora* spp.).

The age of the pebbles shows a more complex pattern. In some instances, as with the rafts, their age is the same as that of the stratigraphic unit in which they occur. For example, sample 17, a pebble from the Hughes Beds at Trehir Quarry (locality 5, Fig. 2) contains a miospore

assemblage indicative of Westphalian D (*Triqitrites spinosus*, Table 2) and sample 28, a pebble from the Brithdir Beds at Llantrisant (locality 1, Fig. 2), has an upper Westphalian C miospore assemblage (*Punctatosporites* spp. and *Densosporites sphaerotriangularis*, Table 2). In other instances the pebbles contain miospore assemblages indicating a much older age than the enclosing stratigraphic unit. For example, samples 2 and c are pebbles from the Rhondda Beds at Nant Garw (locality 2, Fig. 2) in which the miospore assemblages suggest a Westphalian A age (highlighted by *Savitrisporites nux* and *Radiizonates aligerens* respectively, Table 2).

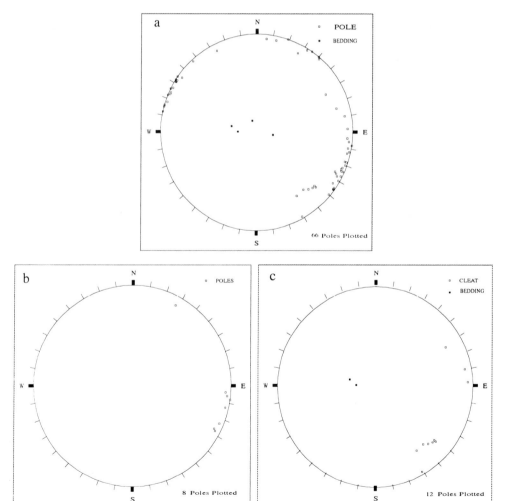

**Fig. 11.** Lower hemisphere stereographic plots of poles to cleat in coal rafts from Craig Ogwr (locality 10, Fig. 2). (**a**) Poles to cleat from all measured rafts, showing broad range in strike from NNW to NE; *n* = 66. (**b**) Poles to cleat from raft A; *n* = 8. (**c**) Poles to cleat from raft B; *n* = 12. See text for explanation.

**Fig. 12.** Photograph of rafts from Craig Ogwr (locality 10, Fig. 2), showing 'fish-tail' termination to large raft and adjoining rafts A and B for which the cleat orientations are shown in Fig. 10b and 10c.

## Interpretation and discussion

### Origin of coal clasts

The rafts appear to have been derived from two distinct sources. The rafts that show irregular terminations and evidence for compaction in the overlying sandstone were probably derived from partly lithified and compacted peat, an interpretation that is consistent with their ages being the same, within the limits of the palynology, as the enclosing strata. The cleat in these rafts shows evidence of being developed during the compaction of the raft *in situ* within the enclosing sandstones (Gayer & Pesek 1992). This is implied by the stronger cleat development in the more completely compacted central regions of the raft compared with the less compacted 'fish-tail' terminations (Fig. 5b), and also by the fact that cleat was developed in these rafts parallel to the regional far-field maximum principal stress (see later). It is likely that the source for these rafts was partly matured peat or lignite that was eroded during channel down-cutting. The other rafts show no evidence for differential compaction, have sharp terminations defined by cleat fractures and contain cleat, which although having a consistent strike orientation within each raft shows a

disorientation between rafts. This second group of rafts is thought to have been derived from already compacted and cleated coal. The fact that the ages of these rafts are the same, within the limits of the palynology, as the enclosing strata suggests that compaction and cleat formation was a very rapid process.

The lack of differential compaction shown in the pebbles implies that they were derived from fully compacted coal seams. The lack of a consistent orientation of the cleat between pebbles, together with the observation that cleat frequently defines the shape of the pebbles, strongly supports an early development of the cleat before erosion from the parent coal seam. The palynologically derived ages for the pebbles indicates that some of the pebbles had a common origin with the second group of rafts and indicate a similar rapid compaction and cleat formation. Some of the pebbles, however, were derived from much older coals. These pebbles imply that the region to the south and east of the present outcrop of the South Wales coalfield must have been uplifted and eroded after deposition, compaction and cleat formation within the Lower Coal Measures, but before deposition in the Upper Coal Measures; a time span of less than 5 Ma. Gayer & Pesek (1992) argued that progressive forelandward

**Table 2.** *Summary of relevant miospore data for a selection of pebbles and rafts used in this study. Left- and right-hand columns show clast sample numbers and localities (in parentheses). Localities refer to those in Table 1 and Fig. 2. Shading represents stratigraphic age as indicated at the top centre of table: shading in boxes in left- and right-hand columns indicates stratigraphic age of clast host rock. See text for explanation*

| PEBBLES (Locality) | MIOSPORES | Pennant | | Coal Measures | | | | | RAFTS (Locality) |
|---|---|---|---|---|---|---|---|---|---|
| | | Upr | Lr | Middle | | Lower | | | |
| | | WESTPHALIAN | | | | | | | |
| | | WD | U.WC | L.WC | WB | U.WA | M.WA | L.WA | |
| 28(1) | Thymospora spp. | | | | | | | | |
| 17(5) | Triquitrites spinosus | | | | | | | | 19(5) |
| 28(1) | Triquitrites sculptilis | | | | | | | | |
| 17(5) | Punctatosporites oculus | | | | | | | | |
| 17(5) | Punctatosporites pygmaeus | | | | | | | | 19(5) |
| 28(1) 17(5) | Punctatosporites spp. | | | | | | | | 19(5) |
| | Torispora spp. | | | | | | | | 30(1) |
| | Vestispora laevigata | | | | | | | | 1(2) |
| 17(5) | Vestispora fenestrata | | | | | | | | 19(5) |
| | Endosprites globiferus | | | | | | | | 1(2) |
| 28(1) 2(2) 60(2) 75(8) | Densosporites sphaerotriangulus | | | | | | | | |
| 2(2) | Densosporites anulatus | | | | | | | | |
| 28(1) 2(2) 60(2) 75(8) | Densosporites spp. | | | | | | | | 1(2) 19(5) |
| 2(2) | Savitrisporites nux | | | | | | | | |
| C(2) | Radizonates aligerens | | | | | | | | |

migration of Variscan deformation could have been the driving mechanism to explain this second pebble source.

Thus the evidence from this study strongly suggests that three groups of coal clast are present in the Pennant Measures of the South Wales coalfield: (i) type 1 rafts, (ii) type 2 rafts and (iii) pebbles. Each clast type appears to have been derived from a distinct source, characterized by differing thicknesses of overburden: type 1 rafts were sourced from beds of peat that had a very thin capping of sediments; type 2 rafts were eroded from seams that were buried by tens to hundreds of metres and were in the form of cleated lignite or sub-bituminous coal; and pebbles were eroded from seams covered by up to 1000 m of sediment and some had the form of cleated high-volatile bituminous coal.

## Origin of cleat

The origin of cleat in coal is not fully understood. Most workers have observed that cleat fractures commonly terminate abruptly against bed boundaries with the overlying and underlying strata (Close 1993), thus making it difficult to relate cleat development to that of normal joint systems within the Coal Measure sequence. Some workers have demonstrated a parallel orientation of joints in associated sediments with the main (face) cleat orientation (Tremain & Whitehead 1990; Tremain et al. 1991), whereas other studies have shown a lack of parallelism (Grout & Verbeek 1986; 1987; Epsman et al. 1988). In the South Wales coalfield the cleat systems, both face and butt cleat, have been shown to have formed before fracture systems associated with Variscan fold and thrust deformation (Frodsham et al. 1993; Harris et al. this volume; Hathaway & Gayer this volume). Gayer (1993) suggested that the cleat developed during dewatering as the coal-forming material compacted, producing overpressured conditions that allowed cleat fractures to form as extensional, mode one cracks parallel to the regional $\sigma_1$ orientation. Such overpressuring in coal has been suggested by Law et al. (1983)

and, according to the mechanical analysis of Lorenz *et al.* (1991), zonal overpressuring of sediments in a weak regional stress field will lead to the reduction of the least principal stress to a level at which regional closely spaced extensional fractures will develop perpendicular to the least principal stress and parallel to the greatest principal stress.

The regional stress field for the Variscan fold and thrust deformation in South Wales has been determined by Gayer & Nemcok (1994). In the south of the South Wales coalfield the $\sigma_1$ stress orientation was shown to be NNW–SSE, parallel to the main cleat direction of the rafts whose cleat was argued earlier to have formed *in situ* in the Pennant Sandstone (Figs 10 and 11). It thus seems possible that the cleat in the South Wales coalfield formed early during coal formation within a regional stress field that continued into the later fold and thrust deformation, a model consistent with the foreland basin setting discussed by Gayer & Pesek (1992).

## Origin of coal rank

Many studies have shown that the maturation of coal is related to the evolution of the associated sedimentary basin and, more particularly, to the burial history of the coal (Teichmüller & Teichmüller 1982). The coals of the South Wales basin received their rank before the main Variscan deformation which folded the coal rank contours (White 1991). White (unpublished data) argued that the coal rank was developed at maximum depositional burial, i.e. after deposition of the Upper Pennant Measures, and that the pattern of coal rank in the coalfield (Fig. 2) reflected changes in sediment thickness and hence differential subsidence. However, Gayer *et al.* (1991) found no correlation between the maximum preserved stratigraphic thickness and maximum rank, and argued that hot fluids migrating into the coal basin along deep faults and thrusts were responsible for increasing the heat flow and hence coal rank. It might be expected that heat flow through a foreland basin produced by down-flexure of the lithosphere would be low, giving geothermal gradients of approximately $25°C\,km^{-1}$ (Allen & Allen 1990). However, Alderton & Bevins (in press) have suggested a geothermal gradient of $\approx60°C\,km^{-1}$ on the basis of fluid inclusions within quartz crystals within ironstone nodules in the Middle Coal Measures close to locality 8 (Fig. 2). This elevated heat flow could be due to perturbation of a low foreland basin heat flow by hot fluids driven into

the basin by the 'rolling pin' action of foreland-ward migrating thrusts.

The evidence from the coal clasts that bears on the origin of coal rank is equivocal. The $R_m$ determinations reported here suggest that the coal rafts received their present rank *in situ* within the Pennant Sandstone, although it cannot be ruled out that a lower $R_m$ value had been acquired before the rafts had been eroded from their source seam. In the case of the pebbles there is slightly more indication that the older source coals had reached a rank at least as high as the present rank of Pennant Measures coals before erosion. Such an argument finds support in the presence of cleat in the source coals, as several studies have indicated that cleat development starts at bituminous coal rank and progressively intensifies with increasing rank (Levine 1993). If the cleated source coals for both the pebbles and the fully compacted and cleated rafts had acquired the rank of high-volatile bituminous coal before erosion, a temperature of at least $100°C$ would be required. This would imply burial to at least 3 km with a normal foreland basin geothermal gradient of $25°C\,km^{-1}$, but only 1 km with a geothermal gradient of $60°C\,km^{-1}$.

We suggest that the evidence from the coal clasts in the Pennant Sandstone of the South Wales coalfield favours the model outlined in the following.

(1)  The South Wales coal basin developed as a foreland basin as a result of downflexure of the lithosphere in response to a Variscan thrust sheet load in southern Britain.

(2)  Coals, formed within the basin, were rapidly buried by sediment shed from the tectonic load to the south. At the same time, the heat flow was increased by hot fluids migrating along thrusts into the basin from the orogenic hinterland. The combination of rapid burial and heating to $\approx100°C$ produced cleated bituminous coal, with the cleat oriented parallel to the NNW–SSE Variscan compressive stress.

(3)  Migration of Variscan deformation north-wards uplifted coal areas to the south and at the same time increased subsidence in areas to the north. This resulted in rapid erosion of the already cleated and partly matured coal to form clasts that were transported into the basin to the north and deposited in channel lags within the Pennant Sandstone.

(4)  The above sequence must have been repeated several times to produce coal clasts at successive levels throughout the

Pennant Measures and it appears that the southern rim of the South Wales coal basin must have been in permanent uplift during the Westphalian C and D.

The implications of this study for CBM are that gas may be generated in coals formed in a foreland basin setting very soon after deposition, depending on the rates of burial and levels of fluid-enhanced heat flow. At the same time, cleat, oriented parallel to the orogenic convergence direction, will be developed in the coals. Provided that the coal is not uplifted by thrusting and eroded, but remains buried, CBM may migrate out of the coal into adjacent reservoirs, or remain trapped in the coal depending on local stratigraphic, structural and hydrological conditions. It is thus important to consider the very early, post depositional history of the coal to determine the relative timing of structural development, maturation and cleat formation to assess CBM potential.

## Conclusions

(1) Coal clasts, deposited in channel lags at several horizons within the Upper Coal Measures of the South Wales coal-bearing foreland basin, can be divided by shape into (a) large, elongate rafts and (b) small, near-equidimensional pebbles.

(2) Miospore assemblages indicate that the rafts and some of the pebbles were eroded from coals belonging to the same Westphalian stage as the sediments in which the clasts are deposited. Other pebbles were derived from older, Lower Coal Measure coals.

(3) Some of the rafts show evidence for differential compaction and the development of cleat *in situ* within the enclosing sediment. Other rafts and all the pebbles were already fully compacted and had cleat developed before erosion from the source coal seam.

(4) Maceral group analysis of the clasts shows that the Upper Coal Measure coal seams and the rafts have a similar range of composition, dominated by vitrinite group macerals. The pebbles show a wider range of maceral group composition with some pebbles containing high percentages of inertinite and liptinite group macerals.

(5) $R_m$ values for the rafts are mostly 0.3% lower than those of the adjacent coal seams and show a similar rank pattern to that of *in situ* coals in the South Wales coalfield. It is argued that the rafts acquired their maximum $R_m$% *in situ* in the Upper Coal

Measures. The pebbles show a less ordered pattern of $R_m$ values, with some higher than those of the adjacent coal seams. It is argued that most of the pebbles may have been derived from coal seams that had acquired their rank before erosion.

(6) Cleat in South Wales coals developed early during compaction to form coal, as a result of extensional fracturing of overpressured coal in response to NNW–SSE oriented Variscan compression.

(7) Burial to ≈1 km within an elevated geothermal gradient, enhanced by hot fluid flow along thrusts, resulted in rapid maturation to bituminous coal rank.

(8) The development of rank and cleat in South Wales coals is related to the evolution of a foreland basin.

(9) The intimate association of deposition, coal maturation, cleat development and deformation have important implications for the timing of CBM generation and retention in coals, which require detailed geological analysis of a region to assess CBM potential.

## References

ALLEN, P. A. & ALLEN, J. R. 1990. *Basin Analysis— Principles and Applications*. Blackwell Scientific, Oxford.

ALDERTON, D. H. M. & BEVINS, R. E. P–T conditions during formation of quartz in the South Wales coalfield: evidence from co-existing hydrocarbon and aqueous fluid inclusions, in press.

CLOSE, J. C. 1993. Natural fractures in coal. *In*: LAW, B. E. & RICE, D. D. (eds) *Hydrocarbons from Coal*. American Association of Petroleum Geologists, Studies in Geology, **38**, 119–132.

EPSMAN, M. L., WILSON, & 8 others 1988. *Geologic Evaluation of Critical Production Parameters for Coalbed Methane Resources Part II, Black Warrior Basin*. Gas Research Institute Annual Report, **GRI-88/0332.2**, 178pp.

FRODSHAM, K., GAYER, R. A., JAMES, J. E. & PRYCE, R. 1993. Variscan thrust deformation in the South Wales Coalfield—a case study from Ffos-Las Opencast Coal Site. *In*: GAYER, R. A., GREILING, R. O. & VOGEL, A. (eds) *The Rhenohercynian and Subvariscan Fold Belts*. Earth Evolution Science Series. Vieweg, Braunschweig.

GAYER, R. A. 1993. The effect of fluid over-pressuring on deformation, mineralisation and gas migration in coal-bearing strata. *In*: PARNELL, J., RUFFELL, A. H. & MOLES, N. R. (eds) *Contributions to an International Conference on Fluid Evolution, Migration and Interaction in Rocks. Geofluids '93 Extended Abstracts, Torquay*, 186–189.

—— & JONES, J. 1989. The Variscan foreland in South Wales. *Proceedings of the Ussher Society*, **7**, 177–179.

—— & NEMCOK, M. 1994. Transpressionally driven rotation in the external Variscides of southwest Britain. *Proceedings of the Ussher Society*, **8**, 317–320.

—— & PESEK, J. 1992. Cannibalisation of Coal Measures in the South Wales Coalfield—significance for foreland basin evolution. *Proceedings of the Ussher Society*, **8**, 44–49.

——, COLE, J., FRODSHAM, K., HARTLEY, A. J., HILLIER, B. MILIORIZOS, M. & WHITE, S. 1991. The role of fluids in the evolution of the South Wales Coalfield foreland basin. *Proceedings of the Ussher Society*, **7**, 380–384.

GROUT, M. A. & VERBEEK, E. R. 1986. Prediction of joint patterns at depth—examples from the Piceance basin, northwestern Colorado. *Geological Society of America, Abstracts with Programs*, **18**(5), 358.

—— & —— 1987. Regional joint sets unrelated to major folds: example from the Piceance basin, northern Colorado Plateau. *Geological Society of America, Abstracts with Programs*, **19**(5), 279.

HARTLEY, A. J. 1993. A depositional model for the Mid-Westphalian A to late Westphalian B Coal Measures of South Wales. *Journal of the Geological Society, London*, **150**, 1121–1136.

JONES, J. 1989. The influence of contemporaneous tectonic activity on Westphalian sedimentation in the South Wales coalfield. *In*: GUTHRIDGE, P., ARTHURTON, R. S. & NOLAN, S. C. (eds) *The Role of Tectonics in Devonian and Carboniferous Sedimentation in the British Isles*. Occasional Publication of the Yorkshire Geological Society, **6**, 243–253.

KELLING, G. 1974. Upper Carboniferous sedimentation in South Wales. *In*: OWEN, T. R. (ed.) *The Upper Palaeozoic and Post-Palaeozoic Rocks of Wales*. University of Wales Press, Cardiff, 185–224.

——1988. Silesian sedimentation and tectonics in the South Wales Basin: a brief review. *In*: BESLY, B. & KELLING, G. (eds) *Sedimentation in a Syn-orogenic Basin Complex: the Upper Carboniferous of Northwest Europe*. Blackie, Glagow and London, 38–42.

LAW, B. E., HATCH, J. R., KUKAL, G. C. & KEIGHIN, C. W. 1983. Geological implications of coal dewatering. *American Association of Petroleum Geologists Bulletin*, **67**, 2255–2260.

LEVINE, J. R. 1993. Coalification: the evolution of coal as source rock and reservoir rock for oil and gas. *In*: Law, B. E. & RICE, D. D. (eds) *Hydrocarbons from Coal*. American Association of Petroleum Geologists, Studies in Geology, **38**, 39–77.

LORENZ, J. C., TEUFEL, L. W. & WARPINSKI, N. R. 1991. Regional fractures 1: a mechanism for the formation of regional fractures at depth in flat-lying reservoirs. *American Association of Petroleum Geologists Bulletin*, **75**, 1714–1737.

OWEN, T. R. 1974. *In*: OWEN, T. R. (ed.) *The Upper Palaeozoic and Post-Palaeozoic rocks of Wales*. University of Wales Press, Cardiff.

SCOTT, A. C. 1989. Observations on the nature and origin of fusain. *International Journal of Coal Geology*, **12**, 443–475.

STACH, E., MACKOWSKY, M.-TH., TEICHMÜLLER, M., TAYLOR, G. H., CHANDRA, D. & TEICHMÜLLER, R. 1982. *Stach's Textbook of Coal Petrology*, 3rd edn. Gebrüder Borntraeger, Stuttgart and Berlin, 535pp.

TEICHMÜLLER, M. & TEICHMÜLLER, R. 1982. The geological basis of coal formation. *In*: STACH, E., MACKOWSKY, M.-TH., TEICHMÜLLER, M., TAYLOR, G. H., CHANDRA, D. & TEICHMÜLLER, R. (eds) *Stach's Textbook of Coal Petrology*, 3rd edn. Gebrüder Borntraeger, Stuttgart and Berlin, 5–86.

TREMAIN, C. M. & WHITEHEAD, N. H. 1990. Natural fracture (cleat and joint) characteristics and patterns in Upper Cretaceous and Teriary rocks of the San Juan basin, New Mexico and Colorado. *In*: AYERS, W. B., KAISER, W. R., AMBROSE, W. A., SWARTZ, T. E. & LAUBACH, S. E. (eds) *Geologic Evaluation of Critical Production Parameters for Coalbed Methane Resources: Part 1 San Juan Basin*. Gas Research Institute, Annual Report, **GRI-90/0014**, 73–98.

——, LAUBACH, S. E. & WHITEHEAD, N. H. 1991. Coal fracture (cleat) patterns in Upper Cretaceous Fruitland Formation, San Juan basin, Colorado and New Mexico: implications for coalbed methane exploration and development. *In*: AYERS, W. B., KAISER, & 12 others (eds) *Geologic and Hydrologic Controls on the Occurrence and Producibility of Coalbed Methane, Fruitland Formation, San Juan Basin*. Gas Research Institute, Topical Report, **GRI-91/0072**, 97–117.

WHITE, S. 1991. Palaeo-geothermal profiling across the South Wales Coalfield. *Proceedings of the Ussher Society*, **7**, 368–374.

WOODLAND, A. W. & EVANS, W. B. 1964. *The Geology of the South Wales Coalfield, Part IV, the Country Around Pontypridd and Maesteg*, 3rd edn. Memoir of the Geological Survey of Great Britain.

# Thrust-related permeability in the South Wales Coalfield

T. M. HATHAWAY & R. A. GAYER

*Laboratory for Strain Analysis, Department of Earth Sciences, University of Wales,*
*PO Box 914 Cardiff CF1 3YE, UK*

**Abstract:** The South Wales Coalfield is a Variscan foreland basin extensively deformed by both linked and isolated thrusts in response to regional NW–SE compressive stress. Thrusts normally strike NE–SW, but transpression has caused a variable dextral rotation of the thrusts to strike E–W and even NW–SE. The range in orientation has allowed thrust-related fractures locally to be opened within the neotectonic stress field in which $\sigma_1$ is oriented NW–SE. It is argued that similar thrust-related permeability should be developed in other coal-bearing foreland basins both associated with the Late Carboniferous Variscan/ Appalachian orogeny and with younger compressional tectonic systems.
    The dominant meso- to major-scale structures formed during the compression of coal-bearing sequences are thrusts and folds. Strains at leading and trailing tip-lines of isolated thrusts and in the immediate hanging wall and footwall generate tension cracks which may act as methane conduits. Unsealed, these allow permeability parallel to thrust strike. Décollements form as bed-parallel detachments within coals, developing as pervasive shear zones, characterized by cleavage duplexes and C-S fabrics in which a penetrative new hinterland-dipping fabric is formed. This fabric, under changing regional stress conditions, may be opened to form a highly effective gas migration pathway. Coal-bearing strata develop chevron folds with flexural slip in fold limbs and tension gashes in competent strata, generating porosity and permeability parallel to the strike of fold limbs. Incompetent coal seams are strongly sheared, producing cleavage duplexes with contrasting vergence in opposing limbs.

Like many coal-bearing basins worldwide, the South Wales Coalfield has a proved potential for coal bed methane (CBM) production (Cornelius *et al.* 1993). The methane content of the included coals is high, with values up to $22 \, m^3/t$ in the anthracite field to the northwest of the coalfield (Creedy 1988). The recorded incidence of outburst coal conditions in the more strongly deformed seams in the anthracite region of the coalfield is also thought to be associated with high pressures of sorbed gas (Barker-Read unpublished data; Cross unpublished data)
    The development of rank and therefore of CBM in the coalfield occurred before the onset of thrusting (White 1991), when the methane would have been held within the micropore structure of the coal, in spaces between the matrix constituents. Provided there is sufficient permeability and pressure within the reservoir, the methane will flow and escape towards the surface or other areas of low pressure through fractures within the rock (Creedy 1991; Gamson this volume). It is thought that, at this early stage, reservoir permeability was provided by cleats. The coalfield has been affected subsequently by progressively developed fold and thrust deformation (Cole *et al.* 1991; Gayer *et al.* 1991; Jones 1991), which generated additional fracture systems within the coals. Provided that these compressionally derived fractures are open

within the neotectonic stress regime, they should provide additional fracture permeability within the coalbed reservoir during commercial methane production. A detailed investigation of methane desorption related to tectonically induced fracture spacing in South Wales anthracites is given by Harris *et al.* (this volume).
    This paper presents recent observations and interpretations of stress-related fracturing associated with thrust faults and folds. Field examples of these structures in the South Wales coalfield are given and an assessment of CBM potential in the coalfield and applications of the thrust fracture model to other foreland thrust systems are discussed.

## Thrust and fold-related fracture systems in the South Wales Coalfield

The South Wales Coalfield forms an E–W oriented basin of Silesian age in which the basin-fill comprises Namurian to Westphalian siliciclastic sediments (Fig. 1) (Kelling 1974; Hartley 1993). The sedimentary sequence consists of a series of coal cyclothems of seatearth, coal, mudstone and sandstone which exhibit contrasting rheological properties, giving the basin a characteristic mechanical structure. It is this sedimentary structure which exerts the

*From* Gayer, R. & Harris, I. (eds) 1996, *Coalbed Methane and Coal Geology,*
Geological Society Special Publication No 109, pp 121–132.

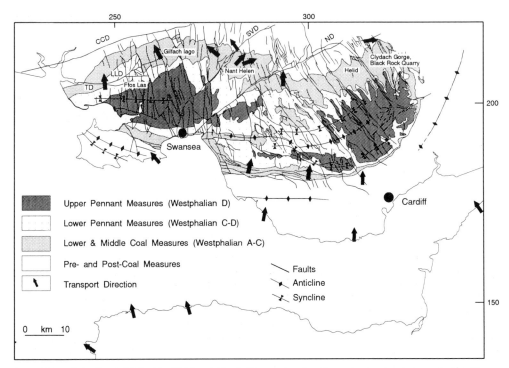

**Fig. 1.** Map of the South Wales Coalfield and North Devon showing the main structural elements, thrust transport directions and the locations of opencast coal sites and Black Rock Quarry. Thrust transport directions on the south crop of the coalfield represent a zone of under thrusting. CCD, Carreg Cennen Disturbance; LLD, Llannon Disturbance; TD, Trimsaron Disturbance; SVD, Swansea Valley Disturbance; ND, Neath Disturbance.

strongest influence over the geometry of thrusts and folds and associated fractures.

The South Wales coalfield has been substantially tectonized with periods of extension and compression. A framework of NE–SW trending disturbances represent the Variscan (latest Carboniferous) reactivation of Caledonide fault zones within the basement (Owen & Weaver 1983). They are characterized by intense folding and faulting in the Upper Palaeozoic cover strata and have been interpreted as areas where a basal thrust underlying the coalfield may link to Caledonoid structures in the basement (Brooks *et al.* 1994). E–W trending folds and E–W striking thrust faults, compartmentalized by NNW trending strike-slip faults, developed. The latter were reactivated as normal faults during NE–SW Mesozoic extension. Local variations in these trends are the result of transpressional rotation during Variscan convergence (Gayer & Nemčok 1994).

Depths to crystalline basement vary across the coalfield, with a region of shallow (<3 km) depth occurring in the northwest of the coalfield (Mechie & Brooks 1984). Although the basement

does not normally appear significantly to have controlled the structure, it is possible that the E–W striking Trimsaron Disturbance (thrust and fold zone) in the west of the coalfield represents an inversion of an earlier extensional fault that defined the southern edge of the region of shallow basement and that the NE–SW striking Vale of Neath Disturbance is connected with a basement fault controlling its southeastern margin (Hillier unpublished data).

## Isolated thrust geometry

Thrust deformation is widespread throughout the coalfield, but is most intensive in the central and western coalfield and becomes less prevalent to the east of Helid Opencast Site. An increase in thrust deformation towards the northwest is coincident with an increase in coal rank and vitrinite reflectance for the Westphalian A B and lower C Coal Measures (White 1991). It has been shown by Gillespie (unpublished data), Hathaway & Gayer (1994) and Gayer *et al.* (1995) that thrusts in the coalfield occur either as

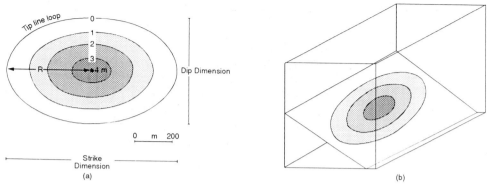

**Fig. 2.** Thrust fault geometry. (**a**) Schematic displacement contour strike-projection of an idealized fault. R is the fault radius. (**b**) Block diagram to show strike projection of the fault. After Walsh and Watterson (1990).

isolated or linked structures. The thrust geometry within linked systems is characterized by extensive flats with dips of <5°, which lie parallel to bedding in lithologies such as shale and coal and have low displacement gradients. These flats are connected by ramps with dips of between 20 and 30°. It is possible that the ramps originate as isolated thrusts and become linked by downwards propagation to join a detachment (Eisenstadt & De Paor 1987). Associated with ramps are fault-propagation folds, discussed later, and high displacement gradients. Here we will consider only strain associated with isolated thrusts and describe the resulting fractures that, if open, may provide permeability for methane within the coalfield.

Isolated faults are analogous to crystal dislocation planes bounded by a dislocation loop (Eshelby 1957). The fault surface is an elliptical plane with the long axis of the ellipse in the strike dimension and the short axis of the ellipse parallel to transport, in the dip direction (Barnett *et al.* 1987) (Fig. 2). Aspect ratios of thrusts range from 3:1 to 18:1 (Gillespie unpublished data). Total displacement and deformation associated with individual faults are contained within an ellipsoidal volume. Displacement varies across each thrust plane, with maximum displacement at the centre of the fault ellipse decreasing systematically to zero displacement at the edges of the ellipse, the tip-line. This variation in displacement across the fault results in strain which is accommodated in the rock adjacent to the fault plane as normal and reverse drag (Fig. 3). Normal drag may have formed as a frictional feature against the

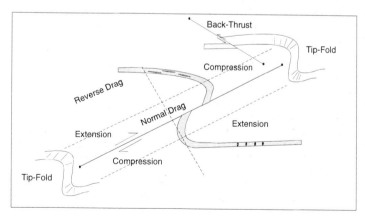

**Fig. 3.** Description of stress orientations around an isolated thrust and the resulting features accommodating strain. The shaded bed represents a competent sandstone which has been stretched parallel to bedding in the extensional field generating subvertical tension gashes, but has been thickened in the compressional field with the production of bed-parallel tension gashes.

thrust or, more realistically, as an original tip strain that has been cut through during thrust propagation (Williams & Chapman 1983) and contributes to the total displacement of the fault whereas reverse drag, where the beds dip away from the opposite cut-off, accommodates the displacement variation on a normal to the fault plane (Gillespie unpublished data).

## Features of strain associated with isolated thrusts

Displacement variation across the fault surface and the opposing slip between the hanging wall and footwall sections is accommodated by ductile (plastic) strains relaxed permanently in the surrounding rock to produce regions of extension and compression around the fault (Barnett et al. 1987; Hyett 1990) (Fig. 3). Fletcher and Pollard (1981) described a similar anticrack model for the propagation of solution surfaces where opposing fields of displacement are identical, but with a changing sign resulting in extension cracks opening on one side of a propagating crack with solution surfaces on the opposite side.

In the vicinity of thrusts ductile strain is manifested as changes in bed thickness and the production of dilatant extensional veins. Dilatant veins around thrusts provide a kinematic indicator and palaeostress orientations. Dilational fractures normally form perpendicular to a minimum effective tensional principal stress in the plane of $\sigma_1$ and $\sigma_2$ with mineral fibres perpendicular to fracture walls and parallel to symbol $\sigma_3$ (Dunne & Hancock 1994). However, it should be noted that extensional veins may also develop in transpression/transtension zones with fibres oblique to the vein wall and McCross (1986) proposed an analysis taking into account such fibre orientations. In regions of compression, beds which are subhorizontal become shortened and thickened, strain being accommodated either by ductile vertical thickening in incompetent rocks or by subhorizontal tension gashes in more competent lithologies. In the regions of extension, subhorizontal beds will be stretched and thinned, the strain being accommodated in competent beds by near-vertical veins with extensional fibres oriented near-horizontal, parallel to the regional $\sigma_1$.

These types of vein structures have been well documented in other areas of thrust faulting. It is these veins, unmineralized, that will provide pathways for methane flow during extraction. However, there are certain constraints on the development of such veins. Parallel extension

fractures and veins are often associated with thrust, normal and wrench faults; they are produced by repeated hydro-fracturing under a shared stress regime and are a low differential stress phenomenon (Nur et al. 1973; Sibson 1981) (Fig. 4). This can only occur in association with thrusts when fluid pressures exceed the lithostatic load. Expressed as an equation for thrust faults, where

$$\sigma_v = \rho g z = \sigma_3$$

$\rho$ is the average rock density, $g$ the gravitational acceleration and $z$ the depth, the conditions for hydrofracturing are

$$\lambda_v = 1 + t/\rho g z$$

where $t$ is the tensile strength and $\lambda_v$ must therefore be greater than or equal to unity; see Sibson (1981) for a full treatment and discussion. Figure 5 shows delimiting curves for hydro-fracturing in the top 10 km of the crust developed by Sibson (1981), who also concluded that repeated fluid pumping may increase fracturing and that stress cycling may be essential to retain fault and fracture permeability. High fluid pressures were seen as a critical factor by Frodsham et al. (1993) in the development of thrusting in the South Wales Basin, resulting in a style of faulting known as progressive easy slip thrusting. The presence of

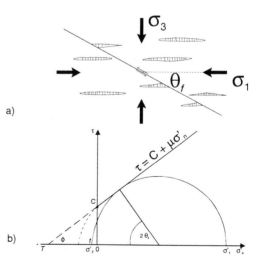

**Fig. 4.** Hydraulic extension fracturing in association with fault. (a) Field relationships and (b) critical stress condition for simultaneous shear failure and reopening of extension fractures. $\sigma_1$ and $\sigma_3$, Effective principal stresses; $\tau$, shear stress; $C$, cohesion; $\mu$, coefficient of friction; $\phi$, angle of friction; and $T$, tensile strength of the intact rock. After Sibson (1981).

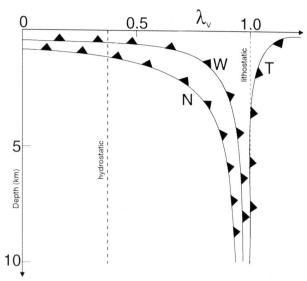

**Fig. 5.** Curves delimiting potential fields of shear failure and hydro-fracturing (barbed side) for different modes (T, thrust; W wrench; N normal), $\lambda_v$ = hydrostatic head. After Sibson (1981).

high fluid pressures during thrust formation therefore increases the probability of there being a well-developed thrust-related fracture system within the South Wales coalfield.

### Thrust-related folds

Various styles of folding occur in association with isolated thrusts and detachments in the coalfield. Jamison (1989) suggested that three distinct processes operated to form such folds. The first occurs when the hanging wall or footwall moves around a bend in the thrust surface, e.g. at a change from ramp to flat. These folds are known as fault bend folds. The second process results from strains introduced into the rock in the immediate vicinity of a propagating thrust tip, producing a fault propagation fold. These may amplify with propagation and later may be cut through to become features of normal drag (see earlier). Finally, the third type is formed when thrusts lie parallel to bedding within the coal seams, which are overlain by competent sandstones. The strata in the sequence overlying the coal become folded as strain is transferred vertically to the overlying sequence, forming a detachment fold. These are common in the Ffos Las opencast site. The result of this folding in competent beds will be to produce fractures in the outer arc of the hinges of the competent units striking parallel to the axial plane of the fold. Fault propagation folding is

common in Nant Helen opencast site where thrusts are more isolated and displacement is lost up and down-dip (Hathaway & Gayer 1994).

### Cleavage duplexes

Bed-parallel detachments are commonly developed within the coals of the South Wales coalfield (Frodsham et al. 1993). It is thought that this association has resulted from fluid overpressuring within the coals during rapid compaction and dewatering of the mudrock and peat sequence (Gayer 1993). The characteristic structures of these coal detachments are cleavage duplexes, which consist of a series of closely spaced, polished, sigmoidal surfaces developed at about 30–45° to the coal bedding and striking parallel to the regional trend of folds and thrusts (Fig. 6a). The surfaces commonly contain slicken-lineations indicating the sense of thrusting. In some instances these surfaces develop as discrete thrust ramps, repeating coal in a true duplex structure (Fig. 6b). More commonly, they occur as a sigmoidal cleavage, giving the typical structure of a ductile shear zone, including asymmetrical folds and S-C fabrics defining the sense of shear (Fig. 6a). The cleavage duplex fabric progressively destroys the earlier bedding and cleat fabrics and eventually produces the dominant fracture system within the coal. Provided this fracture system is open, it will generate pathways for methane flow.

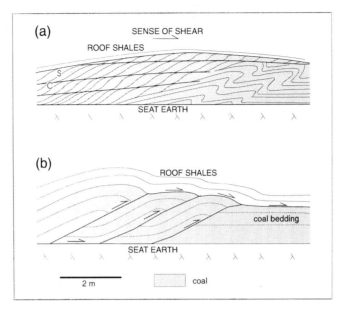

**Fig. 6.** Cleavage duplex. (**a**) S-C fabrics with a coal duplex progressively destroying bedding towards the transport direction and (**b**) a more common duplex where horses occur as sigmoidally shaped packages bounded by a floor and roof thrust.

## Chevron folds

Strong chevron folding is developed in the northwest of the coalfield. The folds are open to closed, inclined structures with subhorizontal hinges. In most instances they verge to the north or northwest and appear to be associated with thrusting, either as fault propagation folds or detachment folds. The folds commonly deform earlier bed-parallel coal detachments so that cleavage duplexes related to the latter are folded around the chevron hinges. Downward facing

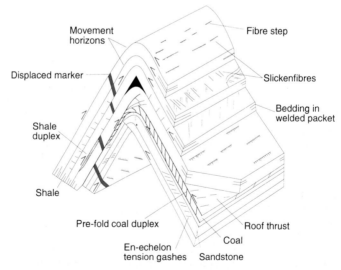

**Fig. 7.** Features found in association with chevron folding in the coalfield. The most important features are duplexes associated with fold limbs and tension cracks on the hinges of folds and in the limbs. Note the presence of a pre-fold duplex in the coal (see also Fig. 9c and 9d). Modified from Tanner (1989).

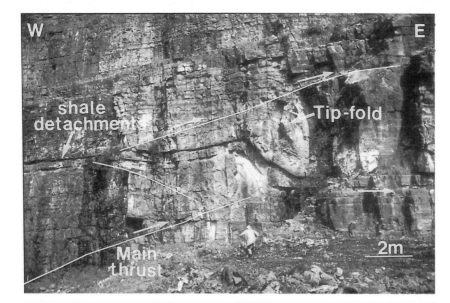

**(a)**

W       E

shale detachments

Tip-fold

Main thrust

2m

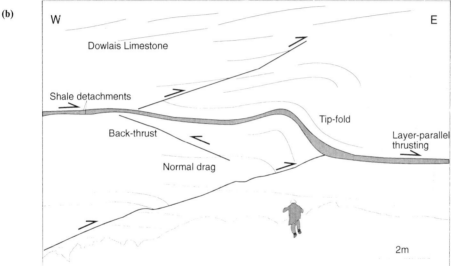

**(b)**

W       E

Dowlais Limestone

Shale detachments

Back-thrust

Normal drag

Tip-fold

Layer-parallel thrusting

2m

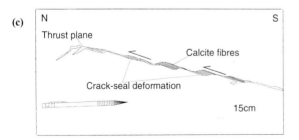

**(c)**

N       S

Thrust plane

Calcite fibres

Crack-seal deformation

15cm

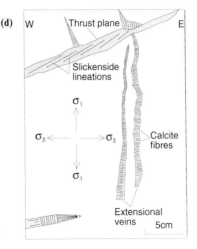

**(d)**

W    Thrust plane    E

Slickenside lineations

$\sigma_1$

$\sigma_3 \longleftarrow \longrightarrow \sigma_3$

$\sigma_1$

Calcite fibres

Extensional veins   5cm

**Fig. 8.** (**a**) Photograph and (**b**) interpretation of Black Rock Thrust within the Lower Carboniferous Dowlais Limestone in the Clydach Gorge. The thrust has a prominent tip-fold and back thrusting in the hanging wall. (**c**) A series of pull-apart calcite veins along the thrust plane with individual calcite fibres orientated along the transport direction. (**d**) These extensional veins are found commonly in the footwall of the thrust where they have been cut through by the propagating tip of the thrust. The calcite fibres are inferred to be oriented parallel with the $\sigma_3$ axis.

thrust detachments thus occur on the steep to overturned northern limbs, in which thrust ramps, associated with cleavage duplexes, dip less steeply north than the coal bedding (Fig. 7).

The folds are formed by a flexural slip process with the development of polished, slickensided bedding planes in the fold limbs. Where packets of competent sandstones are folded, sigmoidal, *en echelon* tension gashes may form within the competent beds, and bedding plane slip is concentrated in the interbedded incompetent coals, producing significant thrust slip within the chevron limbs. Cleavage duplexes, verging towards the fold hinge in both limbs, often develop in this setting (Tanner 1989) (Fig. 7). The fracture systems developed are comparable with those in duplexes related to thrusting and will provide similar migration pathways for methane.

### Field examples

There are many examples within the coal basin of the predicted fractures associated with isolated thrusts. Although thrusting occurs throughout the Coal Measures it is concentrated in the Westphalian A and B Lower and Middle Coal Measures, becoming almost absent in the more competent Upper Pennant Measures. Thrusting is also common in the Carboniferous Limestone below the Coal Measures, and at Black Rock Quarry in the Clydach Gorge (Fig. 1) on the northeastern margin of the coalfield, a thrust and thrust propagation fold are well exposed (Fig. 8a, b). Associated with the thrust are a series of calcite-filled veins.

These veins are of two types, classified by their orientation. The first type of vein occurs in jogs along the thrust plane and consists of slicken-fibres with stretched calcite crystals parallel to symbol $\sigma_1$ (Fig. 8c). The second set of veins is subvertical and occurs in the immediate footwall of the thrust. Stretched calcite fibres parallel to symbol $\sigma_1$ in these veins are subhorizontal (Fig. 8d). The second set of veins is cut by the thrust plane and can be assumed to have formed at the tip of the thrust, which later propagated outwards and cut through the veins. Owing to the height of the thrust in the quarry wall it was not possible to examine the hanging wall closely in the volume where veining would be predicted, but horizontal calcite veins are developed in the immediate hanging wall.

Figure 9a shows an example of an isolated thrust cutting Westphalian B Coal Measures in Nant Helen opencast coal site in the anthracite field on the northern margin of the coalfield (Fig. 1). This is one of a series of uniformly (30–50 m) spaced, isolated thrusts, dipping at about 20–30°SW and restricted to a sequence containing well-defined sandstone interbeds. In the example illustrated, a 30 cm sandstone bed has been thinned and stretched in the footwall of the thrust with the development of subvertical tension gashes partly infilled with quartz. The vein quartz has stretched fibres oriented sub-horizontally, parallel to the regional $\sigma_1$. In the hanging wall, the displaced sandstone has been slightly thickened, presumably by ductile processes without the development of tension gashes.

Examples of cleavage duplexes within detachments parallel to coal seams are extremely common throughout the western and central parts of the coalfield. Figure 9b illustrates a typical example from Nant Helen opencast coal site. The 1.5 m thick coal is deformed by two duplexes separated by a thin shale band within the coal. The slickenside surfaces are sigmoidally curved, dipping towards the southwest, with a spacing of 10–20 cm.

Figure 9c shows an example of a chevron fold within the Lower Coal Measures in the north-western margin of the coalfield at Gilfach Iago opencast coal site (Fig. 1). The fold has an inter-limb angle of 70° and the axial surface dips approximately 60° to the SE, indicating a northwesterly vergence. Bedding planes in the fold limbs are slickensided, but slickenlines plunge almost parallel to the fold hinge rather than down-dip. It is thought that is the result of transpressional deformation, discussed later. Coal seams folded by the structure show a complex pattern of cleavage duplex formation. In the shallow SE dipping limb an apparently single set of slickensided sigmoidal fractures cuts the coal, dipping at between 30° and 50° to the SE (Fig. 9d). In the subvertical northwesterly limb an early cleavage duplex faces downwards and is clearly folded around the fold hinge, but a weaker, second set of sigmoidal fractures, dipping NW, is superimposed on the first duplex. We interpret the first downward facing duplex as a pre-folding layer-parallel detachment folded around the chevron and the second set of fractures to be the result of flexural slip in the fold limb, directed towards the outer arc of the fold hinge. The absence of a second cleavage duplex in the shallow dipping SE limb may be due to the fact that the flexural slip episode would have had the same sense of shear in this fold limb as that of the early thrust detachment, which would have resulted in an intensification of the early fabric and thus would be difficult to distinguish.

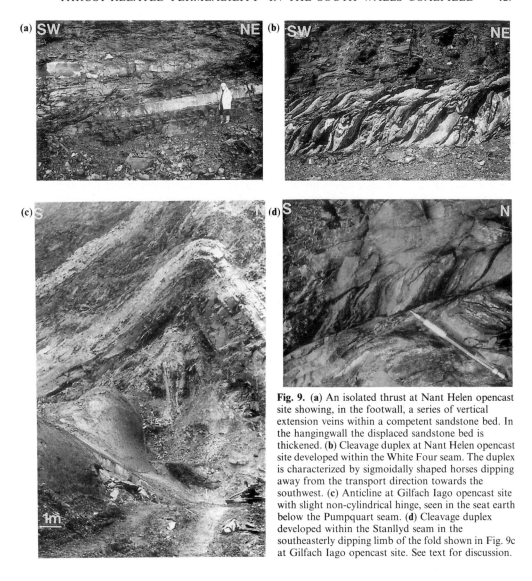

**Fig. 9.** (a) An isolated thrust at Nant Helen opencast site showing, in the footwall, a series of vertical extension veins within a competent sandstone bed. In the hangingwall the displaced sandstone bed is thickened. (b) Cleavage duplex at Nant Helen opencast site developed within the White Four seam. The duplex is characterized by sigmoidally shaped horses dipping away from the transport direction towards the southwest. (c) Anticline at Gilfach Iago opencast site with slight non-cylindrical hinge, seen in the seat earth below the Pumpquart seam. (d) Cleavage duplex developed within the Stanllyd seam in the southeasterly dipping limb of the fold shown in Fig. 9c at Gilfach Iago opencast site. See text for discussion.

## Discussion

### Thrust-related fractures as permeability pathways

Before thrusting, methane would have been adsorbed in the micropore system of the coal and freely held within the larger pores and cleat fractures. Methane will escape where there is either tectonic activity and associated fractures or where mining disturbs deformed strata, but most gas will be held in place by a caprock of impermeable mudrocks and sandstone. Free gas escape can be seen in most opencast pits, where it bubbles through collected water and is associated with fault planes. For the thrust-related fracture systems described earlier to have a significant influence on the permeability of the coal, these fractures must be open at the time of methane extraction, but not during methane generation when the coal would have acted as both a source and reservoir. The fractures generated by extensional stresses associated with reverse drag must not be completely mineralized and the fractures generated directly by compressive stresses would need to have been opened within the neotectonic stress field. It is thus neccssary to know the relationship between the orientation of the thrust-related fractures and the neotectonic stress field. Where cleat fractures are incompletely mineralized they may link with strike-parallel fractures in cleavage duplexes, facilitating the migration of methane into tectonically generated fractures.

A significant variation in transport direction of the thrusts occurs along the northern margin of the coalfield. The three localities described in the previous section record transport directions ranging from eastwards in the Clydach Gorge in the NE, through northeastwards at Nant Helen in the north-centre, to northwestwards in the NW. The regional direction of Variscan convergence is thought to be towards the NW, in common with Variscan tectonics in SW England and farther E in Europe (Andrews 1993; Gayer & Nemčok 1994). Palaeostress analysis of the Variscan structures in South Wales (Gayer & Nemčok 1994) indicated that the variation in Variscan transport direction along the north outcrop of the coalfield was due to a combination of transpressional block rotation and stress deflection, resulting from oblique dextral convergence against the E–W Wales Basement Massif.

The effect of this transpressional rotation is that thrust-related fractures will have a variety of orientations, depending on the geometry of transpressional tectonics at any particular locality.

The regional neotectonic stress field is constant across much of Europe north of the Alps. It appears to have been in existence since the Early Miocene and is oriented with $\sigma_1$ NNW–SSE (Bergerat 1987). Thus it is likely that in regions of the coalfield where thrusts and folds trend ENE–WSW, any thrust-related fractures will have remained tight. On the other hand, where the Variscan structures trend NW–SE, as they do, for instance, in the vicinity of Nant Helen opencast coal site, it is likely that the thrust-related fracture systems will have been opened. In such areas they would be expected to enhance significantly the coal permeability.

Linked thrusts have until now not been discussed and little is known about the state of stress and stress orientations in their vicinity. Observations from the NW of the South Wales coalfield at Ffos Las opencast coal site have shown that linked thrusts are often associated with areas of friable coal (Frodsham et al. 1993; Harris 1995; Harris et al. this volume). Friable coal has developed powdery properties resulting from intense thrust deformation and crushing. Associated with this friability are cleavage duplexes, asymmetrical folds and other high strain indicators. Harris et al. (this volume) have

demonstrated that such high strain increases the rate and amount of methane desorption from the coal, but it is also possible that the friable coal may block pore and fracture permeability unless suitable precautions are taken.

## Applicability to other coal basins

The examples used for this study have all been taken from the South Wales coal basin. However, we believe that many of these features are not restricted to this basin. The other Variscan/Appalachian coal-bearing foreland basins in which CBM is currently produced or is being actively considered are the Ruhr coalfield in Germany, the Upper Silesian coalfield in Poland and the Czech Republic and the Pennsylvanian and Black Warrior basins in the USA. Pashin & Carroll (1993) have documented fractures related to the main thrust adjacent to the Cahaba Synclinorium of Alabama. The fractures they have identified are associated with low displacement thrusts that are contained within coal seams. The fractures occur as inclined surfaces parallel to the strike of bedding, but dipping 60° with slickensided surfaces and normal displacement where they cut coal. Similar slip fractures are also reported from the South Wales Coalfield, but these do not appear to displace bedding and are thought to be the result of regional compression. Pashin & Carroll (1983) also attribute their fractures to compression, but suggest that they are formed by flexural slip on the limbs of folds, where strata are thinned.

In addition, there are many basins developed in similar tectonic environments worldwide which could be expected to show thrust-related fracture systems. The model developed here has been shown to be applicable to both Coal Measures sequences and limestones and thus it is reasonable to assume that thrusts in these other basins will show a similar relationship to the thrust regime and that a comparable geometry will be shown in the strains developed.

## Conclusions

Several categories of fractures may form in association with thrusts. Firstly, extensional fissures formed by strains associated with reverse drag developed in the volume surrounding isolated thrusts; secondly, a range of fabrics formed as a result of thrust detachments within coal seams; and, thirdly, fractures and fabrics developed in association with chevron folds formed as thrust detachment or thrust propagation folds. The first category is likely to provide permeability pathways for methane provided the fractures have remained unmineralized. The second and third categories result from compressive strains and are likely to remain tight unless opened in a later stress field. The permeability formed by all the above processes will be strongly aligned with the strike of the thrusts and the trend of associated fold hinges.

In the South Wales coalfield, thrust detachments, with associated fabrics, are present in many of the coal seams in the centre and west of the coalfield. As a result of Variscan transpressional rotation, the thrust-related fracture systems in the coalfield are locally variable in orientation, so that in some areas they have been opened within the NNW–SSE oriented neotectonic stress field. In these areas they are likely to enhance fracture permeability in the coals.

To determine the CBM potential of areas affected by foreland folding and thrusting it is essential to undertake studies of the regional structural geology and to ascertain the relationship between the neotectonic stress regime and the trends of earlier compressive structures.

The authors thank G. Williams for his review comments and especially P. Gillespie for help and advice in improving the text. T.M.H. is in receipt of an NERC studentship sponsored by British Coal Opencast. We are grateful to British Coal Opencast for allowing access to working open pit mines.

# References

ANDREWS, J. R. 1993. Evidence for Variscan dextral transpression in the Pilton Shales, Croyde Bay, north Devon. *Proceedings of the Ussher Society*, **8**, 198–199.

BARNETT, J. A. M., MORTIMER, J., RIPPON, J. H., WALSH, J. J. & WATTERSON, J. 1987. Displacement geometry in the volume containing a single normal fault. *Bulletin of the American Association of Petroleum Geologists*, **71**, 925–937

BERGERAT, F. 1987. Stress fields in the European platform at the time of the Africa–Eurasia collision. *Tectonics*, 6 99–132.

BROOKS, M., MILIORIZOS, M. & HILLIER, B. V. 1994. Deep structure of the Vale of Glamorgan, South Wales, UK. *Journal of the Geological Society, London*, **151**, 909–917.

COLE, J. E., MILIORIZOS, M. FRODSHAM, K., GAYER, R. A., GILLESPIE, P. A., HARTLEY, A. J. & WHITE, S. C. 1991. Variscan structures in the opencast coal sites of the South Wales Coalfield. *Proceedings of the Ussher Society*, 7, 375–379.

CORNELIUS, C. T., HARTLEY, A., GAYER, R. & ROSS, C. 1993. Coal deposition and tectonic history of the South Wales Coalfield, U. K.: implications for coalbed methane resource development. *In: Proceedings of the 1993 International Coalbed Methane Symposium, University of Alabama/Tuscaloosa*, 161–172.

CREEDY, D. P. 1988. Geological controls on the formation and distribution of gas in British Coal Measures Strata. *International Journal of Coal Geology*, **10**, 1–31.

——1991. An introduction to geological aspects of methane occurrence and control in British deep coal mines. *Quarterly Journal of Engineering Geology*, **24**, 209–220.

DUNNE, W. M. & HANCOCK, P. L. 1994. Palaeostress analysis of small-scale brittle structures. *In*: HANCOCK, P. L. (ed.) *Continental Deformation*. Pergamon Press, Oxford, 101–120.

EISENSTADT, G. & DEPAOR, D. G. 1987. Alternative model of thrust fault propagation. *Geology*, **15**, 630–633.

ELLIOTT, D. 1976. The energy balance and deformation mechanisms of thrust sheets. *Philosophical Transactions of the Royal Society of London, Series A*, **283**, 289–312.

ESHELBY, J. D. 1957. The determination of the elastic field of an ellipsoidal inclusion and related problems. *Proceedings of the Royal Society of London A*, **241**, 376–396.

FLETCHER, R. C. & POLLARD, D. D. 1981. Anticrack model for pressure solution surfaces. *Geology*, **9**, 419–424.

FRODSHAM, K., GAYER, R. A., JAMES, E. & PRYCE, R. 1993. Variscan thrust deformation in the South Wales Coalfield—a case study from Ffos-Las Opencast Coal Site. *In*: GAYER, R. A., GREILING, R. O. & VOGEL, A. K. (eds) *Rhenohercynian and Subvariscan Fold Belts*. Earth Evolution Science, **6**, 315–348.

GAYER, R. 1993. The effect of fluid over-pressuring on deformation, mineralisation and gas migration in coal-bearing strata. *In*: PARNELL, J. RUFFELL, A. H. & MOLES, N. R. (eds) *Contributions to an International Conference on Fluid Evolution, Migration and Interaction in Rocks. Geofluids '93 Extended Abstracts, Torquay*, 186–189.

—— & NEMČOK, M. 1994. Transpressionally driven rotation in the Variscides of South Wales. *Proceedings of the Ussher Society*, **8**, 317–320.

——, COLE, J., FRODSHAM, K., HARTLEY, A. J., HILLIER, B., MILIORIZOS, M. & WHITE, S. C. 1991. The role of fluids in the evolution of the South Wales Coalfield foreland basin. *Proceedings of the Ussher Society*, 7, 380–384.

——, HATHAWAY, T. M. & DAVIS, J. 1995. Structural geological factors in open pit coal mine design, with special reference to thrusting: case study from the Ffyndaff sites in the South Wales coalfield. *In*: SPEARS, D. A. & WHATELEY, M. K. G. (eds) *European Coal Geology*. Geological Society, London, Special Publication, **82**, 233–249.

HARRIS, I. H. 1995. Newly developed techniques to determine proportions of undersized (friable) coal during prospective site investigations. *In*: SPEARS, D. A. & WHATELEY, M. K. G. (eds) *European Coal Geology*. Geological Society, London, Special Publication, **82**, 99–114.

——, DAVIES, G. A., GAYER, R. A. & WILLIAMS, K. 1996. Enhanced methane desorption characteristics from South Wales anthracites affected by tectonically induced fracture sets. *This volume*.

HARTLEY, A. J. 1993. A depositional model for the Mid-Westphalian A to late Westphalian B Coal Measures of South Wales. *Journal of the Geological Society, London*, **150**, 1121–1136.

HATHAWAY, T. M. & GAYER, R. A. 1994. Variations in the style of thrust faulting in the South Wales Coalfield and mechanisms of thrust development. *Proceedings of the Ussher Society*, **8**, 279–284.

HYETT, A. J. 1990. Deformation around a thrust tip in Carboniferous limestone at Tutt Head, near Swansea, South Wales. *Journal of Structural Geology*, **12**, 47–58.

JAMISON, W. R. 1989. Geometric analysis of fold development in overthrust terranes. *Journal of Structural Geology*, **9**, 207–219.

JONES, J. A. 1991. A mountain front model for the Variscan deformation of the South Wales coalfield. *Journal of the Geological Society, London*, **148**, 881–891.

KELLING, G. 1974. Upper Carboniferous sedimentation in South Wales. *In*: OWEN, T. R. (ed.) *The Upper Carboniferous and Post-Palaeozoic and Post-Palaeozoic Rocks of Wales*. University of Wales Press, Cardiff, 185–224.

MCCROSS, A. M. 1986. Simple construction for deformation in transpression/transtension zones. *Journal of Structural Geology*, **8**, 715–718.

MECHIE, J. & BROOKS, M. 1984. A seismic study of deep geological structure in the Bristol Channel area, SW Britain. Geophysical *Journal of the Royal Astronomical Society, London*, **78**, 661–689.

NUR, A., BELL, M. L. & TALWANI, P. 1973. *Proceedings of the Conference on Tectonic Problems of the San Andreas Fault System*, **13**, 391–404.

OWEN, T. R. & WEAVER, J. D. 1983. The structure of the main South Wales Coalfield and its margins. *In*: HANCOCK, P. L. (ed.) *The Variscan Fold Belt in the British Isles*. Adam Hilger, Bristol, 74–87.

PASHIN, J. C & CARROLL, R. E. 1993. Origin of the Pottsville Formation (Lower Pennsylvanian) in the Cahaba Synclinorium of Alabama: genesis of coalbed reservoirs in a synsedimentary foreland thrust system. *In*: *Proceedings of the 1993 International Coalbed Methane Symposium, University of Alabama/Tuscaloosa*, 623–631.

SIBSON, R. H. 1974. Frictional constraints on thrust, wrench and normal faults. *Nature*, **249**, 542–544.

——1980. Power dissipation and stress levels in the upper crust. *Journal of Geophysical Research*, **85**, 6239–6247.

——1981. Controls on low-stress hydro-fracture dilatancy in thrust, wrench and normal fault terrains. *Nature*, **289**, 665–667.

TANNER, P. W. G. 1989. The flexural-slip mechanism. *Journal of Structural Geology*, **11**, 635–655.

WALSH, J. J. & WATTERSON, J. 1990. New methods of fault projection for coalmine planning. *Proceedings of the Yorkshire Geological Society*, **48**, 209–219.

WHITE, S. 1991. Palaeo-geothermal profiling across the South Wales Coalfield. *Proceedings of the Ussher Society*, **7**, 368–374.

WILLIAMS, G. D. & CHAPMAN, T. J. 1983. Strains developed in the hangingwalls of thrusts due to their slip/propagation rate: a dislocation model. *Journal of Structural Geology*, **5**, 563–572.

# Nature and origin of fractures in Permian coals from the Bowen Basin, Queensland, Australia

C. I. PATTISON[1], C. R. FIELDING[2], R. H. MCWATTERS[1]
& L. H. HAMILTON[3]

[1] CSIRO, Division of Petroleum Resources, PO Box 3000, Glen Waverley,
Victoria 3150, Australia
[2] Department of Earth Sciences, University of Queensland, St Lucia,
Queensland 4072, Australia
[3] School of Geology, Queensland University of Technology, GPO Box 2434,
Queensland 4001, Australia

**Abstract:** A field investigation into the development of cleat and other fractures within Permian coal measures was undertaken in the Bowen Basin of eastern Queensland, Australia to provide predictive information on coal permeability for use in coalbed methane exploration. Data were collected from five open-cut and three underground mines in the southern and central Bowen Basin, covering the Upper Permian German Creek Formation and Rangal Coal Measures. Fractures noted in the coals fall readily into the four categories of faults and shear zones, extension and compression-related joint sets, mining-induced fractures and coal cleats. Four geometric varieties of coal cleats were identified: class A orthogonal sets, class B sinusoidal sets, class C polygonal sets and class D chaotic sets. Superimposed or overprinted type A sets were also observed locally. On a regional scale the cleats typically parallel dip (face cleats) and strike (butt cleats), but local departures from this pattern occur in proximity to faults, where face and butt cleat directions were often reversed, rotated or multimodal. Field data suggest that coal cleats of classes A B and C were formed by brittle fracturing of the coal during burial, wheras class D and superimposed cleats were formed during later compressional events. Constraints on the timing of deformation events affecting the Bowen Basin, and geochemical data provided by other workers indicate that the cleats were formed within 5 Ma of formation of precursor peats, but were in many instances modified by later structural events.

The considerable black coal deposits of the Permo-Triassic Bowen Basin in east-central Queensland, Australia are mined extensively by open-cut and underground methods. In recent times there has been an active interest in the potential for commercial coalbed methane extraction from subsurface coal measures in the basin. As part of a broader research effort into the controls on methane generation, storage and production potential of Bowen Basin coals, this paper presents results from a regional investigation into the characteristics and origin of cleats and their relationship with other structures and sedimentary features associated with two major Late Permian coal measure units (German Creek Formation and Rangal Coal Measures). As such, it forms a companion to the paper by Faraj *et al.* on cleat mineralization in Bowen Basin coals.

The aims of the project were to determine the local and regional distribution and orientation of fractures occurring naturally within coal, to investigate the relationship between cleats and other fractures (joints, shears, faults), interpret the origin of cleats and establish a regional model for their development.

## Regional geology

The Bowen Basin is a large, elongate Permo-Triassic sedimentary basin extending north–south for approximately 900 km in eastern Queensland and northern New South Wales, Australia (Fig. 1). It contains up to 10 km of variably deformed shallow marine and fluvio-deltaic siliciclastic sedimentary rocks, which are unconformably overlain in the south by relatively flat-lying Jurassic and Cretaceous rocks of the Surat Basin. The region hosts significant economic accumulations of coal, natural gas and precious metals, and is one of the most intensively explored basins in Australia.

In gross structural terms, the Bowen Basin can be readily divided into distinct units that reflect the underlying basement morphology and the effects of subsequent tectonic events on the basin sequence (Fig. 2). The main depocentre of the

*From* Gayer, R. & Harris, I. (eds) 1996, *Coalbed Methane and Coal Geology,*
Geological Society Special Publication No 109, pp 133–150.

Bowen Basin, known as the Taroom Trough, occurs along the eastern margin of the basin and lies adjacent to the Connors and Auburn Arches, which collectively form part of the extensive New England Fold Belt. Subsequent erosion of these arches has exposed Late Carboniferous batholiths that represent the core of a former Andean-style continental margin mountain range. The western flank of the basin comprises the Collinsville Shelf and Comet Ridge, which together represent a basement terrain that was originally cratonized to the Australian plate during the Early Palaeozoic (Kirkegaard 1974). The remaining significant tectonic element is the Denison Trough, which along with recently identified smaller marginal rift basins (Fielding *et al.* 1994), formed during the initiation of the Bowen Basin.

On the larger tectonic scale, the Bowen Basin forms part of the Bowen–Gunnedah–Sydney foreland system of eastern Australia, which began forming in response to convergence of the palaeo-Australian and palaeo-Pacific plates during the Early Permian (Murray *et al.* 1987; Murray 1990; Caritat & Braun 1992; Elliott 1993). Initial sediment accumulation occurred on a basement of Siluro-Devonian and Carboniferous volcanic and marine sedimentary rocks and comprised thick piles of predominantly basaltic volcanic rocks and volcaniclastic sediments (Dear 1994). Continued downwarping of the basin during the Early Permian, or a marine transgression, subsequently initiated widespread marine sedimentation in the eastern part of the basin and mixed marine, deltaic and fluvial sedimentation in the north and west, where coal measures periodically accumulated. Towards the end of this activity, in earliest Late Permian times, extensive coal-forming environments developed on deltaic and coastal plains occurring along the western margin of the basin that are now preserved as the German Creek Formation and equivalents (Fielding & McLoughlin 1992; Falkner & Fielding 1993; John & Fielding 1993).

This period was later followed by renewed uplift and associated volcanic activity to the east of the Bowen Basin, which initiated a period of dominantly terrestrial sedimentation and the effective reorganization of dispersal systems. The marine basin was progressively filled by successive pulses of coarse sediment supply derived from the east and feeding into major southward-flowing axial river systems. Lowland alluvial and coastal plain environments were rapidly established across the entire basin, leading to the accumulation of the

Rangal Coal Measures and equivalents at the end of the Permian period (Fielding *et al.* 1993). Continued uplift along the eastern margin of the basin ultimately led to the development of southwesterly divergent thrust fronts within sedimentary rocks occurring along the eastern margin of the basin and adjacent to major faults developed at the boundary with shelf regions to the west, including the German Creek–Oaky Creek area on the Collinsville Shelf and the South Blackwater, Cook and Curragh areas on the Comet Ridge. This event is also recorded by a major thermal event in the form of radiometric dates from S-type granites and associated alteration systems within the adjacent New England Fold Belt. Most ages in this suite cluster around the interval 245–240 Ma.

Uplift along the eastern margin of the basin and associated compression of the basin sequence continued into Triassic times, culminating in further major episodes of crustal shortening in the Early to early Late Triassic (Ziolkowski & Taylor 1985; Elliott 1993). This period also corresponds with the maximum burial of Bowen Basin sediments determined from subsidence modelling (Baker & Caritat 1992). Subsequent geological events of potential relevance to this study include a widespread Cretaceous thermo-magmatic event within the Bowen Basin and adjacent regions (Marshallsea *et al.* 1985; Pattison 1990) and a Middle Cretaceous compressive deformation (Dickens & Malone 1973; Elliott 1993) that affected much of eastern Queensland.

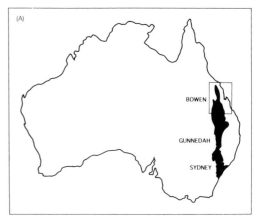

**Fig. 1.** (A) Position of the Permo-Triassic Bowen, Gunnedah and Sydney basins in eastern Australia. (B) Extent of coal-bearing formations in the Bowen Basin and the location of past and present coal mines.

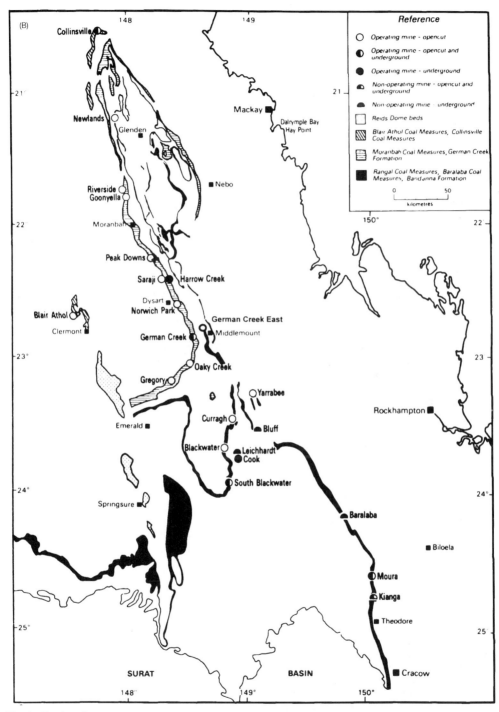

**Fig. 1.** Continued.

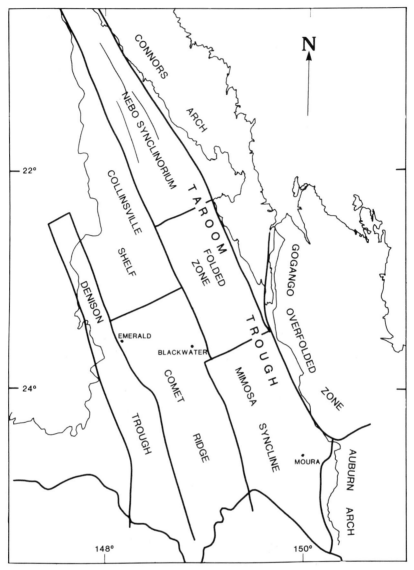

**Fig. 2.** Major structural elements of the Bowen Basin.

## Methodology

Fracture measurements were undertaken at five mine sites in the central and southern Bowen Basin which exploit the Late Permian German Creek (Oaky and German Creek) and Rangal Coal Measures (Moura, South Blackwater, Cook and German Creek East). All accessible open-cut pits (totalling 14) and underground workings (three) were visited and exposed coal faces studied at regular intervals (50–100 m). In some areas, data were collected from more closely spaced

stations, typically 5 or 10 m. At each location the orientation and spacing of cleats and any other coal fractures were recorded at regular intervals from the top to bottom of the exposed seams.

In cases where fracture orientations were consistent throughout the vertical profile of the seam, 10 readings were recorded at each station. However, where variations were evident, the seam was divided into identifiable structural units, with each unit being analysed separately. The physical nature of the fractures was also recorded at each sampling locality, with particular attention paid to any curvature in the vertical and horizontal planes, as well as the presence and composition of infilling minerals. The prevailing angle between the predominant cleat sets was measured periodically and detailed coal profiles attempted at a few sites only.

Wherever possible, the traverses incorporated larger scale faults, which were examined for kinematic indicators. Detailed measurements of the strike, dip and intensity of jointing in overburden strata were also taken, along with sedimentological descriptions and related palaeo-current data. Additional cleat directions for deep-level Rangal Coal Measure coals, as determined from remnant magnetism data for selected MIM Holdings Ltd coalbed methane drillholes in ATP 524P, are included for comparative purposes.

## Coal fracture systems

The natural fractures found in coals examined during this study fall readily into four distinct classes, which in descending order of magnitude are (1) faults and shear zones, (2) extension and compression-related joint sets, (3) mining-induced fractures and (4) coal cleat (Fig 3). Although fractures induced by mining will not be considered here, it is important to note that drilling, blasting and the de-stressing effects associated with overburden (open-cut) or coal removal (underground) can produce fracture systems that can be either misinterpreted as part of the natural fracture system or effectively overprint the pre-existing network (e.g. Hanes & Shepherd 1981). Consequently, every attempt was made to isolate and exclude from the analysis any fractures associated with blasting and those mirroring the surrounding workings.

Of the three remaining classes, only the joint and cleat sets require further definition here. The complex nature of fracturing in the coals examined made this a particularly difficult task and ultimately involved revisiting mine sites to clarify interpretations that developed as our experience grew. Nonetheless, consistent observations were identified that did not conform with conventional descriptions reported elsewhere. This may in large be due to the relative preponderance of publications related to cleating in Laurasian coals, as opposed to Gondwana coals, and ultimately to compositional differences between the two. As such, the definitions provided should be specific to the Permian coals of eastern Australia.

### Cleats in coal

The use of the word cleat as a mining term dates back to the late 19th century (Kendall & Briggs 1933) and has since been adopted by miners,

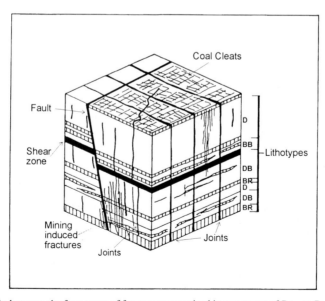

**Fig. 3.** Relationship between the four types of fractures recognized in exposures of Bowen Basin coals: faults and shear zones, extension- and compression-related joints, mining-induced fractures and coal cleats. BR, bright coal; BB, bright banded; DB, dull banded; and D, dull.

geologists and engineers to describe a variety of fractures commonly found in coal. Most workers regard cleats in coal as equivalents to joints in clastic rock (e.g. Ver Steeg 1942; Ramano & Moiz 1968, McCulloch *et al.* 1974*a, b*, 1976; Diamond *et al.* 1976; Grout 1981; Ward 1984; Spears & Caswell 1986; Condon 1988), or as closely spaced, pervasive fractures (Heidecker 1979; Laubach *et al.* 1991) that originated by imperceptible movement associated with mode 1 extensional opening (Close 1991; Close & Mavor 1991; Tremain *et al.* 1991). Consequently, cleats in coal have been variously described as being restricted to individual coal bands or, in some mine exposures, to encompass the vertical extent of the seam (Tremain *et al.* 1991).

The more generalized definition can be misleading, especially when applied to the Permian coals examined during the present study. Of all the observed patterns to emerge from our work, the most consistent relates to the vertical extent of cleat development. In all instances, cleating was found to be well developed within bright coal bands, but was rarely well-developed in dull bands. Such a tendency has been noted previously by numerous workers (e.g. Ting 1977; Hucka 1989; Law 1991; 1993; Tremain *et al.* 1991; Close & Mavor 1991; Close 1991; Mavor *et al.* 1991), and is regarded here as of fundamental significance.

It was also noted that cleats rarely crossed lithotype boundaries unless they were linked through the duller lithotypes by through-going fractures. As vertically extensive fracture sets were not universally developed, and where they existed only linked a small percentage of available cleats, this suggested both a genetic and temporal difference between these two fracture systems (i.e. lithotype controlled and vertically unrestricted fractures).

For the purposes of this study, a cleat is defined as an extensional fracture occurring within coal that is confined to a particular lithotype or microlithotype. This definition may not be universally applicable, but it is interesting to note that lithotype controlled cleat height has also been described for coals of the Uinta (Bunnell 1987; Hucka, 1989; 1991) and Raton (Close 1988) basins in the USA.

## Joints in coal

Given that cleats are defined here as being lithotype-bound, then the definition of a joint in coal refers to any other extensional fracture that is confined to or transects a coal seam. Although the definition of a joint automatically assumes an extensional origin, no specific contemporaneous tectonic stress regime or structural mechanism is implied in the term.

The coal joints noted during this study generally extended vertically through only a small number of lithotypes, but in some instances extended throughout the entire seam, this being especially notable in mine exposures of German Creek coals. This latter type of joint has been documented elsewhere, where they are often referred to as master cleats (Henkle *et al.* 1978; Laubach *et al.* 1991). However, based on our field observations, it is likely that these seam-height joints represent late-stage tensile fractures associated with unloading and destressing, and would not be present at any great depth. Primary evidence for this is the lack of any visible infilling mineralization associated with these major fractures, even when minerals are evident within the cleat system, and the noted lack of vertically extensive fractures found in coal cores derived from depths greater than 300 m. Although not entirely conclusive, it does agree with the interpretation forwarded by Tremain *et al.* (1991) that so-called 'cleats' found in some immature Tertiary coals in the USA are nothing more than recently developed joints.

The further distinction between differing generations of joints, and perhaps differing modes of formation, could be important to our understanding of the structural history of a coal seam and hence its potential permeability to fluids and gas. However, in situations where seam-height joints existed, the intensity of fracturing commonly made the identification of any original macroscopic fabric impossible. As a consequence, no attempt has been made here to distinguish between the differing hierarchies of coal joints.

## Mineralization in cleats and joints

Cleats examined were typically filled by one or more mineral species, wheras joints in coal and clastic rocks showed more erratic mineralization. Cleats within coals of the German Creek Coal Measures were noticeably less mineralized than their counterparts in the Rangal Coal Measures, where the presence of reddish brown clays was noted at all exposures throughout the southern and central Bowen Basin. Within these latter coals, clays were ubiquitous in face cleats, whereas they were only occasionally found in butt cleats, with illite and kaolinite the dominant phases (see Faraj *et al.* this volume). Butt cleats and joints, on the other hand, were dominated by calcite and other carbonate minerals. In all instances noted, carbonates were clearly formed after the clay minerals and occupied the centres of cleat and joint apertures.

## Fracture geometries

### Cleats

As described elsewhere, cleats in coal generally form an orthogonal set of fractures that are essentially perpendicular to bedding surfaces. These cleat sets are well developed and continuous in bright coal bands and poorly developed and discontinuous in the dull bands. For historical reasons the predominant or primary cleat is referred to as the face cleat and the secondary or end cleat is known as the butt cleat. In plan view, face cleat geometries typically range from linear to curvilinear and form parallel to subparallel sets that can be continuous beyond the open-cut pit scale ($\approx 50$ m). Butt cleats generally form parallel sets that are aligned normal to the face cleats. However, there are many exceptions to this idealized geometry. In some areas the face and butt cleat are largely indistinguishable or similar in appearance, whereas in other areas the cleat pattern is chaotic or resembles that of polygonal shrinkage cracks (Fig. 4).

Four general classes of cleat geometries were identified during the course of this study: class A, orthogonal sets; class B, sinusoidal sets; class C, polygonal sets; and class D, chaotic arrangements (Fig. 4). The distinction between these classes was sometimes complicated by the presence of a second or superimposed cleat set (Fig. 4) and geometries with some elements of classes A B and C were also observed. Nonetheless, of the cleat geometries described, the only one obviously structurally related is the chaotic class which can be directly linked to the proximity of in-seam shearing. It is difficult to estimate the relative proportions of each class, due primarily to limited plan exposure, but it appears that classes A and B predominate, even within seams affected by shearing. Polygonal cleats were noted in areas where no other cleat was evident, a relationship also noted by Tremain *et al.* (1991) within some US coals.

Such cleat development was particularly evident in the dull coals. The superimposed cleat sets were largely recognized only in situations where their orientation differed significantly from previous generations, although a difference in the presence or composition of infilling minerals was commonly evident.

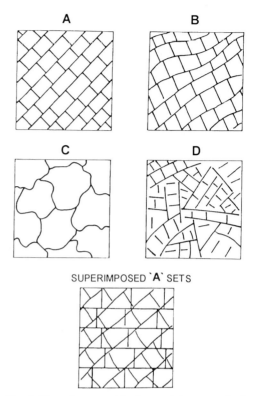

**Fig. 4.** Four plan geometric classes of cleat recognized in Bowen Basin coals: (A) orthogonal sets; (B) sinusoidal sets; (C) polygonal sets; and (D) chaotic arrangements. Additionally, type A cleats were found locally in superimposed sets indicating overprinting.

Both face and butt cleat spacings for class A and B geometries ranged from 1 to 100 mm, dependent on the host lithotype, with 1–10 mm being typical in bright lithotypes and 5–20 mm in duller bands. In the vertical dimension both cleat sets generally occurred as linear fractures occurring perpendicular to bedding. However, some cleats were noted that curved towards the horizontal plane, especially in close proximity to shear horizons. No systematic pattern in spacing was evident for the polygonal class of cleat geometries, but of the examples recorded the diameter of individual polygons rarely exceeded 50 mm.

### Joints

The coal joints defined in this study extended over greater vertical distances than the cleats, but were significantly less frequent, especially in cored coal samples. These joints in many instances occurred as roughly orthogonal sets that conformed with the pre-existing cleat fabric, dipping at high angles to the bedding. However, this was not always so and joint sets dipping at low to moderate angles to bedding were noted in several localities. These obliquely dipping joints were particularly evident in highly faulted areas (e.g. Cook Colliery), where the joints were aligned parallel to the ENE dip of the numerous reverse and thrust faults that cut the coal seam.

The coal joints commonly found in mine site exposures appear to occur at regular intervals ranging from less than 10 cm to greater than 100 cm. In some instances the joints found in coal were aligned parallel to joints in the overburden strata, whereas in others no clear relationship existed. It was also noticeable that joints within a given coal seam rarely extended any great distance into the immediate overburden strata.

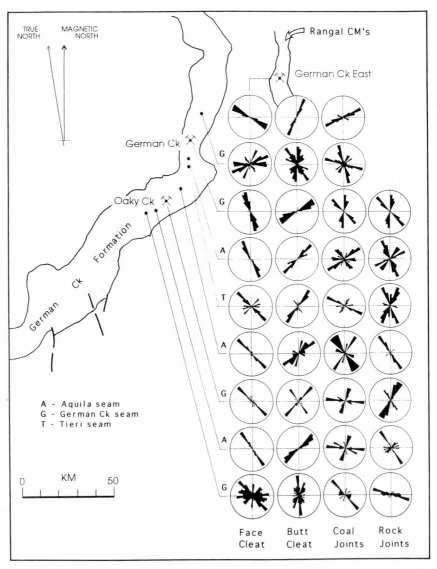

**Fig. 5.** Regional cleat and joint orientations in the German Creek–Oaky Creek area of the Bowen Basin. Note the general consistency in cleat orientations between seams of the German Creek Formation and the similarity with coals of the Rangal Coal Measures (German Creek East).

Joints in clastic overburden and interburden strata were generally more variable in both orientation and distribution than their coal counterparts. Joints were best developed in sandstones, where as many as three sets were common, and within close proximity to faults. It was generally found that one of the joint sets was aligned roughly parallel to either the face or butt cleat direction in adjacent coal, although significant departures did occur.

## Regional trends

Fracture orientation data collected at regular intervals along exposed highwalls and in underground headings were combined for the various study sites and compared to assess regional patterns. Data for different formations were treated separately and, where possible, the data for individual seams within a given formation were combined independently.

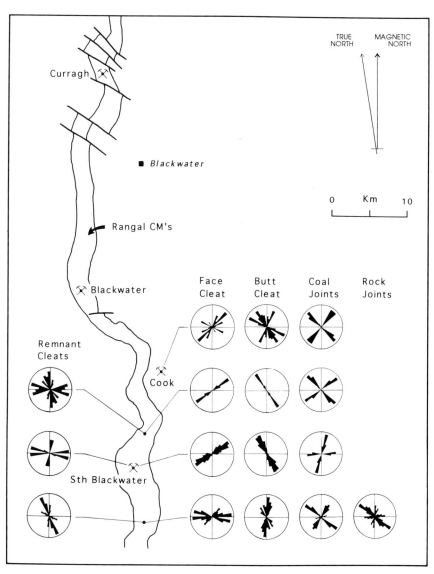

**Fig. 6.** Map showing regional cleat and joint orientations in the Blackwater area of the Bowen Basin. Note the common presence of remnant cleat sets in the region.

On a regional scale, face and butt cleats were found to align consistently parallel with structural dip and strike, respectively. This pattern prevailed in the German-Oaky Creek (Fig. 5), South Blackwater-Cook (Fig. 6), and Moura (Fig. 7) areas, even where gross strike variations occur. In relatively undisturbed areas, the orientation of face and butt cleats remained constant from seam to seam within a single formation and even between different coal measure sequences exposed in the same area (Fig. 5). The determination of cleat directions for coal core from the Moura area also indicates that face and butt cleat trends remain fairly constant at depth over a large area.

This general regional consistency is significant in that likely cleat orientations at depth could be predicted from outcrop patterns or from stratigraphically higher seams in the same sequence. However, caution must be exercised

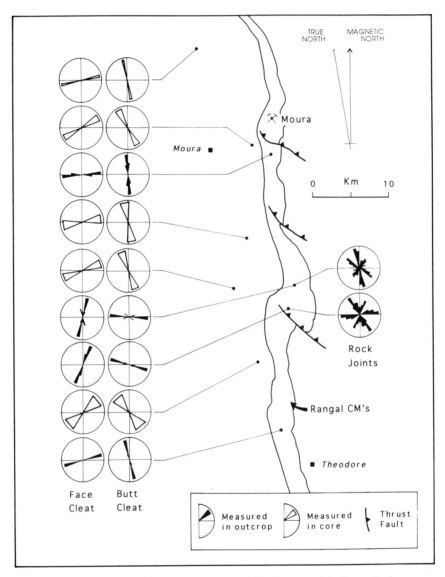

**Fig. 7.** Map showing regional cleat orientations in the Moura–Theodore area of the Bowen Basin, as measured in both outcrop and from core samples derived from depths greater than 300 m. Measurements from outcrop include data of Bassett (1984) and Miller (1992).

as variations in general cleat orientations were found to exist at both the regional and local scales, particularly in areas adjacent to faults or where in-seam shearing is severe. The best examples of such regional variations occurred in the north of the German Creek Mine, where faulting is intense (Fig. 5), and in the Moura area, where cleat orientations are markedly different in proximity to a northwest-trending thrust fault (Fig. 7). Regionally consistent joint orientations in both coal and clastic rocks were less cvident, but some consistency was apparent over distances in the range 1–10 km. Cleats recognized as being remnant sets, based on the presence of infilling minerals, were variable from site to site, although a general north–south and east–west trend was evident in the data (Fig. 6).

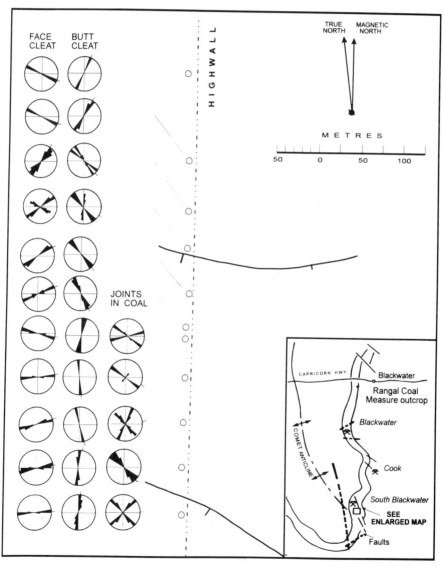

**Fig. 8.** Orientation of cleats and coal joints in the Argo seam exposed in Ramp 14 North, South Blackwater Mine. Note the bimodal orientation of cleats adjacent to the northerly of the two normal faults intersected by the highwall, and the rotation of cleats into this fault. Inset map showing regional context covers an area of approximately 50 by 35 km.

## Relationship to structure

Cleat orientations were found to vary most commonly in proximity to fault zones. In these zones, either the cleat set which was closest to the fault orientation tended to be best developed, regardless of whether it was the locally consistent face or butt, or each cleat set exhibited a bimodal orientation (Fig. 8). In some intensely faulted areas the face and butt cleat sets were interchangeable on a metre by metre scale. A further noticeable trend associated with some normal faults was the tendency for cleat orientations to show a progressive rotation into the plane of the fault, becoming parallel and perpendicular to the exposed fault trace. In instances where this was evident, it strongly implies that either the faults existed before cleat formation or that both formed contemporaneously.

Within coal seams affected by intense in-seam shearing a gradual or abrupt variation in cleat orientations was also noted to occur vertically through the seam (Fig. 9). This was particularly evident in the German-Oaky Creek area and is largely responsible for the complex face and butt cleat patterns seen in the composite rose diagrams in Fig. 5. As chaotic cleat sets are also more evident in coals affected by shearing, it is apparent that significant crushing and rotation accompanied shortening of the coal measure sequence. The implication of this finding is that although the generation of crushed zones associated with shearing could provide ideal avenues for fluid and gas flow, it is also likely that the overall vertical connectivity of the seam is adversely affected. Less significant variations in cleat orientations and frequency were also noted across major facies changes in overlying clastic strata.

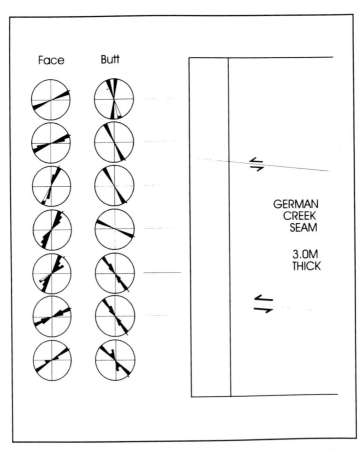

**Fig. 9.** Variations in cleat orientations in a vertical section of the German Creek seam affected by in-seam shearing at the German Creek Mine.

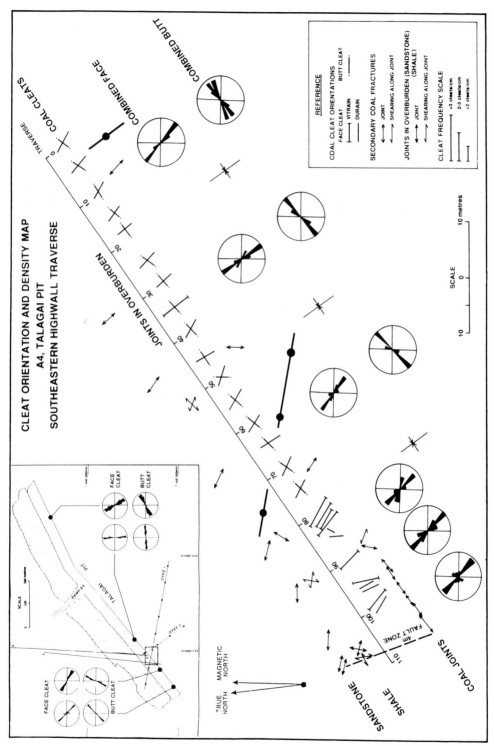

**Fig. 10.** Cleat and joint data gathered from close-spaced stations in the Talagai Pit (A4), Oaky Creek Mine. Note the increase in coal joints and lack of development of butt cleats adjacent to the fault which intersects the highwall at 108 m.

Cleat spacings were generally found to show similar lateral consistency to orientation, with similar degrees of variation adjacent to zones of structural disturbance. Cleat frequency was not typically found to increase near fault zones, although the incidence of pervasive fractures (joints) within coal seams and overburden strata did (Fig. 10). As noted previously, cleat frequency also varied systematically with lithotype, the highest frequency being associated with vitrain layers. Joints in overburden strata tended to show more complex patterns than those shown by cleats. However, in areas of faulting joints through both coal and overburden were often aligned with and perpendicular to the fault trend, and showed a good match with cleat directions.

The regional consistency in cleat orientations and relationships with strike and dip trends evident within the Permian coal measures of the Bowen Basin has also been noted in various other parts of the world (Close 1993). Cleat orientations that progressively curve into the plane of normal faults (i.e. face perpendicular to fault) have been reported from the USA (Laubach *et al.* 1991), as has the tendency for fault trends to be superimposed onto cleat patterns (Tremain *et al.* 1991). The data set summarized in this paper, however, combining regional scale information with detailed observations on a much finer scale, allows an understanding of the controls on cleat formation and characteristics which has hitherto been unavailable for most coal-bearing basins. This theme is developed in the next section.

## Cleat formation

Theories proposed to explain the formation of cleats in coal have been categorized by Close (1993) into three groups: (1) endogenetic, involving soft sediment deformation processes such as differential compaction and shrinkage during coalification; (2) exogenetic, related to tectonic stresses imposed during burial and/or uplift; and (3) duogenetic, involving some combination of the preceding. Various workers have pointed to the regional consistency in cleat orientation and relationship with dip and strike as evidence of a tectonic origin. Close (1991) and Close & Mavor (1991) surmised from structural and other data that cleat formation in Cretaceous coals of the San Juan and Raton basins (USA) occurred during burial and before folding of the strata and occurred within a few million years of surface peat accumulation. In the present study, detailed structural observations are combined with direct mineralogical and

geochemical evidence for the timing of cleat formation (Faraj *et al.* this volume), allowing a more precise and detailed interpretation to be made.

In a similar manner to Close (1991), we interpret the perpendicular nature of cleats with respect to bedding in Bowen Basin coals as evidence that cleats were formed before folding. Equally, the parallelism with strike trends suggests a relationship with the burial process, as discussed earlier, wheras modifications to the regional pattern suggest that cleats were modified by pre-existing structures, or in some instances overprinted by later structural events. The generally recognized orthogonal geometry of cleats strongly indicate that they formed by brittle failure of lithified material. Consequently, the evidence suggests that most cleats are of tectonic origin and formed under the influence of a regional stress pattern. However, as cleat geometries can be variable, it is likely that their orientation was controlled to some degree by local stress fields, with the generation of classes A, B and C sets reflecting differing degrees of stress anisotropy.

As recognized during previous studies, butt cleats were always confined between face cleats, which has in the past been used as evidence to suggest that face cleats were the first to form. However, the presence of polygonal cleat geometries and the similar mineral assemblages infilling face and some butt cleats suggests that the two cleat sets may have formed contemporaneously or butt cleat initiation closely followed face cleat generation. Given the common directional divergence between cleat sets and other fracture sets within the coal, it can be further concluded that coal joints formed separately from cleats, even though the initial cleat anisotropy obviously influenced later generations of fractures. Some of these coal joints are clearly related to local compressional structures, whereas others have orientations that mimic joint sets in overlying clastic rocks, suggesting they formed as a result of unloading and de-stressing.

The Late Permian coal measures of the Bowen Basin accumulated immediately before and during a period of flexural loading and progressive contractional deformation. The WSW advance of thrust sheets into the eastern part of the basin simultaneously led to the release of vast amounts of sediment into the basin, which initiated flexural subsidence and the rapid burial of sediment. The orientation of the thrust vector (and facies patterns evident in outcrop) imply that the contractional deformation affected the northern part of the basin earlier than the southern part and the relative intensity of compression indicates that shortening was more severe in the north. The foreland phase of the Bowen Basin began around 255 Ma ago and the youngest strata preserved were formed around 235 Ma ago at about the time of maximum burial (Baker & Caritat 1992). The climax of compression (Hunter–Bowen Orogeny) and uplift of the Bowen Basin succession occurred between 235 and 230 Ma ago, by which time it is argued that cleat sets had been formed. The modification of cleat patterns around the thrust and other faults of probable Hunter–Bowen origin implies that cleats formed in a stress environment locally modified by these structures.

Radiometric dates from illites filling face cleats (the earliest phase of cleat mineralization recognized) cluster around 245, 232 (two samples) and 225–215 Ma ago (Faraj et al. this volume), indicating that the cleats had formed by 245 Ma. According to subsidence analysis, the German Creek and Rangal Coal Measures would have been buried to at least 1500 and 1000 m respectively, by this time and would have been under the influence of a WSW orientated compressive stress regime. The period 245–240 Ma was also evidently a time of significant intrusive and hydrothermal activity along the eastern margin of the southern Bowen Basin, as evidenced by radiometric dates from granites and alteration systems associated with precious and base metal mineralization.

All evidence cited points towards cleat formation during the Permo-Triassic burial event, which preceded uplift and deformation of the succession. Face cleats, which were aligned parallel with the axis of principal compressive stress and hence dilational, were mineralized early, whereas butt cleats remained predominantly closed. The latest period of illite formation in face cleats corresponds to a post-deformational period of crustal extension and high heat flow during the Late Triassic. This was followed by carbonate-dominated mineralization which infiltrated face, butt and other fracture systems within the coal. The age of this mineralizing event is poorly constrained, but may correspond with an Early to mid-Cretaceous period of magmatic and structural activity within the Bowen Basin and partially overlying Great Artesian Basin.

## Conclusions

Coal cleats in the Bowen Basin of eastern Queensland, Australia are interpreted to have formed by the brittle failure of macerals during burial in the Permo-Triassic. Age dates on

minerals infilling the cleat system indicate that they formed before 245 Ma, which is less than 10 Ma after surface peat accumulation. Evidence also indicates that face and butt cleats either formed contemporaneously, or butt cleat formation closely followed face cleat generation. Cleat orientations were largely dictated by the prevailing regional stress pattern, with face cleats aligned parallel to the principal maximum horizontal compressive stress. However, localized influences on the existing stress field, predominantly associated with earlier or contemporaneous faults, modified cleat orientations. These influences resulted in the formation of cleat sets with different geometries, each of which reflects a variable state of stress anisotropy.

Subsequent uplift and deformation of the Bowen Basin succession during the Middle to Late Triassic further modified the cleat pattern and initiated additional fracture systems within the coal, many of which parallel contemporaneous thrust-related structures. A second phase of clay mineralization occurred in the Late Triassic (225–220 Ma BP) and was followed by a later phase of extensive carbonate mineralization, potentially during compression in the Early to mid-Cretaceous. It is also possible that an additional cleat set was superimposed on former sets during this compressive regime. More recent uplift and unloading, combined with mining activities, resulted in the formation of vertically extensive joint systems within coals located at or near the surface. Together, these fracture systems provide a record of the structural history of the Bowen Basin.

The relationships established during this study have some predictive value in the exploration for economic coalbed methane targets. The regional parallelism between face cleats and structural dip provide a basis from which to predict cleat orientations in structurally uncomplicated areas. Cleat orientations may also be predicted in zones of faulting by assessing the nature and orientation of faults and comparing them with the detailed case histories presented herein.

It is clear from the data gathered during the present study that previous structural deformation may not in itself improve fracture permeability in Bowen Basin coals. No increase in cleat frequency was noted adjacent to younger faults exposed in the mines and it was often found that cleat systems were preferentially mineralized in such areas. Equally, crushed zones generated adjacent to in-seam shear zones could provide permeable avenues for fluids and gas, but the overall effect on the seam was the generation of chaotic cleat sets and the disturbance of any vertical fracture continuity. Although the intensity of vertical to subvertical joints was greatest near fault zones, the typical spacing was such that they would be unlikely to significantly improve gas flow to the well bore. Vertically extensive joint sets interpreted as forming during recent destressing could have a significant impact on coal seam permeability, especially where the average spacing is of the order of 2 to 5 cm. However, no such extensive fracture sets have been recorded in coal cores recovered from depths greater than 300 m.

The presence and distribution of minerals in Bowen Basin coal cleats is also predictable to some degree. Face cleats are characterized by the presence of clays and calcite, whereas butt cleats are generally either filled by calcite or are essentially unmineralized. This information, combined with an understanding of present stress conditions, could be utilized to design the most effective stimulation strategy for a given situation. Clearly, coal seam permeability to gas is the most significant concern to coalbed methane exploration in the Bowen Basin. It is evident from downhole tests that major areas of the basin are under a net compressive stress regime and that this anisotropy is affecting well productivity. To generate effective future exploration targets, cleat orientation and mineralization data must be combined with an understanding of directional permeability determined from downhole stress measurements.

This project was financially supported by the CSIRO Division of Petroleum Resources, MIM Holdings Ltd and by CSIRO/University of Queensland and CSIRO/Queensland University of Technology collaborative grants. B. Douglas, S. Edgar, C. Larkin, C. Huddy and J. Esterle assisted in field data collection. Coal Resources of Queensland Pty Ltd, South Blackwater Mines, BHP-Mitsui Moura Mine, Curragh Mining, Oaky Creek Coal and Capricorn Coal Management are thanked for allowing access to exposures and other data, and in particular mine geologists K. Rixon, D. Somer, P. Cottam, A. Kovago, P. Ledger, E. Dahl, N. LeFevbre, R. Phillips and D. Moelle are thanked for their help. E. Burdin drafted some of the diagrams at short notice, R. Holcombe provided useful comments on the original draft and access to his GEORIENT program, and R. Ellison and an anonymous colleague reviewed the submitted manuscript.

# References

BAKER, J. C. & CARITAT, P. DE 1992. Postdepositional history of the Permian sequence in the Denison Trough, eastern Australia. *AAPG Bulletin*, **76**, 1224–1249

BASSETT, K. G. 1984. Geological investigations at Moura. *In*: SOUVAN, A. J. *et al.* (eds) *Gas, Coal, Stress and Stability Investigations at Thiess Dampier Mitsui Coal Pty Ltd, Moura, Queensland*. National Energy Research, Development and Demonstration Program, End of Grant Report Number 359, Department of Resources and Energy, Canberra.

BUNNELL, M. D. 1987. *Roof Geology and Coal Seam Characteristics of the No. 3 Mine, Hardscrabble Canyon, Carbon County, Utah*. Utah Geological and Mineral Survey, Special Studies, **69**.

CARITAT, P. DE & BRAUN, J. 1992. Cyclic development of sedimentary basins at convergent plate margins: 1. Structural and tectono-thermal evolution of some Gondwana basins of eastern Australia. *Journal of Geodynamics*, **16**, 241–282.

CLOSE, J. C. 1988. *Coalbed methane potential of the Raton Basin, Colorado and New Mexico*. PhD Thesis, Southern Illinois University at Carbondale.

——1991. *Western Cretaceous Coal Seam Project—Natural Fractures in Bituminous Coal Gas Reservoirs*. Gas Research Institute, Chicago, Illinois, Report GRI-91/0337, 1–140.

——1993. Natural fractures in coal. *In*: LAW, B. E. & RICE, D. D. (eds) *Hydrocarbons From Coal*. American Association of Petroleum Geologists, Studies in Geology Series, **38**, 119–132.

—— & MAVOR, M. J. 1991. Influence of coal composition and rank on fracture development in Fruitland coalbed natural gas reservoirs of the San Juan Basin. *In: Coalbed Methane of Western North America Conference*. Rocky Mountain Association of Geologists, 1–29.

CONDON, S. M. 1988. *Joint Patterns on the Northwest Side of the San Juan Basin (Southern Ute Indian Reservation), Southwest Colorado*. Rocky Mountain Association of Geologists, 61–68.

DEAR, J. F. 1994. Major cycles of volcanism in the southern Connors Arch. *In*: HOLCOMBE, R. J., STEPHENS, C. J. & FIELDING. C. R. (eds) *GSA Queensland 1994 Field Conference, Capricorn Region, Central Coast, Queensland*. Geological Society of Australia, Queensland Division, Brisbane, 31–45.

DIAMOND, W. P., MCCULLOCH, C. M. & BENCH, B. M. 1976. *Use of Joint and Photolinear Data for Predicting Subsurface Coal Cleat Orientation*. US Bureau of Mines, Report of Investigation, **8120**.

DICKENS, J. M. & MALONE, E. J. 1973. Geology of the Bowen Basin, Queensland. *Bureau of Mineral Resources, Geology & Geophysics, Australia, Bulletin*, **130**.

ELLIOTT, L. G. 1993. Post-Carboniferous tectonic evolution of eastern Australia. *APEA Journal*, **33**, 215–236.

FALKNER, A. J. & FIELDING, C. R. 1993. Geometrical facies analysis of a mixed-influence deltaic system—the Late Permian German Creek Formation, Bowen Basin, Australia. *In*: MARZO, M. & PUIGDEFABREGAS, C. (eds) *Alluvial Sedimentation*. International Association of Sedimentologists, Special Publications, **17**, 195–209.

FIELDING, C. R. & MCLOUGHLIN, S. 1992. Sedimentology and palynostratigraphy of Permian rocks exposed at Fairburn Dam, central Queensland. *Australian Journal of Earth Sciences*, **39**, 631–649.

——, —— & SCOTT, S. G., 1993. Fluvial response to foreland basin overfilling: the Late Permian Rangal Coal Measures in the Bowen Basin, Queensland, Australia. *Sedimentary Geology*, **85**, 475–497.

——, HOLCOMBE, R. J. & STEPHENS, C. J. 1994. A critical evaluation of the Grantleigh Trough, east-central Queensland. *In*: HOLCOMBE, R. J., STEPHENS, C. J. & FIELDING, C. R. (eds) *1994 Field Conference, Capricorn Region, Central Coastal Queensland*. Geological Society of Australia (Queensland Division), Brisbane, 17–30.

GROUT, M. A. 1991. *Cleats in Coalbeds of the Southern Piceance Basin, Colorado–Correlation with Regional and Local Fracture Sets in Associated Clastic Rocks*. Rocky Mountain Association of Geologists Field Conference Guidebook, 35–47.

HANES, J. & SHEPHERD, J. 1981. Mining induced cleavage, cleats and instantaneous outbursts in the Gemini seam at Leichardt Colliery, Blackwater, Queensland. *Proceedings of the Australasian Institute of Mining and Metallurgy*, **277**, 17–26.

HEIDECKER, E. J. 1979. Photostructural detection of concealed cleat and faults influencing coal mine planning. *Proceedings of the Australasian Institute of Mining and Metallurgy*, **270**, 39–45.

HENCKLE, W. R. JR, MUHM, J. R. & DEBUYL, M. H. F. 1978. Cleat orientation in some sub-bituminous coals of the Powder River and Hanna Basins, Wyoming. *In*: HODGSON, H. E. (ed.) *Proceedings of the Second Symposium on the Geology of Rocky Mountain coal—1977*. Colorado Geological Survey, Denver, 129–141.

HUCKA, B. P. 1989. *Analysis of Cleats in Utah Coal Seams*. Utah Geological and Mineral Survey, Open File Report, **154**.

HUCKA, B. P. 1991. *Analysis and Regional Implication of Cleat and Joint Systems in Selected Coal Seams, Carbon, Emery, Sanpete, Sevier and Summit Counties, Utah*. Utah Geological and Mineral Survey, Special Study, **74**.

JOHN, B. H. & FIELDING, C. R. 1993. Reservoir potential of the Catherine Sandstone, Denison Trough, east-central Queensland. *Australian Petroleum Exploration Association Journal*, **33**, 176–187.

KENDALL, P. F. & BRIGGS, H. 1933. The formation of rock joints and the cleat of coal. *Proceedings of the Royal Society of Edinburgh*, **53**, 164–187.

KIRKEGAARD, A. G., 1974: Structural elements of the northern part of the Tasman Geosyncline. *In*: DENMEAD, A. K., TWEEDALE, G. W. & WILSON, A. F. (eds) *The Tasman Geosyncline—a Symposium*. Geological Society of Australia (Queensland Division), Brisbane, 47–62.

LAUBACH, S. E., TREMAIN, C. H. & AYERS, W. B. JR, 1991. Coal fracture studies: guides for coalbed methane exploration and development. *Journal of Coal Quality*, **10**(3), 81–88.

LAW, B. E. 1991. The relationship between coal rank and cleat density—a preliminary report [abstract]. *AAPG Bulletin*, **75**, 1131.

——1993. The relationship between coal rank and cleat density: implications for the prediction of permeability in coal. *In*: *1993 International Coalbed Methane Symposium*, Birmingham, Alabama, Paper 9341.

MARSHALLSEA, S. J., GREEN, P. F., DUDDY, I. R. & GLEADOW, A. J. W. 1985. The thermal history of the southern Bowen Basin. *In*: *Bowen Basin Coal Symposium*. Geological Society of Australia, Abstracts, **17**, 237–241.

MCCULLOCH, C. M., DEUL, M. & JERAN, P. W. 1974a. *Cleat in Bituminous Coalbeds*. US Bureau of Mines, Report of Investigation, **7910**.

——, —— & —— 1974b. Structural control of cleat in bituminous coalbeds. Geological Society of America. *Abstract of Programs*, **6**(7), 862–863.

——, LAMBERT, S. W. & WHITE, J. R. 1976. *Determining Cleat Orientation of Deeper Coalbeds from Overlying Coals*. US Bureau of Mines, Report of Investigation, **8116**.

MAVOR, M. J., CLOSE, J. C. & PRATT, T. J. 1991. *Western Cretaceous Coal Seam Project Summary of the Completion Optimisation and Assessment Laboratory (COAL) Site*. Gas Research Institute, Topical Report, **GRI-91/0377**.

MILLER, D. E. 1992. *Geology of the Southern Moura Mine and Malakoff Range, Southeast Bowen Basin, Central Queensland*. Honours Thesis, University of Queensland, Brisbane.

MURRAY, C. G. 1990. Tectonic evolution and metallogenesis of the Bowen Basin. *In*: *Bowen Basin Symposium 1990. Proceedings*. Geological Society of Australia (Queensland Division), Brisbane, 201–212.

——, FERGUSSON, C. L., FLOOD, P. G., WHITAKER, W. G. & KORSCH, R. J. 1987. Plate tectonic model for the Carboniferous evolution of the New England Fold Belt. *Australian Journal of Earth Science*, **34**, 213–236.

PATTISON, C. I. 1990. *Igneous Intrusions in the Bowen Basin*. MSc Thesis, Queensland University of Technology.

RAMANO, R. N. & MOIZ, A. A. 1968. Study of the cleat in north and south Godavari Valley coalfields of Andhra Pradesh. *Journal of Mines, Metals and Fuels, India, Calcutta*, **16**, 154–160.

SPEARS, D. A. & CASWELL, S. A. 1986. Mineral matter in coals: cleat minerals and their origin in some coals from the English Midlands. *International Journal of Coal Geology*, **6**, 107–125.

TING, F. T. C. 1977. Cleating in coals; its origin, spacing, and implications in mining and preparation. *Geological Society of America, Abstracts*, **9**, 1201–1202.

TREMAIN, C. M., LAUBACH, S. E. & Whitehead, N. H. 1991. Coal fracture (cleat) patterns in Upper Cretaceous Fruitland Formation, San Juan Basin, Colorado and New Mexico: Implications for coalbed methane exploration and development. *In*: SCHWOCHOW, S. (ed.) *Coalbed Methane of*

*Western North America.* Rocky Mountain Association of Geologists Field Conference Guidebook, 49–59.

VER STEEG, K., 1942: Jointing in the coal beds of Ohio. *Economic Geology*, **37**, 503–509.

WARD, W. E., DRAHOVZAL, J. A. & EVANS F. E. JR 1984. *Fracture Analysis in Selected Areas of the Warrior Coal Basin.* Alabama Geological Survey, Alabama, Circular, **111**.

ZIOLKOWSKI, V. & TAYLOR, R., 1985: Regional structure of the northern Denison Trough. *In*: *Bowen Basin Symposium 1990, Proceedings.* Geological Society of Australia (Queensland Division), Brisbane, 129–135.

# Cleat mineralization of Upper Permian Baralaba/Rangal Coal Measures, Bowen Basin, Australia

BASIM S. M. FARAJ[1,2], CHRIS R. FIELDING[1] & IAN D. R. MACKINNON[2]

[1] *Department of Earth Sciences, The University of Queensland, St Lucia, Brisbane, Queensland 4072, Australia*
[2] *Centre for Microscopy and Microanalysis, The University of Queensland, St Lucia, Brisbane, Queensland 4072, Australia*

**Abstract:** Coals from the Permo-Triassic Bowen Basin have been investigated using a variety of complementary analytical techniques. Face cleat minerals in the studied samples are dominated by authigenic clays, notably pure illite or illite–chlorite mixtures. Butt cleats and joints are dominated by carbonates, mostly calcite with less abundant ferroan calcite, ankerite and siderite. K/Ar ages of cleat-fill illites and one fracture illite–smectite fill indicate three phases of illite formation during the Triassic: (1) an early phase with ages clustered around 244 Ma ago; (2) a second phase about 232 Ma ago; and (3) the latest phase about 219 Ma ago. Thermal modelling of selected boreholes from the study area and other geological data indicate that the first phase of illitization occurred when the coal seams were about 1000 m below the surface and at a temperature of between 70 and 80°C, at a time when the basin was rapidly subsiding. These early illites were stable and retained their $1M$ polytype even after being exposed to temperatures of 150–190°C during maximum burial in the southeastern Bowen Basin. The second phase of illitization (at $232 \pm 3$) occurred at about the time of maximum burial. The latest phase took place between $223 \pm 3$ and $212 \pm 3$ Ma ago and occurred in a wider temperature range (between 170 and 100°C) during a rapid uplift in the Late Triassic. The Bowen Basin area experienced a second cycle of subsidence that commenced in the Early Jurassic with the formation of the Surat Basin, which overlies the Bowen Basin and forms part of the Great Artesian Basin system. Regional uplift and erosion in middle Cretaceous times terminated sediment accumulation in the Surat Basin. During the second cycle of burial widespread carbonate mineralization in butt-cleats and joints took place. Calcite fluid inclusions indicate that calcite was precipitated from meteoric water at about 80°C. The widespread carbonate mineralization in butt cleats and the near-absence of carbonates in the face cleats is here attributed to permeability anisotropy, caused by a change in the direction of lateral compressive stress during Jurassic–Cretaceous times relative to that during the Triassic Hunter–Bowen Orogeny.

Coal seams are both source and reservoir rocks for coalbed methane (Levine 1993; Rice 1993). They are excellent source rocks due to their extreme richness in organic matter and their efficiency in retaining much of the initial hydrogen until high maturity levels are reached ($\approx 1.0\% R_o$) (Tissot & Welte 1984; Rightmire 1991; Littke & Leythaeuser 1993). Methane gas is present as an adsorbed phase within the coal matrix and in cleats and fractures in the coal seams as a free gas and/or dissolved in water, where its flow is governed by Darcy's flow (Kim 1977; Diamond 1993). Furthermore, the distribution patterns, frequency and connectivity of fractures and cleats in coal seams are major factors affecting the rate and direction of flow of gases and water in coal (Kim 1977; Stone & Snoeberger 1978; Tremain et al. 1991).

Mineralization of cleats and fractures impedes the movement of gases and water in coal seams and has been found to negatively influence methane producibility from coal (Tremain et al. 1991). This phenomenon is analogous to the precipitation of porosity reducing cements in conventional hydrocarbon reservoirs. Therefore, a knowledge of the cleat minerals, their spatial distribution and origin are important aspects of the reservoir characterization of coal seams. This study provides details of the nature and spatial distribution of cleat minerals in Upper Permian coal measures from the Bowen Basin of eastern Australia. Our results provide an insight into the controls on fluid flow and mineralization within the Bowen Basin. It is hoped that results from this work and those from the regional investigation into the characteristics and origin of cleats (Pattison et al. this volume), will be useful in the characterization of coal seams as reservoir rocks for coalbed methane in the Bowen Basin, and indeed in other basins.

The Bowen Basin is located in central-east Queensland, Australia (Fig. 1) and is one of the

*From* Gayer, R. & Harris, I. (eds) 1996, *Coalbed Methane and Coal Geology*,
Geological Society Special Publication No 109, pp 151–164.

**Fig. 1.** Location map for the Bowen Basin and its relationship with the coeval Gunnedah and Sydney basins. Note the MIM Ltd coalbed methane lease boundaries (in black). Modified after Baker *et al.* (1993).

largest coal-producing basins in the world with a total area of about $70\,000\,\text{km}^2$ (Huleatt 1991). Over 20 coal mines are operational in the Bowen Basin (Fig. 1) and produce the bulk of Queensland's black coal. The north–south elongate Bowen Basin (Fig. 1) formed in Early Permian times as a series of fault-bounded graben and half-graben during a phase of limited crustal extension. A middle Permian phase of mainly passive thermal subsidence was followed by a period of thrust loading from the east, during which time the Bowen Basin became a foreland basin. From the Late Permian until the Middle

Triassic, the eastern half of the basin was periodically affected by pulsed, WSW propagation of thrust sheets from the rising New England Fold Belt to the east. Sediment accumulation stopped in late Middle Triassic times (about 235 Ma BP), when a major basin-wide thrusting event was followed by uplift and erosion of the Permo-Triassic succession. For further details, see Pattison *et al.* (this volume) and references cited therein.

The Upper Permian Baralaba/Rangal Coal Measures are widespread over the basin and show regular rank increase from sub-bituminous at the basin's margins to anthracite towards the depocentre (Shibaoka & Bennett 1976; Beeston 1986). The vitrinite content of coal seams ranges between 30 and 70% (Staines & Koppe 1979). Thus Bowen Basin coal is suitable for many applications, including steam generation and coke-making. Since 1976 an interest in the commercial development of coalbed methane resources in the Bowen Basin has gained momentum (Oldroyd & Brotherton 1992). An exploration programme was undertaken by MIM Holdings Ltd in which 21 boreholes were drilled into the Rangal/Baralaba and earlier Upper Permian German Creek Coal Measures (Fig. 2). Core from these drillholes forms the primary database for this study.

## Materials and methods

Mineralized core samples from the Rangal/Baralaba and German Creek Coal Measures were selected for this study. Borehole samples ranging in depth (below surface) from 300 to 989 m were taken from eight different seams in 13 boreholes (Fig. 2). The rank of these coal measures varies from high-volatile bituminous coal ($0.87\%\ R_\text{o}$) in samples from Moura area to semi-anthracite ($2.36\%\ R_\text{o}$) in samples from Bluff area (Fig. 2). Textural, chemical and mineralogical characterization of the cleat minerals was carried out utilizing a number of complementary analytical techniques, including transmitted and reflected light optical microscopy, scanning electron microscopy with energy dispersive spectrometry (SEM/EDS), analytical electron microscopy (AEM), electron microprobe and X-ray diffraction (XRD) analysis. All coal samples were examined under a stereoscopic microscope and the cleat minerals were carefully separated using fine forceps. Whenever possible, the minerals separated were related to their habitat. Bedding-discordant discontinuities in coal samples examined have been classified into (1) 'systematic' cleat sets (face, butt and

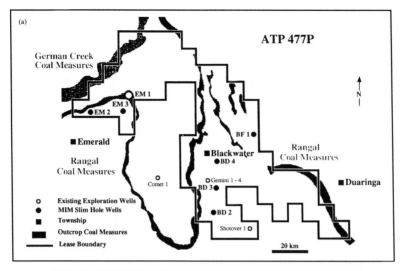

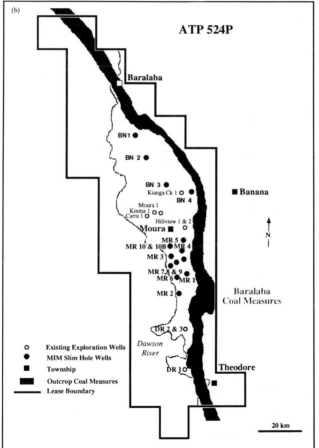

**Fig. 2. (a)** Location map of the MIM's coalbed methane lease ATP 477P. Samples were collected from the Rangal Coal Measures intersected by MIM Slim Hole Wells BF1, BD4, BD3 and BD2. Also, samples were collected from both the Rangal and German Creek Coal Measures from the MIM Slim Hole Wells EM1, EM2 and EM3 northeast of Emerald. **(b)** Location map of MIM's coalbed methane ATP 524P adjoining south of ATP 477. Samples were collected from the Baralaba Coal Measures (lateral equivalent of the Rangal Coal Measures) intersected by MIM Slim Hole Wells, BN1, BN2, BN3 and BN4 in the Banana area, and MR5, MR7 and MR2 from the Moura area. Map modified after MIM Ltd.

superimposed), (2) 'non-systematic' cleat sets and (3) joints as detailed in the companion paper by Pattison *et al.* (this volume). K/Ar radiogenic isotope dating was also performed on nine illite samples from cleats in the Rangal/Baralaba coal seams.

## Results

The overall variety in cleat minerals across the studied area is limited, with most of the minerals found belonging to two groups: clay minerals dominated by illite with minor occurrences of chlorite and kaolinite, and carbonates dominated by calcite with minor phases such as ferroan calcite, ankerite and siderite. Rare accessory minerals (strontianite, wollastonite and whewellite) have been also identified. In addition, limited occurrences of barite and pyrite were noted.

Face cleat minerals of the Rangal/Baralaba Coal Measures are dominated by clays. The major mineral found is illite, with minor occurrences of ammonium illite, chlorite and kaolinite. In contrast, butt cleat and joint minerals are dominated by carbonates, of which calcite is the dominant phase, with minor occurrences of ferroan calcite and siderite. Figure 3 shows the distribution of illite, chlorite and calcite in face, butt cleats and joints from the study area.

### Illites

Cleat-filling illites encountered in this study are brown or black in hand specimen. They form flakes within the cleat space, up to several centimetres in length and typically between 50 and 200 $\mu$m thick (Fig. 4a). Under SEM, illite fills display smooth flat surfaces parallel to the cleat aperture, often displaying casts or moulds of the internal cleat surfaces on which the illite grew (Fig. 4b). In cross-section, the illite grains appear as curved, densely packed, irregular

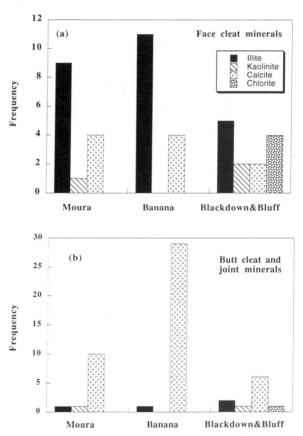

**Fig. 3.** Histograms showing the occurrence frequency of minerals found in (**a**) face cleats and (**b**) butt cleats and joints from the study area.

(a)

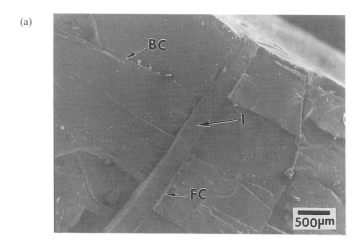

(b)

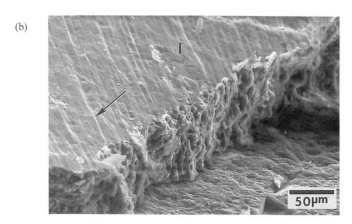

**Fig. 4.** Scanning electron photomicrographs illustrating cleat-fill illite from the study area. (**a**) Illite-filled face cleat from a coal sample in BN2 well (seam A Upper) at depth 879.9 m. Note the illite (denoted I) presence only in face cleats (denoted FC), whereas butt cleats (denoted BC) are not mineralized. (**b**) Close up photomicrograph of face cleat illite from seam X Lower at 554.9 m depth, from a coal sample in MR5 well. Note the moulds (arrowed) on the surface of the fill. This type of morphology is typical of cleat-filling illite found in this study.

plates, 5–10 $\mu$m across. Most of the cleat-fill illites examined display similar morphology and texture.

X-Ray diffraction traces of oriented, air-dried, ethylene glycolated and heated preparations of cleat-fill illites show them to be composed of almost pure illite intermixed with traces ($\approx$5%) of expandable smectite layers (Reynolds 1980; Moore & Reynolds 1989). An example is shown in Fig. 5a. The face cleat-fill illites are of similar chemistry. Table 1 shows electron microprobe analyses of nine illite samples. Illite flakes were separated and mounted flat onto glass slides (without any polishing) using carbon dag. Up to

10 individual analyses were carried out on each sample and then averaged and presented as a single analysis in Table 1. Analyses show that these illites are aluminium-rich, with variations between 0.5 and 1.1 atoms of Al/unit cell substituted for Si in the tetrahedral layer. In the octahedral layer, variations between 1.8 and 1.0 Al atoms/unit cell are evident. However, the total Si/Al ratio is less variable, with ratios between 1.3 and 1.4 except for samples MIM 016 and the ammonium rich sample bn4-2-1 (Table 1). In addition, Fe and Mg are present in more or less similar proportions, whereas traces of Ti are also noted. The dominant interlayer cation is K$^+$

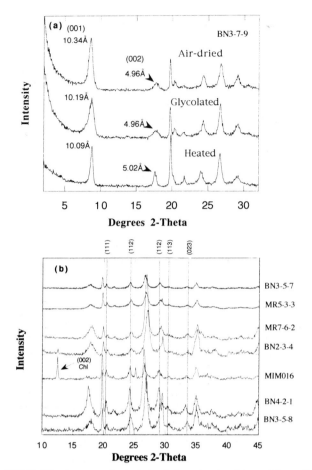

**Fig. 5.** (**a**) Powdered XRD diffractograms of oriented, ethylene glycolated (overnight at 60°C) and heated (550°C for 2 h) samples of face cleat illite-fill (sample BN3-7-9 from seam C at 827 m depth from BN3 well). Note the slight decrease in *d*-spacing of 001 illite peak on glycolation and heating, and the near-absence of change in 002 peak, indicative of a small amount (≈5%) of smectite layers associated with illite. (**b**) Random X-ray diffractograms of face cleat illite-fills from seven samples, including a highly illitic (≈90%) illite/smectite mixed-layer sample (MIM 016), showing the presence of 1*M* diagnostic peaks (vertical lines) in these samples. Also note the chlorite (002) peak in MIM 016 sample. All diffractograms obtained using copper Kα radiation.

with values ranging between 0.70 and 0.77. The only exception to this range of values is sample MIM 016 (from Bluff-1 well), which is a mixture of mixed-layer illite/smectite (≈90% illite), chlorite and minor kaolinite. This mixture is present as a co-precipitate in the face cleat samples from Bluff-1 well. Similar material is also noted in face cleats from Blackdown samples.

*Illite polytypes.* Based on random-powdered XRD runs, all cleat-fill illites, including the ammonium illite, are a 1*M* polytype. This identification is based on several characteristic *hkl* peaks present on either side of the (003) basal

reflection (Bailey 1980; Moore & Reynolds 1989). Only one sample (mr5-3-3) showed the presence of a small amount (<10%) of 2*M*1 polytype (Bailey 1980).

*Distribution of cleat-fill-illite.* Illite is second to calcite in overall frequency of occurrence in the samples studied (Fig. 3). Unlike calcite, illite dominates face cleat minerals. Illite has been identified in cleats of the Rangal/Baralaba coal seams in the Moura and Banana areas, but is absent or minor from butt cleats (Fig. 3). Illite is also dominant in face cleat minerals identified from the Bluff and Blackdown coal samples (Fig. 3a). Illite from these areas is not found as a single

**Table 1.** *Microprobe analyses and structural formulae for nine cleat-fill illite samples from the Bowen Basin. Sample numbers and well names (in parentheses) are also shown*

| Wt.% oxide | mr7-2-1 (MR7) | mr5-1-2 (MR5) | bn3-7-9 (BN3) | bn2-3-4 (BN2) | mr2-B Lr (MR2) | mim016§ (BF1) | mr5-3-3 (MR5) | bn1-2-3 (BN1) | bn4-2-1‡ (BN4) |
|---|---|---|---|---|---|---|---|---|---|
| $Na_2O$ | 0.212 | 0.212 | 0.393 | 0.262 | 0.759 | 0.193 | 0.329 | 0.198 | 0.374 |
| MgO | 0.929 | 0.929 | 0.948 | 1.400 | 1.024 | 0.151 | 0.682 | 0.793 | 0.206 |
| $Al_2O_3$ | 32.391 | 32.391 | 31.190 | 28.393 | 29.014 | 35.540 | 32.930 | 27.608 | 32.268 |
| $SiO_2$ | 50.347 | 50.347 | 48.661 | 47.839 | 47.275 | 42.295 | 50.394 | 43.832 | 45.445 |
| $K_2O$ | 8.805 | 8.805 | 8.890 | 8.130 | 8.127 | 0.472 | 8.709 | 7.498 | 6.738 |
| CaO | 0.052 | 0.052 | 0.086 | 0.011 | 0.040 | 0.188 | 0.021 | 0.079 | 0.116 |
| $TiO_2$ | 0.034 | 0.034 | 0.053 | 0.025 | 0.027 | 0.012 | 0.027 | 0.026 | 0.014 |
| MnO | 0.010 | 0.010 | 0.005 | 0.005 | 0.010 | 0.013 | 0.006 | 0.006 | 0.011 |
| FeO† | 1.411 | 1.411 | 1.570 | 1.355 | 1.212 | 0.888 | 1.294 | 1.220 | 0.924 |
| Total wt.% oxides* | 94.190 | 94.702 | 91.796 | 87.421 | 87.489 | 79.752 | 94.393 | 81.258 | 86.096 |

Number of cations on the basis of $[O10(OH)_2 - 1H_2O]$

| | | | | | | | | | |
|---|---|---|---|---|---|---|---|---|---|
| Si | 3.327 | 3.470 | 3.216 | 3.161 | 3.124 | 2.795 | 3.330 | 2.896 | 3.003 |
| Al | 0.673 | 0.530 | 0.784 | 0.839 | 0.876 | 1.205 | 0.670 | 1.104 | 0.997 |
| Σtetrahedral | 4.000 | 4.000 | 4.000 | 4.000 | 4.000 | 4.000 | 4.000 | 4.000 | 4.000 |
| Al | 1.850 | 1.817 | 1.645 | 1.373 | 1.384 | 1.563 | 1.895 | 1.047 | 1.516 |
| Fe | 0.078 | 0.090 | 0.087 | 0.075 | 0.067 | 0.049 | 0.072 | 0.067 | 0.051 |
| Mg | 0.092 | 0.153 | 0.093 | 0.138 | 0.101 | 0.015 | 0.067 | 0.078 | 0.020 |
| Ti | 0.002 | 0.001 | 0.003 | 0.001 | 0.001 | 0.001 | 0.001 | 0.001 | 0.001 |
| ΣOctahedral | 2.021 | 2.061 | 1.827 | 1.587 | 1.553 | 1.627 | 2.035 | 1.193 | 1.588 |
| K | 0.742 | 0.709 | 0.749 | 0.685 | 0.685 | 0.040 | 0.734 | 0.632 | 0.568 |
| Na | 0.027 | 0.038 | 0.050 | 0.034 | 0.097 | 0.025 | 0.042 | 0.025 | 0.048 |
| Ca | 0.008 | 0.020 | 0.012 | 0.002 | 0.006 | 0.026 | 0.004 | 0.012 | 0.016 |
| ΣInterlayer | 0.777 | 0.762 | 0.837 | 0.767 | 0.850 | 0.102 | 0.778 | 0.773 | 0.721 |
| Si/Al ratio | 1.319 | 1.478 | 1.324 | 1.430 | 1.382 | 1.010 | 1.298 | 1.347 | 1.195 |
| Si/K | 4.48 | 4.89 | 4.29 | 4.61 | 4.56 | 7.09 | 4.53 | 4.58 | 4.99 |
| Tetrahedral charge | −0.673 | −0.530 | −0.684 | −0.596 | −0.632 | −0.880 | −0.682 | −0.647 | −0.750 |
| Octahedral charge | −0.105 | −0.204 | −0.154 | −0.180 | −0.219 | +0.776 | −0.096 | −0.128 | −0.065 |
| Total layer charge | −0.778 | −−0.762 | −0.838 | −0.776 | −0.851 | −0.103 | −0.777 | −0.775 | −0.686 |
| Interlayer charge | +0.777 | +0.762 | +0.837 | +0.776 | +0.850 | +0.102 | +0.778 | +0.773 | +0.721 |

* Total wt.% oxide is reported without the structural $H_2O$.
† Total Fe is reported as FeO.
‡ Ammonium-rich illite.
§ This sample is a mixture of illite/smectite, kaolinite and chlorite.

phase, but rather as an intimate mixture with chlorite and/or kaolinite, whereas butt cleats are either not mineralized or contain calcite (Fig. 3b). Joints are well developed in the Bluff samples and contain an illite/chlorite mixture, chlorite, calcite and quartz. Furthermore, coal samples examined from both the Rangal and German Creek Coal Measures at Emerald are devoid of illite. Kaolinite is the dominant clay mineral found in joints and superimposed cleats in the Emerald area, whereas face and butt cleats are either absent or not well developed in samples studied from that region.

*Radiogenic isotope (K/Ar) dating.* The purity of cleat-fill illites found in this study (based on XRD and SEM/EDS) and the ease with which they can be separated from cleats has rendered them ideal for K/Ar age determinations, one of the most difficult aspects of K/Ar dating being contamination by other K-bearing minerals such as potassium feldspars, detrital illites and mica (Hamilton *et al.* 1989; Liewig *et al.* 1987; Lundegard 1989; Emery & Robinson 1993). Furthermore, the problem of multiple generations of illites which tend to yield mixtures of K/Ar age (commonly varying with the illite size

**Table 2.** *List of K/Ar ages (Ma BP) of nine cleat-fill illites, including an illite/smectite (sample MIM 017b) from the Bowen Basin. All the illite samples are 1M polytype except sample MR5-3-3, which contains an additional small amount (≈10%) of 2M1 polytype. Well name, present day depth and sample number of each sample are also shown*

| Well name | Sample number | Depth (m) | Illite polytype | K/Ar age (Ma BP) |
|-----------|---------------|-----------|-----------------|------------------|
| BN-1 | BN1-2-3 | 554.9 | 1*M* | 220 ± 3 |
| BN-2 | BN2-3-4 | 879.9 | 1*M* | 249 ± 3 |
| BN-3 | BN3-7-9 | 827.0 | 1*M* | 242 ± 3 |
| BN-4 | BN4-2-1 | 563.9 | 1*M* | 241 ± 4 |
| MR-5 | MR5-1-2 | 329.0 | 1*M* | 229 ± 3 |
| MR-5 | MR5-3-3 | 421.4 | 1*M* + 2*M*1 | 220 ± 3 |
| MR-7 | MIM 017b | 815.1 | I/S (roof-fill) | 234 ± 3 |
| MR-7 | MR7-6-2 | 734.1 | 1*M* | 223 ± 3 |
| BF-1 | MIM 016 | 837.7 | 1*M* | 212 ± 3 |

fraction) is also eliminated. This dependence on size fraction can produce different ages in sandstone reservoir rock illites (Hamilton *et al.* 1989). In this study, all radiogenic dating was performed on whole illites separated from cleats, except for one sample, where highly illitic (≈90%) illite/smectite mixed-layer clay occurs as a mudrock fracture-fill from the Moura-7 borehole (sample MIM 017b, Table 2).

Table 2 shows K/Ar ages of nine samples. The ages range between 249 ± 3 and 212 ± 3 Ma BP Figure 6 shows the K/Ar ages plotted against present day depth. The oldest samples cluster around 244 ± 3 Ma BP, whereas the K/Ar ages about 232 ± 3 Ma BP are isolated from those both (Table 2) older and younger. The younger values show a spread of ages between 223 ± 3 and 212 ± 3 Ma BP in samples trending from

Moura to the south, up to Bluff to the north, a distance of over 200 km (see Fig. 2).

*Stable isotopes of illite.* Table 3 shows oxygen and hydrogen isotopes of three face cleat-fill illites from Moura-2, Moura-5 and Banana-3 wells (Fig. 2). The three samples are depleted in $^{18}O$ (relative to the PDB standard) with $\delta^{18}O$ values ranging between −16.0 and −23.3‰. A wide range of $\delta D$ values was noted (between −80.1 and −123‰).

## Calcite

Calcite was found to occur most commonly as transparent flakes lining cleat space. Figure 3 shows that calcite occurs most frequently in butt

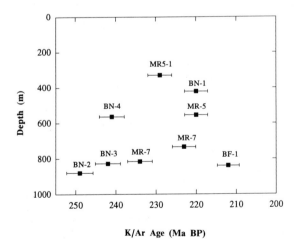

**Fig. 6.** Plot of K/Ar illite ages (shown in Table 2) against present day depth. Note the clustering of the early ages about 244 Ma BP, two intermediate ages (234, 229 Ma BP) the spread of the later ages (between 225 ± 3 and 216 ± 3 Ma BP).

**Table 3.** *Oxygen and hydrogen stable isotope values of three face cleat-fill illites from the Bowen Basin*

| Borehole | Sample depth (m) | $\delta^{18}O$ (‰ SMOW) | $\delta^{18}O*$ (‰ PDB) | $\delta D$ (‰ SMOW) | Fill type |
|---|---|---|---|---|---|
| Moura-2 | 568 | +14.4 | −16.0 | −80.1 | Face cleat |
| Moura-5 | 421 | +6.9 | −23.3 | −123 | Face cleat |
| Banana-3 | 827 | +10.0 | −20.2 | −85.5 | Face cleat |

* Conversion from SMOW and/or PDB $\delta$-values is carried out according to the following formulas:
(i) $\delta_{SMOW} = 1.03086\delta_{PDB} + 30.86$, and (ii) $\delta_{PDB} = 0.97006\delta_{SMOW} - 29.94$ (Friedman & O'Neil 1977).

cleats and joints, but is rarely found in face cleats as a single phase. However, calcite is common in non-coal, roof and floor mudrock fractures from the study area. Some of these fracture fills are up to 1.3 cm thick and three samples have been utilized for fluid inclusion analysis. Calcite-hosted inclusions from the samples studied are generally two-phase, but with variable liquid to vapour ratios. Most appear to be secondary in origin, but examples of primary types have been noted. The homogenization temperatures ($T_h$) range between 82.3 and 158.1°C. The temperatures show a wide scatter, the high $T_h$ values most probably due to stretching as calcite is prone to this phenomenon (Roedder 1984; Rankin 1989). Accordingly, the minimum values (close to 80°C) of $T_h$ are interpreted to indicate minimum trapping temperatures of inclusions in the samples studied (Shepherd *et al.* 1985; McNeil 1995, pers. comm.). Temperatures of last melting ($T_{lm}$) of fluid inclusions from calcite samples range between −3.5 and 1.6°C (mean value 0.44°C). The low $T_{lm}$ values indicate that the fluid in the inclusions is of very low salinity, suggesting a meteoric origin (Roedder 1984).

## Other carbonates

Ankerite is distributed mainly within non-systematic cleats across the entire study area. Textural relationships suggest that its formation post-dated that of illite in the samples examined, but that it preceded calcite formation. Siderite is present as a minor phase in cleats and fractures in some areas, where it can be seen to have pre-dated calcite formation. Rare ferroan calcite was found in a similar habitat to calcite.

## Discussion

### Early illite

All the cleat minerals are authigenic phases that formed from solutions or fluids circulating in the subsurface. K/Ar ages of cleat-fill illites indicate episodic formation during the Triassic (Fig. 6). The earliest illitization phase took place around 244 Ma BP in the Moura–Banana areas. Thermal modelling of selected boreholes from the study area indicates that the first phase of illitization occurred when the coal seams were about 1000 m below the surface about 6 Ma after deposition, at a temperature of between 70 and 80°C and inferred vitrinite reflectance between 0.5 and 0.55% $R_o$. At that time, the Bowen Basin was rapidly subsiding during a period of foreland thrust loading. By this stage the coal seams were evidently capable of brittle failure and cleat formation (cf. Pattison *et al.* this volume). Also during this time period, precious and base metal mineralization formed at several localities along the eastern edge of the Basin (Golding *et al.* 1994).

All the illites formed before maximum burial (about 235 Ma BP) were stable at the maximum temperatures associated with the increase in burial depth. The estimated maximum temperatures attained from the samples studied vary between 150°C in the Moura area and about 190°C in the Banana area. The stability of illite is indicated by retention of the 1*M* polytype. Yoder & Eugster's (1955) experimental work indicates conversion of 1*M* muscovite to 2*M* polytype occurs over a temperature range 200–350°C and at 15 000 psi water pressure. Other workers have also concluded that the conversion of a 1*M* to 2*M*1 polytype is a high temperature transition (Weaver & Brockstra 1984; Srodon & Eberl 1984).

### Late illite

The second phase of illitization at 232 Ma BP occurred at about the time of maximum burial (Pattison *et al.* this volume) at a temperature of about 170°C. The latest phase of illitization recognized took place between about 223 and 212 Ma BP. This pulse of illitization of face cleats occurred during a period of rapid uplift of the

Bowen Basin at the end of the Hunter–Bowen Orogeny (Pattison *et al.* this volume). The illitization of face cleats occurred over a period of 11 Ma (between $223 \pm 3$ and $212 \pm 3$ Ma BP), as shown in Table 2. Thermal modelling indicates that this phase of illitization occurred under a temperature range between 170 and 100°C.

## Origin of illite and possible mechanism of formation

Illite phase equilibria show that illite can form and be stable over a temperature range from 50 to about 200°C, the controlling factors on the temperature of formation being pH, quartz saturation and activity of potassium ions (Aja *et al.* 1991*a*). Furthermore, experimental work on phase relationships suggest illite, chlorite and kaolinite are stable as a mixture in the presence of quartz at temperature of about 150°C (Aja *et al.* 1991*b*).

The chemical similarity of the early and late illites (from XRD, EDS and microprobe results) on such a large scale (at least 200 km front from south to north) suggests a large-scale illitization event, which given the distribution of K/Ar dates may have been pulsed. Nevertheless, there

appears to be some geographical variation in the chemistry of the fluids responsible for face cleat mineralization—for example, in the Bluff and Blackdown areas where mixtures of chlorite and illite co-precipitated in the cleats rather than illite, as is the case in the Banana and Moura samples. This may reflect a change in the chemistry and temperature of the water responsible for precipitation. Indeed, it appears that the illitization episode was confined to the eastern part of the Bowen Basin, as no illites or chlorites were found in samples from the Rangal or the German Creek Coal Measures in the western part of the basin at Emerald (60–80 km west of Bluff and Blackdown areas) (Fig. 2b).

Stable isotope results indicate that the illite-forming solutions were depleted in $^{18}O$, and of meteoric origin (Craig 1961; Longstaffe 1989). Illite–water fractionation curves, based on Savin & Lee's (1988) illite–water fractionation formula, are generated for three illite samples and shown in Fig. 7, together with the range of the most likely temperature values under which they precipitated. The corresponding water composition of two samples indicate fluids depleted in $^{18}O$ with values ranging between $-9$ and $-3‰$ relative to SMOW. One sample shows a slight enrichment of $^{18}O$ with $\delta^{18}O$ water values between 1 and 4‰ relative to SMOW, possibly

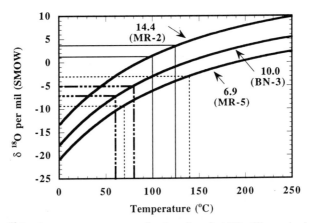

**Fig. 7.** Plot of water $\delta^{18}O$ values versus temperature for three face cleat illite-fill samples (see Table 3). The three thick solid curves represent water–illite fractionation curves of the three illite samples (arrowed). $\delta^{18}O$ value and borehole name (in parentheses) of each sample are also indicated (see Fig. 2 for location). The water–illite fractionation formula used is $1000 \, Ln\alpha(\text{illite–water}) = 2.39 \times 10^{6} \, T^{-2} - 4.19$, where $T$ is temperature in Kelvin (after Savin & Lee 1988). The similar pairs of vertical lines that connect each fractionation curve represent the most likely 'temperature window' under which each illite sample has precipitated, obtained independently from thermal modelling and burial curves. Corresponding horizontal lines drawn from the point of intersection of the vertical lines and each fractionation curve which intersects the water $\delta^{18}O$ coordinate enable a most likely $\delta^{18}O$ water values range for each illite sample to be deduced. For example, illite of sample BN-3 most likely precipitated at a temperature between 60 and 80°C, the corresponding $\delta^{18}O$ water values are negative (between $-9$ and $-7‰$), indicating precipitation from meteoric water.

due to mixing with formation water (Schwartz & Longstaffe 1988; Longstaffe, 1993) or water–rock interaction given the observation that Triassic surface waters in the Bowen Basin were highly depleted in $^{18}O$ down to $-15‰$ (Baker et al. in press).

Cook (1975) and Beeston (1986) noted that the coalification gradient in the Bowen Basin was 'very steep'. In this study, palaeogeothermal gradients in the Bowen Basin were estimated by converting vitrinite reflectance values measured from several boreholes to maximum palaeotemperatures using Barker & Pawlewicz's (1986) empirical formula. Assuming a mean annual surface temperature of 15°C during the Triassic, based on the latitudinal position of Australia in Triassic times (Embleton 1984; Baker & Caritat 1992), the eroded thickness from each area can be estimated using the method outlined by Shibaoka et al. (1973) and Shibaoka & Bennett (1976), using Beeston's (1986) coalification curve as a baseline. The averaged palaeogeothermal gradient for the Bowen Basin during the Triassic is 6.3°C/100 m.

The high geothermal gradients in the Triassic are basin-wide, evident from the coalification gradient, which drops markedly in Jurassic–Cretaceous times (Beeston 1986). The presence of illites in face cleats with K/Ar dates that coincide with the timing of the thrusting, subsidence and uplift related to the Hunter–Bowen Orogeny, the similar chemistry of illites from a wide geographical area in the eastern part of the Bowen Basin and the stable isotope values which infer meteoric water origin for the illites, suggest that the cause for this anomaly may have been the meteoric water recharge from topographically high areas that were generated by the advance of the Hunter–Bowen Orogeny. In such a model, the gravity-driven flow system creates a high potential hydraulic head that is capable of driving hot fluids basinward from the core of the orogeny (Bethke 1986; Oliver 1986). These advective fluids are an efficient medium for the transport of solutes and heat which may then precipitate authigenic minerals in permeable zones. The conduits for fluid movement are faults and permeable zones in the deeper part of the basin. The build-up of basinal fluid pressure as thrust fronts advance causes intermittent ingress of basinal fluids into coal seams and associated rocks (Oliver 1986). Temperature gradients of between 5°C/100 m and 9.5°C/100 m are possible in a gravity-driven groundwater flow in large systems (Garven & Freeze 1984b). Such fluids circulating at depth are capable of forming tension fractures and joints (Lee & Farmer

1993) or opening previously formed fractures (Secor 1965).

Compaction-driven, cross-formational fluid flow from overpressured zones within basins provides another possible fluid flow mechanism (cf. Garven & Freeze 1984a; Schwartz & Longstaffe 1988). Such a mechanism has been argued to explain ankerite formation during burial diagenesis in the Denison Trough, part of the Bowen Basin (Caritat & Baker 1992). However, such systems are believed to be short-lived and stop operating after the release of overpressure (Schwartz & Longstaffe 1988). Such a model cannot explain large-scale fluid flow capable of perturbing geothermal gradients basin-wide and forming large-scale mineralization, due to the slowness of the compaction-driven flow and cooling of the fluids by conduction as they move upwards (Bethke 1986; Oliver 1986). In a study from the Western Canada Basin, Longstaffe (1993) found no significant contribution of $^{18}O$-rich shale pore water to Mesozoic sandstone reservoirs and suggested large-scale, gravity-driven meteoric water as the dominant mechanism in deep burial diagenesis of most sandstones. Furthermore, computer simulations (Bethke 1986), theoretical analysis (Toth 1980; Garven & Freeze 1984a) and present-day fluid flow patterns in sedimentary basins (Hitchon 1984) generally support the large-scale gravity-driven fluid flow concept. Such systems are also advocated to account for some precious and base metal mineral deposits (Hitchon 1980; Garven & Freeze 1984a; Bethke 1986; Longstaffe 1989). The Bowen Basin may therefore have considerable potential for discovery of further precious and base metal mineralization (see Murray 1990; Golding et al. 1994).

## Calcite

Several lines of evidence point to markedly different geological conditions controlling butt cleat mineralization compared with the face cleats in the Bowen Basin. For example, although calcite fluid inclusions from the Banana area indicate precipitation from meteoric water at temperatures of about 80°C, the meteoric water circulating at the time must have been rich in dissolved carbon dioxide, unlike Triassic water. Furthermore, carbonate mineralization, particularly calcite, is widespread over a large area that extends at least from Bluff, Blackdown, Banana and Moura to the east, across to Emerald in the western part of the Basin (Fig. 2). The dominance of calcite in butt cleats and the limited carbonate mineralization in the face cleats imply a regional permeability anisotropy in the basin, which is significantly different in direction and milder in magnitude than that of the Triassic. All this points towards a distinct, regional event responsible for this later phase of mineralization.

Regional considerations support the idea of the Jurassic–Cretaceous period of subsidence associated with the formation of the Surat Basin (Exon 1976; Elliott 1989) as being responsible for carbonate mineralization. Regional uplift and erosion in middle Cretaceous times terminated sediment accumulation in the Surat Basin (Exon 1976; Elliott 1989; Baker & Caritat, 1992; Elliott 1993). During this second episode of subsidence, the estimated palaeogeothermal gradient of the Bowen Basin was significantly less than that of the Triassic ($2.5°C/100$ m, calculated from Beeston's (1986) coalification gradient of the Surat Basin). The dominance of calcite butt cleat mineralization and the near-absence from face cleats was most probably caused by permeability anisotropy linked to this second cycle of subsidence and uplift during Jurassic–Cretaceous times. A lack of mineral phases suitable for radiogenic age dating preclude a more precise diagnosis at this time.

## Conclusions

(1) Cleat minerals provide a valuable tool to assess burial diagenesis in the context of tectonic processes. Cleats can yield pure phases of diagenetic minerals that are ideal for stable and radiogenic isotope geochemical analyses.

(2) Face and butt cleats in the Upper Permian Rangal/Baralaba Coal Measures of the Bowen Basin, Australia were formed during the latest Permian and Triassic as a result of rapid burial and stresses related to the Hunter–Bowen Orogeny.

(3) Illite and illite/chlorite face cleat mineralization formed under high geothermal gradient (average $6.3°C/100$ m) in at least three discrete pulses during the Triassic. The mineralization reflects a permeability anisotropy caused by WSW thrust propagation during the Hunter–Bowen Orogeny.

(4) The driving force for the introduction of mineralizing fluids during the Triassic was possibly a large-scale gravity-driven flow system, capable of displacing hot fluids from the deeper parts of the basin during the Hunter–Bowen Orogeny.

Such a model suggests that the Bowen Basin may have considerable potential for discovery of further precious and base metal mineralization.

(5) Butt cleat mineralization most probably occurred during the Jurassic–Cretaceous burial cycle of the Surat Basin. Again, permeability anisotropy of different direction and milder than that of the Triassic is inferred.

This study represents part of the senior author's doctoral research, supported by MIM Ltd through an Australian Postgraduate Research Industry Award. We are grateful to J. Baker, P. Talbot, L. Elliott, J. Kassan and J. Beeston for their constructive criticism and stimulating discussions during the course of this study. J. Esterle and G. O'Brien from CSIRO are thanked for making available their Zeiss microscope for vitrinite reflectance measurements. D. Cousins and R. Rasch are acknowledged for their help in the microprobe analysis. Much appreciation is due to B. McNeil for performing the fluid inclusion analysis. Acknowledgments are also due to N. Suzuki of Hokkaido University, Japan for his help and advice with the thermal modelling of boreholes. R. Grinan's efforts in redrafting Fig. 2 are appreciated. MIM Ltd are thanked for permission to publish this paper. The coalbed methane team of MIM Ltd, especially R. Brotherton and B. Lowe-Young, are thanked for their valuable suggestions, many stimulating discussions and continuous support during the course of this study. Reviews of the submitted manuscript were provided by B. Beamish and M. Miliorizos.

## References

AJA, S. U., ROSENBERG, P. E. & KITTRICK, J. A. 1991a. Illite equilibria in solutions: I. Phase relationships in the system $K_2O$–$Al_2O_3$–$SiO_2$–$H_2O$ between 25 and 250°C. *Geochimica et Cosmochimica Acta*, **55**, 1353–1364.

——, —— & ——1991b. Illite equilibria in solutions: II. Phase relationships in the system $K_2O$–$MgO$–$Al_2O_3$–$SiO_2$–$H_2O$. *Geochimica et Cosmochimica Acta*, **55**, 1365–1374.

BAILEY, S. W. 1980. Structures of layer silicates. *In*: BRINDLEY, G. W. & BROWN, G. (eds) *Crystal Structure of Clay Minerals and their X-ray Identification*. Mineralogical Society, London, Monograph, **5**, 1–123.

BAKER, J. C. & CARITAT, P. DE 1992. Post-depositional history of the Permian sequence in the Denison Trough, eastern Australia. *AAPG Bulletin*, **76**, 1224–1249.

——, KASSAN, J. & HAMILTON, J. P. Early diagenetic siderite as an indicator of depositional environment in the Triassic Rewan Group, southern Bowen Basin, eastern Australia. *Sedimentology*, in press.

——, FIELDING, C. R., CARITAT, P. DE & WILKINSON, M. M. 1993. Permian evolution of sandstone composition in a complex back-arc extensional to foreland basin: the Bowen Basin, eastern Australia. *Journal of Sedimentary Petrology*, **63**, 881–893.

BARKER, CH. E. & PAWLEWICZ, M. J. 1986. The correlation of vitrinite reflectance with maximum temperature in organic matter. *In*: BUNTEBARTH, G. and STEGENA, L. (eds) *Paleogeothermics*. Lecture Notes in Earth Sciences, **5**, Springer-Verlag, Berlin, 79–93.

BEESTON, J. W. 1986. Coal rank variation in the Bowen Basin, Queensland. *International Journal of Coal Geology*, **6**, 163–180.

BETHKE, C. M. 1986. Hydrologic constraints on the genesis of the Upper Mississippi Valley Mineral District from Illinois Basin brines. *Economic Geology*, **81**, 233–249.

CARITAT, P. DE & BAKER, J. C. 1992. Overpressure release, cross-formational porewater flow and diagenesis. *In*: KHARAKA, Y. K. & MAEST, A. S. (eds) *Proceedings of the 7th International Symposium on Water-Rock Interactions*, Park City, Utah, USA, 1161–1164.

COOK, A. C. 1975. The spatial and temporal variation of the type and rank of Australian coals. *In*: COOK, A. C. (ed.) *Australian Black Coal, its Occurrence, Mining, Preparation and Use*. The Australian Institute of Mining and Metallurgy, Illawarra Branch, 63–84.

CRAIG, H. 1961. Standard for reporting concentrations of deuterium and oxygen-18 in natural waters. *Science*, **133**, 1833–1834.

DIAMOND, W. P. 1993. Methane control for underground coal mines. *In*: LAW, B. E. & RICE, D. D. (eds) *Hydrocarbons from Coal*. American Association of Petroleum Geologists, Studies in Geology, **38**, 237–267.

ELLIOTT, L. 1989. The Surat and Bowen Basins. *The Australian Petroleum Exploration Association Journal*, **29**, 398–416.

——1993. Post-Carboniferous tectonic evolution of eastern Australia. *The Australian Petroleum Exploration Association Journal*, **33**, 215–236.

EMBLETON, B. J. J. 1984. Continental palaeomagnetism. *In*: VEEVERS, J. J. (ed.) *Phanerozoic Earth History of Australia*. Oxford Monographs on Geology and Geophysics, **2**, Clarendon Press, Oxford, 11–16.

EMERY, D. & ROBINSON, A. 1993. *Inorganic Geochemistry: Applications to Petroleum Geology*. Blackwell Scientific, Melbourne.

EXON, N. F. 1976. *Geology of the Surat Basin in Queensland*. Australian Bureau of Mineral Resources, **161**.

FRIEDMAN, I. & O'NEIL, J. R. 1977. Compilation of stable fractionation factors of geochemical interest. *In*: FLEISCHER, M. (ed.) *Data of Geochemistry*, 6th ed. USGS Professional Paper, **440-kk**.

GARVEN, G. & FREEZE, R. A. 1984a. Theoretical analysis of the role of groundwater flow in the genesis of stratabound ore deposits.

1. Mathematical and numerical model. *American Journal of Science*, **284**, 1085–1124.

——1984b. Theoretical analysis of the role of groundwater flow in the genesis of stratabound ore deposits. 2. Quantitative results. *American Journal of Science*, **284**, 1125–1174.

GOLDING, S. D., STEPHENS, C. J., VASCONCELOS, P. V., HOLCOMBE, R. J. & FIELDING, C. R. 1994. Metallogeny of the northern New England Fold Belt. *In*: HOLCOMBE, R. J., STEPHENS, C. J. & FIELDING, C. R. (eds) *1994 Field Conference, Capricorn Region Central Coastal Queensland*. Geological Society of Australia Incorporated (Queensland Division), 137–145.

HAMILTON, P. J., KELLEY, S. & FALLICK, A. E. 1989. K–Ar dating of illite in hydrocarbon reservoirs. *Clay Minerals*, **24**, 215–231.

HITCHON, B. 1980. Some economic aspects of water-rock interaction. *In*: ROBERTS, W. H. & CORDELL, R. J. (eds) *Problems of Petroleum Migration*. American Association of Petroleum Geologists, Studies in Geology, **10**, 109–119.

——1984. Geothermal gradients, hydrodynamics, and hydrocarbon occurrences, Alberta, Canada. *AAPG Bulletin*, **68**, 713–743.

HULEATT, M. B. 1991. *Handbook of Australian Black Coals: Geology, Resources, Seam Properties, and Product Specifications*. Bureau of Mineral Resources, Australia, Report, **7**.

KIM, A. 1977. *Estimating Methane Content of Bituminous Coalbeds from Absorption Data*. US Department of Interior, Bureau of Mines Research Investigation, **8245**.

LEE, C. H. & FARMER, I. 1993. *Fluid Flow in Discontinuous Rocks*. Chapman & Hall, Melbourne.

LEVINE, J. R. 1993. Coalification: the evolution of coal as source rock and reservoir rock for oil and gas. *In*: LAW, B. E. & RICE, D. D. (eds) *Hydrocarbons from Coal*. American Association of Petroleum Geologists, Studies in Geology, **38**, 39–77.

LIEWIG, N., CLAUER, N. & SOMMER, F. 1987. Rb–Sr and K–Ar dating of clay diagenesis in Jurassic sandstone oil reservoir, North Sea. *AAPG Bulletin*, **71**, 1467–1474.

LITTKE, R. & LEYTHAEUSER, D. 1993. Migration of oil and gas in coals. *In*: LAW, B. E. & RICE D. D. (eds) *Hydrocarbons from Coal*. American Association of Petroleum Geologists, Studies in Geology, **38**, 219–236.

LONGSTAFFE, F. J. 1989. Stable isotopes as tracers in clastic diagenesis. *In*: HUTCHEON, I. E. (ed) *Short Course in Burial Diagenesis*. Mineralogical Association of Canada, Montreal, 201–277.

——1993. Meteoric water and sandstone diagenesis in the Western Canada Sedimentary Basin. *In*: HORBURY, A. D. & ROBINSON, A. G. (eds) *Diagenesis and Basin Development*. American Association of Petroleum Geologists, Studies in Geology, **36**, 49–68.

LUNDEGARD, P. D. 1989. Temporal reconstruction of sandstone diagenetic histories. *In*: HUTCHEON,

I. E. (ed.) *Short Course in Burial Diagenesis.* Mineralogical Association of Canada, Montreal, 161–200.

MOORE, D. M. & REYNOLDS, R. C. JR 1989. *X-ray Diffraction and the Identification and Analysis of Clay Minerals.* Oxford University Press, New York.

MURRAY, C. G. 1990. Tectonic evolution and metallogenesis of the Bowen Basin. *In*: *Bowen Basin Symposium Proceedings.* Geological Society of Australia, Queensland Division, Brisbane, 201–212.

OLDROYD, G. C. & BROTHERTON, R. A. 1992. The challenge of developing Queensland's coalbed methane resource. *In*: BEAMISH, B. B. & GAMSON, P. D. (eds) *Symposium on Coalbed Methane Research and Development in Australia*, Vol. 5. James Cook University of North Queensland, Townsville, 55–57.

OLIVER, J. 1986. Fluids expelled tectonically from orogenic belts: their role in hydrocarbon migration and other geologic phenomena. *Geology*, **14**, 99–102.

RANKIN, A. H. 1989. Fluid inclusions. *Geology Today*, **21**, 21–24.

REYNOLDS, R. C. JR 1980. Interstratified clay minerals. *In*: BRINDLEY, G. W. & BROWN, G. (eds) *Crystal Structure of Clay Minerals and their X-ray Identification.* Mineralogical Society, London, Monograph, **5**, 249–303.

RICE, D. D. 1993. Composition and origins of coalbed gas. *In*: LAW, B. E. & RICE, D. D. (eds) *Hydrocarbons from Coal.* American Association of Petroleum Geologists, Studies in Geology, **38**, 159–184.

RIGHTMIRE, C. T. 1991. Coalbed methane resource. *In*: RIGHTMIRE, C. T., EDDY, G. E. & KIRR, J. N. (eds) *Coalbed Methane Resources of the United States.* American Association of Petroleum Geologists, Studies in Geology, **17**, 1–13.

ROEDDER, E. 1984. *Fluid Inclusions.* Reviews in Mineralogy, **12**.

SAVIN, S. M. & LEE, M. 1988. Isotopic studies of phyllosilicates. *In*: BAILEY, S. W. (ed.) *Hydrous Phyllosilicates (Exclusive of Micas).* Reviews in Mineralogy, **19**, 189–223.

SCHWARTZ, F. W. & LONGSTAFFE, F. J. 1988. Ground water and clastic diagenesis. *In*: ROSENSHEIN, J. S. & SEABER, P. R. (eds) *The Geology of North America*, Vol. O-2, *Hydrogeology.* Geological Society of America, Boulder, 413–434.

SECOR, D. T. JR 1965. Role of fluid pressure in jointing. *American Journal of Science*, **263**, 633–646.

SHEPHERD, T. J., RANKIN, A. H. & ALDERTON, D. H. M. 1985. *A Practical Guide to Fluid Inclusion Studies.* Blackie, London.

SHIBAOKA, M. & BENNETT, A. J. R. 1976. Effect of depth of burial and tectonic activity on coalification. *Nature*, **259**, 385–386.

——, BENNETT, A. J. R. & GOULD, K. W. 1973. Diagenesis of organic matter and occurrence of hydrocarbons in some Australian sedimentary basins. *The Australian Petroleum Exploration Association Journal*, **23**, 73–80.

SRODON, J. & EBERL, D. D. 1984. Illite. *In*: BAILEY, S. W. (ed.) *Micas.* Reviews in Mineralogy, **13**, 495–544.

STAINES, H. R. & KOPPE, W. H. 1979. The geology of the North Bowen Basin. *Queensland Government Mining Journal*, **80**, 172–191.

STONE, R. & SNOEBERGER, D. F. 1978. Cleat orientation and areal hydraulic anisotropy of a Wyoming coal aquifer. *Ground Water*, **15**, 434–438.

TISSOT, B. P. & WELTE, D. H. 1984. *Petroleum Formation and Occurrence*, 2nd edn. Springer-Verlag, New York.

TOTH, J. 1980. Cross-formational gravity-flow of groundwater: a mechanism of the transport and accummulation of petroleum (the generalized hydraulic theory of petroleum migration). *In*: ROBERTS, III W. H. & CORDELL, R. J. (eds). *Problems of Petroleum Migration.* American Association of Petroleum Geologists, Studies in Geology, **10**, 121–167.

TREMAIN, C. M., LAUBACH, S. E. & WHITEHEAD, N. H. III. 1991. Coal fracture (cleat) patterns in Upper Cretaceous Fruitland Formation, San Juan Basin, Colorado and New Mexico—implications for coalbed methane exploration and development. *In*: AYERS, W. B. JR, KAISER, W. R. & 12 others (eds) *Geologic and Hydrologic Controls on the Occurrence and Producibility of Coalbed Methane, Fruitland Formation, San Juan Basin.* Gas Research Institute Topical Report, **GRI-91/0072**, 97–117.

WEAVER, C. E. & BROCKSTRA, B. R. 1984. Illite–mica. *In*: WEAVER, C. E. & ASSOCIATES (eds) *Shale–Slate Metamorphism in the Southern Appalachians.* Elsevier, Amsterdam, 67–97.

YODER, H. S. & EUGSTER, H. P. 1955. Synthetic and natural muscovites. *Geochimica et Cosmochimica Acta*, **8**, 225–280.

# Coal microstructure and secondary mineralization: their effect on methane recovery

PAUL GAMSON[1], BASIL BEAMISH[2]
& DAVID JOHNSON[3]

[1] *AusSpec International Pty Ltd, P.O. Box 2235, Kew MDC, Melbourne, Victoria 3101, Australia*
[2] *Department of Geology, The University of Auckland, Private Bag 92019, Auckland, New Zealand*
[3] *James Cook University, Townsville, North Queensland, Australia*

**Abstract:** Methane production from coal seams rather than porous sandstone reservoirs is now recognized as a valuable and recoverable energy source in Australia. The Bowen Basin of Australia possesses well defined coal seams that contain major methane resources. However, commercial gas production to date has been hampered by the low permeabilities of the coal seams. Recovery of this valuable resource will be assisted by a fundamental understanding of coal microstructures and presence of mineralization, and their influence on the gas flow behaviour through coal.

This paper examines the relationships between coal type, microstructure, secondary mineralization and gas flow behaviour. The study demonstrates that Bowen Basin coals should not be viewed as simply a dual porosity system of micropores which are surrounded by cleats. Instead, studies using scanning electron microscopy show the Bowen Basin coals have a third porosity system comprising a hierarchy of micron-sized fractures and micron-sized cavities at a level between the micropores and the cleat/macropore system, which vary according to coal type.

To determine the influence such microstructures have on the flow of gas through the coal matrix, sorption experiments were carried out on small solid blocks of coal, using a new gravimetric technique. The results demonstrate that a clear distinction exists between diffusivity of dull and bright coals in response to coal microstructure. The gas sorption data suggests that both dull and bright coals can be divided into two categories: coal which have a rapid sorption behaviour and coals which have a slow sorption behaviour.

The results of the sorption experiments indicate that size, continuity, connectivity of the microstructures, and the extent of minerals infilling the fractures and cavities, play a significant contribution to overall permeability, and are likely to play a major rate-limiting factor in the flow of methane through coal at a level between diffusion at the micropore level and laminar flow at the cleat level. The studies indicate that the flow behaviour of gas through coal seams in the Bowen Basin is unlikely to be solely dependent on the cleat system but rather a combination of the cleat, microstructure and secondary mineralization in coal.

Methane primarily resides in the coal as an adsorbed layer on the internal coal surface or as free gas in large pore and fractures. As noted by Harpalani & Schraufnagel (1990a), however, a 'Knowledge of conventional gas reservoir modelling is of little value in the case of coalbed methane reservoirs, due to the unique mechanism of gas storage in coalbeds and the unusual flow behaviour of gas in coals'. Present models of methane flow through a coal seam indicate that the absorbed methane after desorption into the gaseous phase must diffuse through the micropore structure of the coal matrix until reaches a cleat (King 1985; Harpalani & Schraufnagel 1990a,b), followed by Darcy flow through the cleats to a well (Fig. 1).

The relationship of coal seam structure and gas flow behaviour is usually modelled as a dual porosity system of macropores (cleats) surrounding a matrix of micropores (Fig. 2).

*From* Gayer, R. & Harris, I. (eds) 1996, *Coalbed Methane and Coal Geology*, Geological Society Special Publication No 109, pp 165–179.

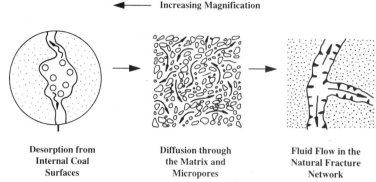

**Fig. 1.** Current model of methane flow through coal showing desorption, diffusion and Darcy flow.

According to this model, the diffusion is usually modelled using Fick's Law and the free flow is modelled using Darcy's Law. Darcy's Law describes the flow of water through porous media that contains pores of uniform cross-section and of uniform packing, so that pore size, shape, distribution and the connectivity of pores are grouped together under one parameter, permeability. As the cleats are regarded as having a uniform pore geometry (size, shape and spacing) that is representative of coal as a whole, Darcy's Law allows the description of the general behaviour of methane flow through macroscopic structures in coal such as the natural cleat network. The cleats are

modelled as defining the size of the matrix blocks, which contain a microporosity, where diffusional flow through the micropores to the cleats is thought to occur. According to this model of gas flow through coals, gas migration is governed by two main factors: (1) the distance methane has to diffuse, which is dependent upon the cleat spacing that delineates the size of the matrix blocks in the coal; (2) the amount of gas flowing through the cleat which is dependent upon the width, length, continuity and permeability of the cleats.

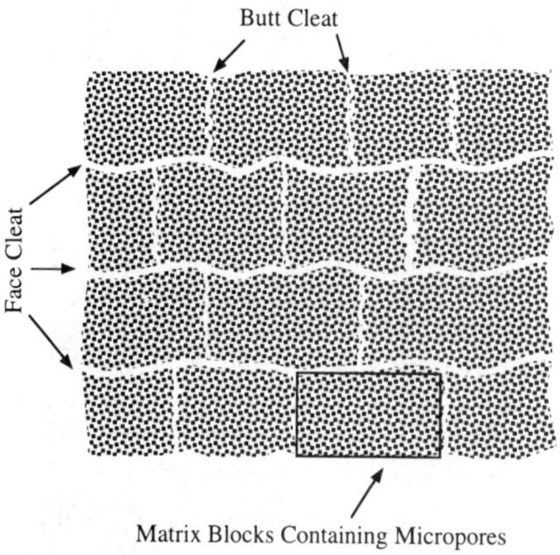

This dual porosity model of gas flow (i.e. gas diffusion through the coal matrix block and laminar flow through the cleats) may well apply to predominantly bright coal seams where well defined, open and unmineralized cleats define uniform microporous blocks. Examples of such coals include the Fruitland Seam Fairway coals of the San Juan Basin and the Blue Creek coals of the Black Warrior Basin. Studies using a scanning electron microscope (SEM) show that these coals comprise closely spaced (0.3–1 mm apart), wide (10–60 mm) cleats with narrow (5–20 mm wide) microcleats in between, all of which are open and non-partially mineralized (Fig 3). Typically the nature of these porosities is reflected in their different permeabilities: cleat permeability $\sim$1–50 md; Matrix permeability $\sim 10^{-5}$ to $10^{-9}$ md.

Given the (micro) structural nature of these coals and the spacing between the cleats and microcleats, it is likely that methane flow through these coals will take place through minute matrix blocks of a size $40\,\mu\text{m} \times 40\,\mu\text{m}$, suggesting that the distance methane has to diffuse is small.

**Fig. 2.** Dual porosity model of macropores (cleat) and micropores.

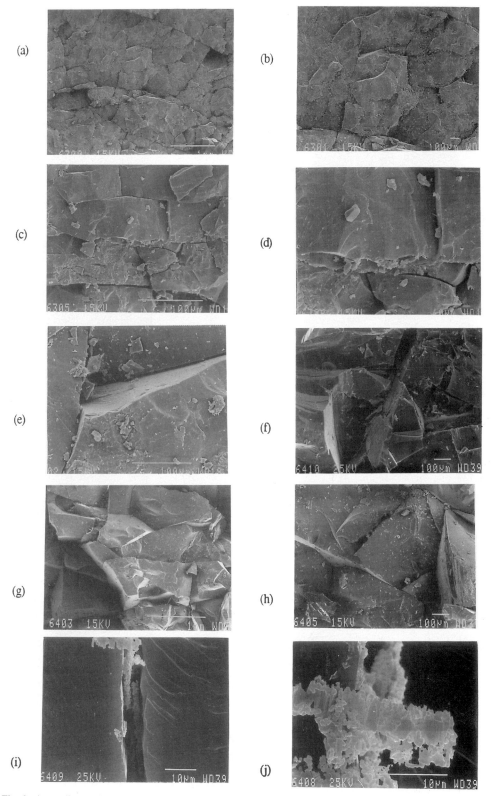

**Fig. 3.** Australian coals exhibiting closely spaced, wide cleats with attendent narrow microcleats.

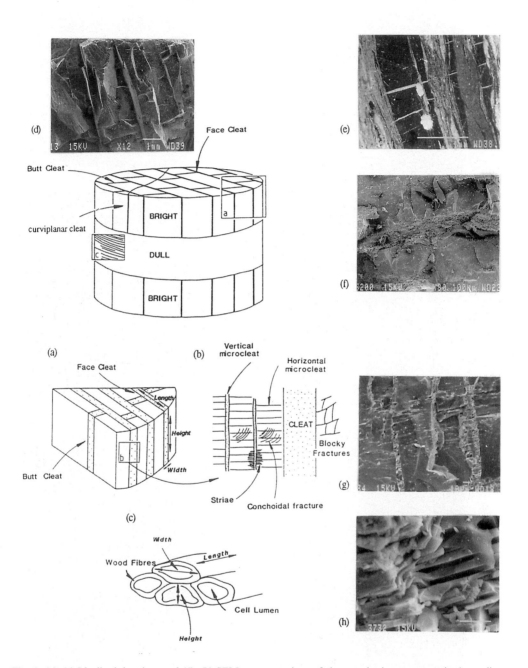

**Fig. 4.** (a)–(c) Idealized drawings and (d)–(h) SEM representations of cleats and microstructures in Australian coals.

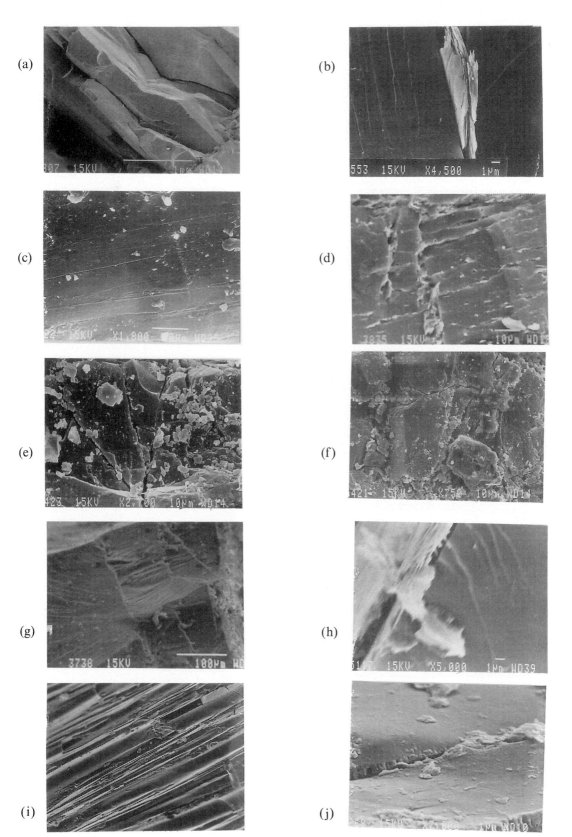

**Fig. 5.** SEM Photomicrographs exhibiting different styles of microstructures in bright coal.

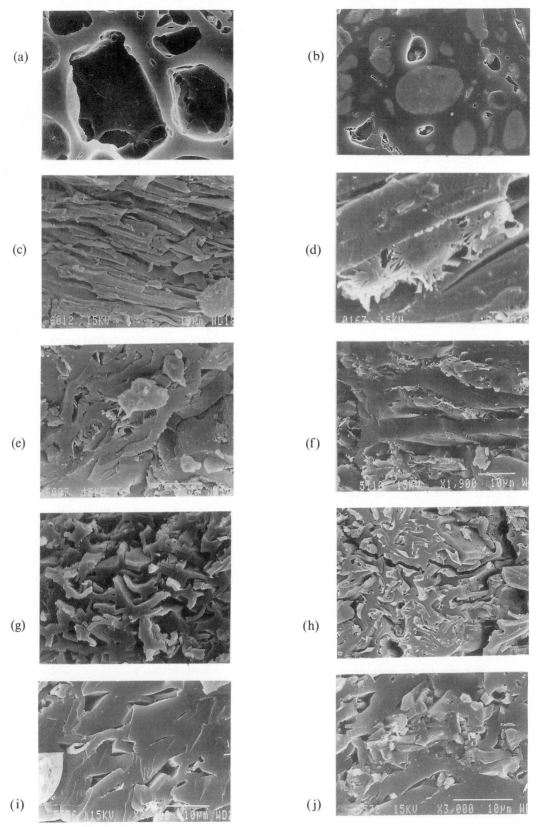

**Fig. 6.** SEM Photomicrographs exhibiting different styles of phytereal structures.

## Structure of Australian coals

The nature of gas flow through Australian coals is perhaps more complex than previously modelled. Examination of coal microstructure using SEM, examples of which are presented here and elsewhere (Gamson & Beamish 1991; Gamson et al. 1993), show that Australian coals of the Bowen Basin are not simply a dual porosity system comprising a matrix of micropores that are surrounded by cleats. Instead, these coals are typically compositionally banded, comprising both bright and dull coal types with non-pervasive cleat systems that are commonly restricted to the bright coal layers and terminate abruptly at the bright/dull coal boundary (Figs 4a, e–f). Moreover, a hierarchy of other micron-sized fractures (microfractures) exist in bright coal at a level between the micropores and the cleat system (Figs 4b, g–h; 5a–b). Typically five different types of microfractures can be recognized (Fig. 5): vertical microcleats (5–20 $\mu$m wide, 50–500 $\mu$m long, and spaced 30–100 $\mu$m apart), horizontal microcleats (0.5–2 $\mu$m wide, 50–300 $\mu$m long and spaced 5–10 $\mu$m apart), blocky fractures (1–15 $\mu$m wide, 50–200 $\mu$m long and spaced less than 100 $\mu$m apart), conchoidal fractures (no regularity in spacing) and striae. Striae are the smallest observed microfractures and comprise a number of closely packed parallel laminations or sheet like layers that are 0.1 $\mu$m wide, 10–100 $\mu$m long and are typically spaced 0.1–0.3 $\mu$m.

Dull coals by contrast, tend not to contain microfractures or cleats, and instead typically contain a phyteral porosity of variously sized and shaped microcavities which are associated with the original plant fragments, rather than microfractures. A common component of these phyteral structures are numerous sheet-like structures arranged in a series of stacked layers parallel to bedding which represent remnants of wood fibres. The sheets are typically 2–4 $\mu$m thick and are separated by long, cylindrical microcavities of cell lumen that are commonly 2–4 $\mu$m high and 10–30 $\mu$m wide (Figs 4g, h).

Although the original cavities were probably rectangular in cross-section forming open sieve structures, many of the structures observed in the SEM appear to have been broken and compressed. This has resulted in various morphological structures: needle (fragmented cell walls resulting in networks of pointed, needle-shaped splinters; Figs 6c–d), compressed (cell walls that have been crushed together almost to lines of compression, but are not broken; Figs 6e–f), bogen (cell walls that have been broken and pushed into one another; Figs 6g–h), bogen-compressed (Figs 6i–j) and highly compressed.

Commonly the various microfractures and microcavities are filled with minerals (Figs 5 & 6a–j). The minerals observed infilling such structures include clays (kaolinite, dickite, illite, illite-smectite, vermiculite, chlorite), carbonates (calcite, ankerite, siderite, dawsonite, strontianite) and quartz. SEM examination of the minerals show that they commonly tightly infill the cleats and microstructure space leaving little pore space for water and gas flow. Minerals in bright coals tend to infill cleats and microfractures and occur as either discrete phases or as mixed mineral phases. In contrast, in dull coals minerals occur in the form of either bands/lenses, as fine particles disseminated through the coal, as cavity infill, and/or clay particles interbedded between maceral fragments.

**Table 1.** *Coal samples measured for their sorption behaviour*

| Sample label | Coal group | Seam name | Rank (VR) | Ash content | Surface area (m²/g, db) | Total porosity | $T_{sorp}$ (minutes) |
|---|---|---|---|---|---|---|---|
| 1A-bright | Fort Cooper | GIRUP | 1.66 | 1.9 | 261 | 2.6 | 148–170 |
| 1B-dull | Fort Cooper | GIRUP | 1.66 | 24.5 | 187 | 12.4 | 5–10 |
| 2A-bright | Baralaba | "0" | 0.81 | 2.1 | 237 | 1.6 | 288–1288 |
| 2B-dull | Baralaba | "0" | 0.81 | 4.2 | 209 | 13.7 | 15–20 |
| 3A-bright | Baralaba | "7" | 0.88 | 3.7 | 277 | 0.9 | 415–1309 |
| 3B-dull | Baralaba | "7" | 0.88 | 13.8 | 175 | 9.0 | 189–210 |

## Coal microstructure and gas flow behaviour

The presence of micron-sized fractures and cavities in both dull and bright coals at scales below the matrix block and between the micropores and the cleat system suggests that the flow behaviour of gas through these coals is unlikely to be solely dependent on the cleat system, but rather a combination of the cleat, microstructure and extent of mineralization in coal. The size, continuity and connectivity of the microstructures suggests that they contribute significantly to the overall (micro)-permeability, and are likely to have a major role in the flow of methane through coal at a level between diffusion at the micropore level and laminar flow at the cleat level. At present, current models of gas flow do not account for these variations.

To understand the relationship between coal type, microstructure and gas flow behaviour, sorption experiments were carried out on selected samples that have been examined in the SEM for their microstructure. Using a new gravimetric technique developed by Beamish *et al.* (1991), sorption experiments were carried out on small (1 g), solid blocks of dried coal rather than crushed samples, using a microbalance. This technique has been developed to test small samples and allows a closer understanding of the influence coal microstructure has on the diffusivity of coal, which controls the gas flow rate through the coal matrix at a level including the cleats. This new approach to gas sorption studies contrasts with previous studies which have measured coal sorption using a bulk crushed sample to measure a coal's maximum gas storage capacity, but tells us little about the time it takes for methane to flow through a solid coal, i.e. the coal's diffusivity.

The individual pieces of dull and bright coal were desorbed from 1.1 mpa to atmospheric pressure, and the amount of gas released measured with time. To understand the effects of microstructure on sorption, parallel samples were analysed in the SEM.

In total thirty coal samples were used to study sorption behaviour and coal microstructure. Of those samples, six are reported on here as representing typical sorption behaviour

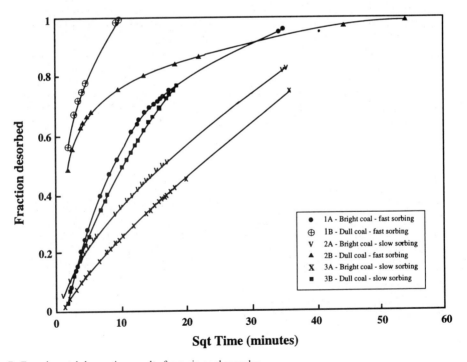

**Fig. 7.** Experimental desorption results from six coal samples.

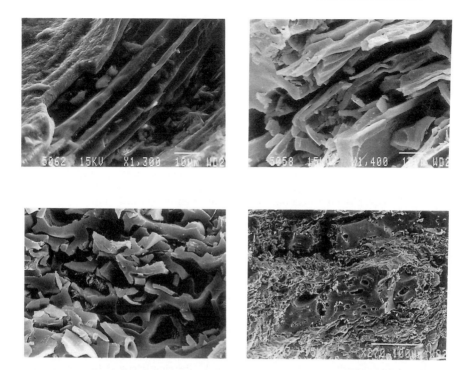

**Fig. 8.** SEM Photomicrographs of dull, fast sorbing coal.

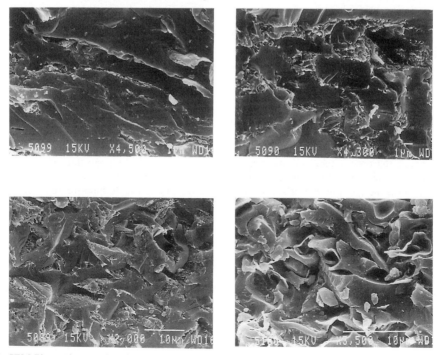

**Fig. 9.** SEM Photomicrographs of dull, slow sorbing coal.

of dull and bright coal (Table 1). The samples consist of three sets of dull and bright coals that were selected from the same seam and the same depth. To reduce the influence of the effect that sample size has on diffusivity, all the samples were cut into similar sized cubic blocks.

## Sorption behaviour of dull and bright coal

The desorption rates of the six samples are shown in Fig. 7, which plots the fraction of gas desorbed versus the square root of time. The desorption data for the six samples (Fig. 7) show that the rate at which methane desorbs from a solid coal differs between coal samples, and coal samples of different coal type. The results show that both the dull and bright coals show two distinct fields of behaviour: fast sorption and slow sorption.

Typically the faster desorbing dull coals have a one-stage sorption process that is characterized by a rapid release of the total gas (Fig. 7). In contrast, the slower desorbing dull coals are characterized by a two-stage sorption process: a first stage during which there is a rapid release of gas, followed by a second stage where sorption is much slower (Fig. 7). Similarly the bright coals are characterized by two types of sorption behaviour. The faster desorbing bright coal is characterized by a two-stage sorption process: a first stage during which there is a rapid release of gas, followed by a second stage where sorption is much slower (Fig. 7). In contrast, the much slower desorbing bright coals are characterized by a one-stage sorption process.

In terms of the time it takes or the dull and bright coals to desorb 63% of their total gas ($T_{sorp}$) the differences in the rate of sorption are marked and vary from a low of 5 minutes to a high of ~22 hours. Typically $T_{sorp}$ for the various coal classes are:

- Dull coal-fast sorbing:        5–50 minutes
- Dull coal-slow sorbing:      150–250 minutes
- Bright coal-fast sorbing:    150–250 minutes
- Bright coal-slow sorbing:   300–1300 minutes

These differences in $T_{sorp}$ suggest that proportions of these coal classes in any seam profile will govern the rate at which methane will be released. The behaviour of these coal classes can be explained in terms of microstructure and extent of mineralization.

The fast sorbing dull coals contain a high proportion of thick inertinite bands (80–90%) which are predominantly composed of open, unmineralized wood fibres that are either: (1) unbroken, (2) broken and pushed into one another, and/or (3) broken and compressed (Fig 8). In contrast, the slow sorbing dull coals contain a lower proportion of inertinite bands (as little as 50%) and a higher proportion of vitrinite bands. The inertinite bands are predominantly composed of wood fibres with closed, mineralized cell lumens that are either: (1) tightly compressed, and/or (2) broken and tightly pushed into one another. In addition, the bands of vitrinite contain micro-fractures which are tightly infilled with minerals (Fig. 9).

The fast sorbing bright coals are characterized by a well defined cleat network. The cleats are moderately infilled with minerals although pore space between the mineral grains is apparent. Microcleats, blocky fractures and smaller micro-fractures are common in the faster sorbing coals and are typically open and unmineralized (Fig. 10). In contrast, the slow sorbing coals bright coals are characterized by tightly infilled face and butt cleats, tightly infilled microcleats and other microfractures, and/or no microfractures (Fig. 11).

The differences in sorption behaviour shown by the two coal classes can best be explained in terms of the macropore and micropore components of the coal. The simplest form of comparison of the desorption data is to apply a spherical unipore model. Of the six coals however, only three of the coals, 1B-dull, 2A- and 3A-bright fit the unipore spherical model. To explain this, the release of methane in such coals may be considered a one-stage sorption process. In the dull coal, 1B, only macropore sorption occurs due to the domination of macropores. In contrast, the one-stage sorption process in the bright coals, 2A and 3A, fit a unipore spherical model due to the domination of micropore diffusion.

The three other coals, 1A-bright, 2B- and 3B-dull, however, do not fit a unipore spherical model due to a distinct curvature in the desorption curve (Fig. 7). To explain this phenomenon, the sorption behaviour is divided into a micro-sphere (micropore) component which is surrounded by a macrosphere (macropore) component (Ruckenstein *et al.* 1971). The release of methane in these coals may be considered a two stage sorption process, where sorption in the

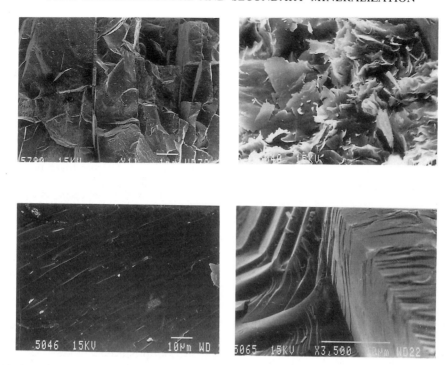

**Fig. 10.** SEM Photomicrographs of bright, fast sorbing coal.

**Fig. 11.** SEM Photomicrographs of bright, slow sorbing coal.

macropores is much faster than in the micropores so that equilibrium is essentially achieved in the macropores before any appreciable release by the micropores is observed. Consequently a first stage is observed during which only macropore sorption occurs, followed by a much slower second stage during which macropore sorption is at equilibrium, and only micropore sorption occurs.

## Gas flow through coals

A combination of scanning electron microscopy and sorption testing using a microbalance has shown that the time taken for methane to travel through the coal matrix varies according to microstructure and is reflected in significant variations in $T_{\text{sorp}}$.

Scanning electron microcopy has shown, that at a scale between the micropores and the cleat system, Australian coals contain a range of microfractures in the bright coals and a range of microcavities in the dull coal, of various pore shapes and sizes which vary according to the degree of mineralization. Consequently, it is likely that methane flow through the matrix will take place not through pores of similar geometry as previously suggested by a dual porosity model, but rather through a complicated network of interconnected microfractures and microcavities, of varying size, shape and cross-section, and minerals filling pore space.

According to current models of gas flow through coals (Fig 1), however, the flow of methane is dependent upon the effective permeability of the coal, i.e. the cleat network. Assuming the diffusion of methane begins and presumably finishes at the micropore level where the majority of gas is stored as an adsorbed layer (and where micropores 4–20 angstroms in diameter are joined by minute passages), as previous models suggest, then methane flow from the micropore system to the cleats in Australian coals must rely upon the effectiveness of the microstructure system to transport the methane.

The various microfractures and microcavities, and the different sorption behaviours exhibited by the dull and bright coals, suggest that microstructures play a rate-limiting role between diffusion at the micropore level and flow at the macropore or cleat level. In terms of modelling, this suggests that additional steps may be involved in the flow of methane through Australian coals (Fig. 12). Four steps are proposed (Fig. 12):

- Step 1. Diffusion from and through the micropores to microfractures in the bright coal and microcavities in the dull coal.
- Step 2. Diffusion and/or flow of methane through microfractures and microcavities partly blocked by diagenetic minerals. Methane flow would be dependent on the size and connectivity of the pore space between the mineral infilling.
- Step 3. Flow through open, unmineralized microfractures in the bright coal and microcavities in the dull coal.
- Step 4. Gas movement through cleats and joints to the well base. Where cleats are generally infilled by minerals, and mostly this infill forms a tight seal, gas movement will be either completely blocked or be by diffusion.

These additional steps are not presented so as to complicate the issue by incorporating more tiers into the gas production process. Instead the extra steps show that microstructures are likely to play an important and probably significant rate-limiting role at scales before gas transport at the cleat or fracture level.

Importantly, the effectiveness of methane flow through the microstructures, as opposed to the cleats, would be ultimately influenced by several microscopic considerations, which include the shape and size of microstructures, microstructure distribution (density, orientation and continuity), connectivity of the microstructures and the cleat system, the amount of fracture infilling with secondary minerals, clay dispersed through the organic matrix, and the change in stress conditions after stimulation. Each of these microscopic considerations will have a different effect on the quantity, rate and direction of gas flow through the coal. For Australian coals the term micropermeability is introduced and refers to the conductivity of the microstructures.

## Implications of microstructures on gas flow models

In terms of gas flow modelling, such variations in coal seams have important implications for present models of gas flow through coal. Currently, models subdivide matrix blocks where diffusion dominates from that involved

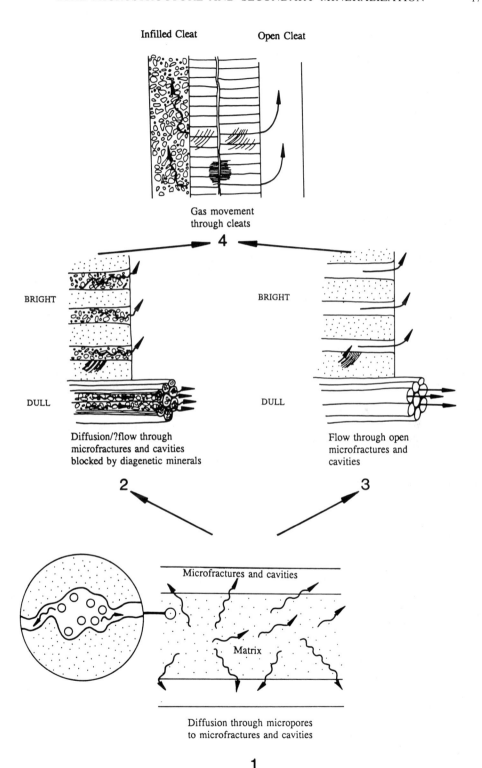

**Fig. 12.** Enhanced desorption model for Australian coals proposing four distinct steps of methane migration.

with laminar flow, on the cleat spacing. Because the effects of microstructure on diffusivity are considered to be real, the presence of open and continuous microstructures in coal suggests that laminar flow is likely to begin at levels smaller than that identified at present by the spacing of cleats. To accommodate the effects of microstructure into models of gas flow, it is necessary to redefine the effective block size within which diffusion dominates.

In bright coals, where there is a hierarchy of open, continuous and connecting microfractures between the cleats, the effective block size is probably not that being defined at present by the cleat spacing, but somewhere between the cleats and the microfractures. In contrast, where cleats and microfractures are infilled with minerals and the pore space for laminar flow is reduced, the effective block size may be larger than previously assumed, and may be more effectively defined by other more widely spaced, open fractures (joints) in the bright coals. Ultimately this could reflect fracture sets beyond the core.

In dull coals however, where there are usually no cleats, it is more difficult to define the effective block size for gas flow. The microstructural and sorption results indicate that where dull coals contain open, unmineralized structures, and macroporous flow dominates, the effective block size in these coals correlates to the size of the phyterals, which could be in the order of $10\,\mu$m. However, where microstructures in the dull coals are heavily mineralized it is probable that the block size relates to a larger composite dimension.

These differences in diffusivity of a coal have important implications for gas drainage. Although a general relationship between coal rank and gas content is well documented (Kim 1977), and that bright coals have a greater storage capacity than dull coals of equivalent rank (Beamish & Gamson 1993), it does not follow, however, that bright coals offer a greater potential for methane flow (i.e. a higher diffusivity) than dull coals of equivalent rank. Instead, it has been shown that dull coals generally have a higher diffusivity than bright coals due to the greater macropore porosity, and associated small block sizes for diffusion, than bright coals of equivalent rank. This is important, for it suggests that areas of low rank coal with low gas contents, due to their low storage capacity, may offer a better gas flow rate than higher rank coals with higher gas contents due to differences in diffusivity between the coals.

Equally significant, it also suggests that coals with high permeability may not offer higher gas flow rates than coals with lower permeability due to low diffusivity.

## Conclusions

Scanning electron microscopy has shown that Australian coals should not be viewed as simply a dual porosity system of micropores which are surrounded by cleats, and instead, viewed as having a third porosity system comprising a hierarchy of micron-sized fractures and micron-sized cavities at a level between the micropores and the cleat/macropore system, which vary according to coal type.

Examination of the sorption behaviour of these coals suggests that bright and dull coals exhibit distinct fields of behaviour: fast sorption and slow sorption. The behaviour of these classes indicate that the size, continuity, connectivity of the microstructures, and the extent of minerals infilling the fractures and cavities play a significant contribution to overall permeability, and are likely to play a major rate limiting factor in the flow of methane through coal at a level between diffusion at the micropore level and laminar flow at the cleat level. A new model is presented to account for these variations in coal type, microstructure and mineralization.

## References

BEAMISH, B. B. & GAMSON, P. D. 1993. Sorption behaviour and microstructure of Bowen Basin coals. *Coalseam Gas Research Institute, James Cook University, Technical Report CGRI TR 92/4 February, 1993.*

——, —— & JOHNSON, D. P. 1991. Investigations of parameters influencing gas storage and release in Bowen Basin coals. *Coalseam Gas Research Institute, James Cook University, Technical Report CGRI TR 91/4 November, 1991.*

GAMSON, P. D. & BEAMISH, B. B. 1991. Characterisation of coal microstructure using scanning electron microscopy. *Proc. of the AusIMM Queensland Coal Symposium, Brisbane, 29–30th August, 1991,* 9–21.

——, —— & JOHNSON, D. P. 1993. Coal microstructure and micropermeability and their effects on natural gas recovery. *Fuel,* **72,** 87–89.

HARPALANI, S. & SCHRAUFNAGEL, R. A. 1990*a.* Shrinkage of coal matrix with release of gas and its impact on permeability of coal. *Fuel,* **69,** 551–556.

—— & —— 1990*b*. Measurement of parameters impacting methane recovery from coal seams. *International Journal of Mining and Geological Engineering*, **8**, 369–384.

KIM, A. G. 1977. Estimating methane content of bituminous coalbeds from adsorption data. *USBM RI 8245.*

KING, G. R. 1985. *Numerical simulation of the simultaneous flow of methane and water through dual porosity coal seams.* PhD Thesis, Pennsylvania State University, P.A., USA.

RUCKENSTEIN, E., VAIDYANATHAN, A. S. & YOUNGQUIST, G. R. 1971. Sorption by solids with bidisperse pore structures. *Chemical Engineering Science*, **26**, 1305–1318.

# Enhanced methane desorption characteristics from South Wales anthracites affected by tectonically induced fracture sets

IAN H. HARRIS[1], GARETH A. DAVIES[1], RODNEY A. GAYER[1]
& KEITH WILLIAMS[2]

[1] *Department of Earth Sciences, University of Wales Cardiff, PO Box 914,
Cardiff CF1 3YE, UK*
[2] *School of Engineering, University of Wales,
Cardiff, PO Box 914, Cardiff CF1 3YE, UK*

**Abstract:** It has generally been assumed that anthracite, despite containing potentially high levels of sorbed methane, is unsuitable for coalbed methane (CBM) production because of low levels of fracture permeability. South Wales anthracites are commonly strongly affected by tectonically developed fracture systems formed during Late Carboniferous Variscan compressional deformation. The fracture systems occur as thrust-related slip, slickenside and feather fractures, and are superimposed on previously formed extensional fracture systems that include cleat, conchoidal and bed-parallel fractures. When these fracture systems are intensely developed, the anthracite becomes incompetent and friable as a result of extremely close fracture spacing. According to current models for gas desorption and migration, methane initially diffuses through the micropore system of the coal matrix before flowing freely through larger pores and fracture systems. The models suggest that the more closely spaced the fractures, the more rapidly will the coal desorb methane.

The tectonically fractured South Wales anthracites were investigated to determine the relationship between the rate of desorption of methane and tectonically induced fracture spacing. The former was established using gravimetric desorption isotherm experiments and *in situ* quantification of the latter was achieved by using a hand-drill penetrometer. The results demonstrate a clear relationship in anthracite between both the rate of methane desorbed and the amount of structural deformation. It is concluded that regions of deformed anthracite, previously considered unsuitable for CBM production, should be re-investigated.

With the demise of the traditional coal mining industry in South Wales, the continued interest in coal as an energy resource has been maintained by the possibility of extracting coalbed methane (CBM). As with many coal regions throughout the UK, considerable interest has been shown in prospective CBM development sites. The South Wales coalfield, with its high values of contained gas (Creedy 1985), has followed this trend. The notable exception in exploration is within the anthracite region, where scientific consensus suggests that anthracite coals do not possess sufficient permeability to allow the rapid flow of methane in commercial amounts. A developing view among South Wales structural geologists suggests that the effect of orogenic deformation on the fracture frequency (and subsequently permeability) within the anthracites is great enough to allow potentially significant amounts of gas to be emitted (e.g. Hathaway & Gayer this volume). This view is supported by methane desorption studies on outburst style coals conducted by Griffin (1978), Barker-Read (1980; 1984) and Creedy (1985), who noted that outburst style coal gave much faster desorption rates than coal from less tectonically affected areas.

This paper describes a series of tests of South Wales anthracites to determine whether the presence of tectonically overprinted fracture sets can affect the rate of desorption of methane from coal.

## South Wales coalfield

The South Wales coalfield is situated within an asymmetrical syncline approximately 96 km E–W and 30 km N–S (Fig. 1). Most deformation recognized within the coalfield dates from the late Carboniferous Variscan orogeny. The level of deformation varies throughout the coalfield. Compressive structures in the eastern half are only locally as well developed as those occurring throughout the western half and especially the northwest of the coalfield. The differences between deformation in the east and west are characterized by the extent of shortening, which varies from 10–40% in the east (Jones 1991) to >60% in the northwest (Frodsham 1990). This stronger deformation has been attributed to the buttressing effect of the Welsh Massif on

*From* Gayer, R. & Harris, I. (eds) 1996, *Coalbed Methane and Coal Geology,*
Geological Society Special Publication No 109, pp 181–196.

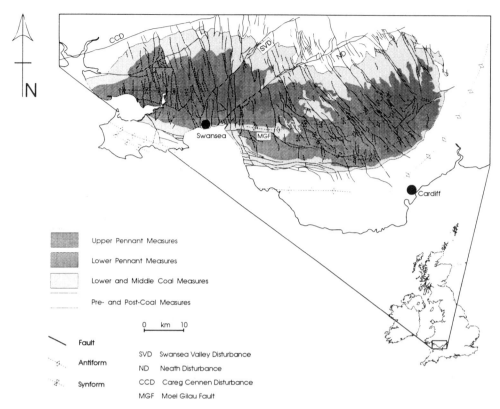

**Fig. 1.** South Wales Coalfield.

Variscan thrust propagation in the northwest of the coalfield (Jones 1991).

Interestingly, the areas of highest deformation comprise the highest coal rank within the coalfield. Although coal rank increases with stratigraphic depth, it also varies across the coalfield independently of the stratigraphic level of the coal seam. Volatile matter content and vitrinite reflectance are two standard methods for determining coal rank (see Stach 1982 for discussion) and comparison between contoured isovol maps and vitrinite reflectance values for coals across the basin show concordant relationships (White 1992). Figure 2 shows vitrinite reflectance values for coals of Westphalian A and B showing the differences in rank from west to east across the coalfield.

There are no conclusive arguments why coal rank is greatest in the northwest of the coalfield. One of the more recent explanations is that the higher rank northwestern coals have been developed by advective heating associated with the migration of fluids from depth through the tectonically permeable coal measure strata (Gayer *et al.* 1991). Recent mineralization studies

within these coals have shown the presence of hot fluids within the coal producing distinct stages of mineral paragenesis (Gayer 1993).

## Fracture styles leading to friable coal

The permeability of coal is strongly affected by the type, continuity and scale of fractures within the coal. Several fracture styles are recognized within coals of the South Wales coalfield. These include the universally present bed-normal cleat fabric, i.e. face and butt cleat, conchoidal fractures recognized within most bright macerals and associated fractures parallel to bedding laminations. However, predominantly in the northwest region, several additional unusual fractures are present. Firstly, slip fractures which are developed at 50–60° to bedding and often show dip slip slicken-lineated polished surfaces; secondly, feather fractures which generally develop in variable orientation upon cones or prisms (Harris 1995a); and thirdly, slickenside fractures. Unlike the cleat, which are generally thought to be associated with

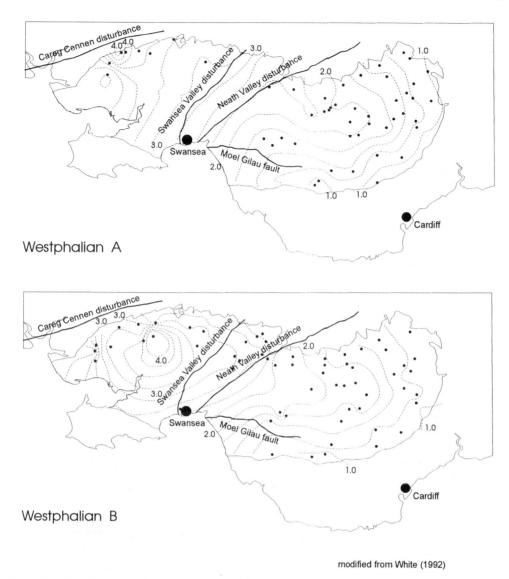

**Fig. 2.** Vitrinite reflectance values for the Westphalian A and B in the South Wales Coalfield.

extensional structures (Close 1993), these over-printed fractures are generally associated with compressive deformation. As well as the macro-scale fractures noted earlier, micro-scale frac-tures are also recognized which generally mirror the orientation of the larger scale macro-fractures. A more comprehensive review and description of the main coal fractures recognized within the South Wales coalfield is given by Harris (1995a).

The most enigmatic tectonically induced fractures are the feather or horsetail fractures. Throughout the tectonically affected coal seams

in the northwest of the coalfield, these are seen to overprint all earlier fracture sets.

Morphologically, feather fractures are conical or prismatic surfaces striated radially from an apex, with the base of the cone ranging in diameter from millimetres to centimeters. The striated surfaces may imply movement, but the conical shape suggests that only minimal slip can have occurred. Individual striations can be seen to bifurcate down the long axis of the cone (Fig. 3) and cones have not been seen to overlap but, rather, to abut each other. A variable orientation of the long axis of the cones has been recorded in some instances associated with later in-seam movement. We do not fully understand the development of feather fractures. However, their remarkable morphological similarity to shatter cones (Roddy & Davis 1977) may suggest an origin associated with the passage of shock waves through the anthracite, perhaps triggered by seismic thrust movements. Feather fractures are rarely present in bituminous grade coals and they may be dependent on the rheological properties of anthracite for their development. They are unlike the other tectonically induced fractures recognized within the northwest of the coalfield, which are recorded throughout the coalfield.

Slickenside fractures are recognized in all coals that have undergone in-seam movement. They are characterized by highly polished surfaces on either side of a fracture discontinuity (Fig. 4). Slickenlines are also sometimes observed generally with oblique-slip attitudes, but dip-slip lineations are not uncommon. Slickenside fractures commonly re-use pre-existing discontinuities, e.g. feather fractures, and movement leads to either a polishing or a complete destruction of the earlier fracture surface.

The movement associated with the development of slickenside fractures results in the comminution of the coal by a process termed nodular encapsulation (Barker-Read 1980). This process degrades the coal by abrasive movements associated with structural deformation.

The development of friable coal is mainly controlled by the extent and distribution of any slickenside fractures. In-seam movements also lead to the opening and comminution of microfractures throughout the coal.

## Methane potential of anthracite

A general rule exists which correlates increased gas adsorption capacity with coal rank. This is due to the increased molecular internal surface area created during the coalification process and is explained by an increase in the amount of micropores within the coal that act as adsorption sites. Table 1 (from Decker *et al.* 1986), shows the percentage of micro-, meso- and macro-pores in different rank coals.

The principal coal seam producers of commercial CBM are of bituminous rank. Coals of

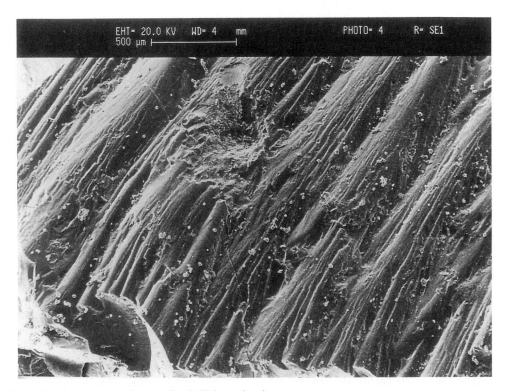

**Fig. 3.** Feather fracture surfaces on South Wales anthracite.

**Fig. 4.** Slickenside fractures on seat earth below the Big Vein anthracite.

**Table 1.** *Distribution of pores within different rank coals*

| Coal rank | Micropores (1.2 nm) | Mesopores (1.2–30 nm) | Macropores (>30 nm) |
|---|---|---|---|
| High volatile bituminous B | 29.9 | 45.1 | 25 |
| High volatile bituminous A | 48 | 0 | 52 |
| Medium volatile bituminous | 61.9 | 0 | 38.1 |
| Low volatile bituminous | 73 | 0 | 27 |
| Anthracite | 75 | 13.1 | 11.9 |

anthracite rank are generally regarded to have insufficient permeability (i.e. to be too tight) to produce CBM in commercial amounts, regardless of the fact that their methane adsorptive capacity is much greater than any other coal rank. The general consensus is that the number of macropores in anthracite is too low at around 11.9% (Table 1) to transmit large volumes of gas by laminar flow. Also, anthracite is perceived to have a wide cleat spacing, thus increasing the mean diffusion distance from the pores to a permeability conduit.

Within the South Wales coalfield the presence of structurally induced fracture permeability characterized by feather and slickenside macro-fractures and related micro-fractures may negate

the lack of non-tectonized permeability. The following section explains how methane is held within the coal seam reservoir and describes the processes by which any methane held is transported through the coal mass into permeable pathways suitable for laminar flow conditions to be initiated.

## Nature of methane sorption, porosity and permeability in coal

The high methane reservoir potential of all coals is due to the inefficient packing of its organic structure during coalification. The density of pure carbon is approximately $2.23 \times 10^3 \, \mathrm{kg\,m^{-3}}$, whereas anthracite has a density around $1.35 \times 10^3 \, \mathrm{kg\,m^{-3}}$. This density contrast is almost totally due to the micro-porosity created by the poor packing structure of the organic coal compounds. The bulk of the pores constituting this porosity are $<0.50 \, \mathrm{nm}$ in diameter (Gan *et al.* 1972). Their small size precludes occupation of these sites by larger molecules, e.g. water, so that only small diameter molecules such as methane (and other alkanes) can occupy this space; methane molecules have an effective diameter of $0.32 \, \mathrm{nm}$ in an uncompressed state.

Methane is stored in these coal reservoirs by two methods. Firstly, gas is held in a free state in the interstitial space between the microfractures. Many authors including Gray (1987) and Barker-Read (1984) have calculated that between 2 and 9% of the total gas within the coal is stored in this manner. The greater the porosity the more methane can be held within the interstitial permeability pathways in a free state. Secondly, the methane is physically adsorbed onto the surface of the pores of the coal. Owing to the large number of these pores within coal the available surface area for methane adsorption is huge and a very large volume of gas can be stored in this manner. This mechanism accounts for the remaining 91–98% of all methane stored in coal.

The two-fold mechanism of methane storage (both free and adsorbed) in coal leads to a two-stage emission behaviour. This was first recorded by Airey (1968) in experiments designed to describe the emission processes from broken coal in a destressed condition. He stated that gas emission was controlled by two processes: (1) gas diffusion following Fick's law; and (2) gas flow following D'Arcy's law.

Free methane is held within the crack structure of coal; as soon as conditions suitable for desorption occur the gas will move in a laminar flow. However, free methane stored in this fracture permeability accounts for less than 10% of the total gas contained in the coal (Barker-Read 1984). The remaining 90% plus of methane must first diffuse through the pore structure of the coal before reaching the fracture permeability, where it can move under laminar flow conditions. In coal with a wide fracture spacing the amount of time required for the diffusing gas to reach the fracture permeability will be much greater than that required in a coal with a closely spaced fracture pattern. The rate of emission of methane from coal is therefore largely controlled by the mean diffusive distance the methane must travel before reaching a suitable fracture for laminar flow.

The mean diffusive distance is defined as half the distance between the fracture permeability, i.e. half the crack spacing.

## Models describing methane desorption

Airey (1968) proposed a two stage model for gas desorption from coals. In stage 1 (solid lump coal with little fracture permeability) methane desorbed from the molecular surface is transported through the coal mass by diffusion following Fick's law, i.e. gas migration follows concentration gradients diffusing through the pore structure

$$Q_t = Q_0 \left( 1 - \frac{6}{\pi^2} \exp -\frac{2\pi^2 D t}{d^2} \right) \qquad (1)$$

where $Q_t =$ quantity of gas lost in time $t$ from a solid of initial gas content $Q_0$; $D =$ diffusion constant; and $d =$ mean diffusion distance.

Stage 2 in coal with a fracture permeability (i.e. a systematic macro- and micro-cleat and/or fracture system), the methane can also be transported by laminar flow, i.e. when the diffused gas reaches an open fracture, movement becomes uninhibited (i.e. free flow) and the transport mechanism follows D'Arcy's law

$$q = \frac{kA}{\mu} \frac{\partial p}{\partial x} \qquad (2)$$

where $q =$ volume flux $(\mathrm{cm^3\,s^{-1}})$; $k =$ permeability constant; $A =$ cross-sectional area $(\mathrm{cm^2})$; $\mu =$ fluid viscosity; $p =$ pressure; and $x =$ length of flow over which pressure drop is measured.

Griffin (1978), Creedy (1985) and Barker-Read (1984) have shown the validity of this two-fold process.

Previous microscopic investigations (Creedy 1985) have shown that on average dull coals have a mean crack spacing of $\approx 3 \, \mathrm{mm}$. Bright coals are more brittle and record a mean crack spacing of $\approx 1 \, \mathrm{mm}$. Thus coal seams with a higher proportion of bright macerals should have faster desorption and emission rates than more dull coals. This statement concerns the larger scale fractures noted within coal, i.e. the cleat fabric and any bedding-parallel fractures. More recent research has shown the greater extent of micro-fractures both parallel and oblique to the larger scale fractures, which would also act as permeability conduits for gas emission.

In an attempt to model gas movement through solid coal Gamson *et al.* (1993) produced a four-stage desorption model. They argued that most coals exhibit a more complex behaviour of gas migration than is modelled by Airey's two-stage system (which was designed to estimate gas movement through destressed, particulate coal). Gamson *et al.* (1993) expanded Airey's two-stage model of diffusion, followed by laminar flow into a four stage model (see Gamson this volume and Gamson *et al.* 1993).

In areas of intense compressive deformation, the condition of the coal seam is severely affected by the superposition of tectonically induced fractures (Harris 1995*b*). In severe examples microscopic investigation of the coal shows a completely granular surface; the frequency of fractures is so great that they cannot be individually identified. The emission rate of methane from coal is principally controlled by the distance the gas has to diffuse before reaching a permeability conduit large enough to allow laminar flow conditions to occur; the presence of tectonically overprinted fracture sets reduces the block size of the coal, thus improving the laminar flow conditions. The high frequency of superimposed fractures in structurally deformed coal seams should result in much lower mean diffusive distances and much more rapid rate of desorption in these tectonically affected coals.

## Relationship of rate of desorption to tectonically disturbed anthracites of the South Wales coalfield

To test the model for the tectonically disturbed, friable, anthracites it was first necessary to design a technique for quantifying the level of disturbance (and hence the micro-fracture spacing).

This is now described using an *in situ* strength indicator. Different levels of disturbed anthracite, using the results of the *in situ* strength tests, were then related to the rate of methane desorption in a gravimetric sorption experiment.

## Fracture spacing in friable (less competent) coal

Research (Harris 1995*b*) has shown that decreases in the *in situ* strength of a coal seam measured using the hand drill penetrometer (HDP) are generally associated with the presence of tectonically overprinted fracture sets and/or the presence of layer-parallel movement within the seam. Variations in the HDP rate indicate the structural condition of the coal area tested (if tested on the same maceral). Faster penetration rates are attributable to an increase in the fracture frequency of any discontinuities; the presence of these fractures makes drilling easier. Similarly, slow penetration rates are commensurate with less fractured, more competent coal. Although the HDP indicates coal with greater fracture porosity, the effect of this increased fracture porosity on rate of gas desorption is dependent on the *in situ* stress conditions at depth. However, given acceptable *in situ* stress conditions at depth the presence of increased fracturing would lead to faster methane emission.

## Sampling procedure

The tests were carried out on a section of anthracite dipping approximately 80°N within an active opencast operation. Along a 20 m transect nine test sites were selected by random number tables. The tests were carried out at right angles to bedding to intersect a standard seam section, traversing several maceral group boundaries.

The observed varied distribution of friability within a coal seam required a system of HDP testing that would both give a reasonable estimate of the *in situ* strength variation and that would not disturb the sample of coal collected for methane desorption analysis. An

**Table 2.** *Mean and standard deviation of the HDP rates*

| Test site | Penetration rates (cm s$^{-1}$) | Standard deviation | Mean penetration rate (cm s$^{-1}$) |
|-----------|--------------------------------|--------------------|-------------------------------------|
| 1 | 1.06, 0.94, 1.04, 0.82 | 0.11 | 0.97 |
| 2 | 0.16, 0.18, 0.41, 0.12 | 0.13 | 0.22 |
| 3 | 1.7, 1.38, 1.5, 1.46 | 0.14 | 1.51 |
| 4 | 0.58, 0.24, 0.51, 0.30 | 0.16 | 0.41 |
| 5 | 0.34, 0.41, 0.61, 1.0 | 0.30 | 0.59 |
| 6 | 0.16, 0.21, 0.15, 0.18 | 0.03 | 0.17 |
| 7 | 0.36, 0.34, 0.27, 0.29 | 0.04 | 0.32 |
| 8 | 1.66, 0.73, 1.33, 0.29 | 0.61 | 1.00 |
| 9 | 0.65, 0.84, 0.70, 0.47 | 0.15 | 0.67 |

arbitrary 10 by 8 cm sampling grid was chosen for this purpose centred on the gas desorption sample site. Four penetration tests were conducted, one at each corner of the rectangle. The average depth of penetration was recorded and coal was collected from the rectangle to this depth for later desorption testing. This procedure allowed the close correlation between the coal tested for penetration rate and that analysed for methane desorption rate. Table 2 shows results of the *in situ* HDP tests. The standard deviations indicate that sites 5 and 8 have above average variances between the four tests. This deviation is generally due to small-scale changes in the friability pattern of the seam and has been shown not to be associated with testing procedure (Harris 1995*b*). Such small-scale friability variations are commonly encountered when *in situ* testing tectonically affected coal and require the use of at least four penetration readings for each site.

## Desorption experimental procedure

The coal samples collected during the HDP analysis were crushed using a tungsten carbide Tema mill and sieved to 200 μm. The crushing was necessary to achieve rapid adsorption of the methane onto the surface of the coal. Crushing to this size eradicates all permeability pathways spaced at intervals greater than 200 μm. This crushing process invariably creates new fractures, generally oblique to the principal fracture (cleat) (Gamson pers. comm.). However, it is difficult to differentiate these fractures from existing fractures in coal, which does not possess fracture mineralization and therefore their effect is difficult to quantify. To maintain consistency between the samples they were crushed for an equal period of time. Given their similar maceral composition (Table 4), this should minimize variations in the rate of desorption introduced by the crushing process.

A set volume of coal (calculated by multiplying the bulk density of the sample by the 20 cm$^{-3}$ volume chosen for testing) was weighed into a canister of known internal volume and attached to the gravimetric sorption apparatus. After ensuring a complete seal, the samples were gradually evacuated to a high vacuum over a period of 18 h to ensure the removal of any residual gas and/or moisture. After this time the canisters were removed from the apparatus and weighed in this evacuated state.

After weighing, the canisters were reconnected to the gravimetric apparatus and laboratory-grade methane was admitted to the canisters and

allowed to equilibrate over 2 h to the pressures chosen for the desorption experiment, in this case 1 and 15 bar methane (absolute pressure). At the end of each pressure equilibration the canisters were removed from the apparatus and weighed. Any increase in the weight of the canisters must be attributable to the presence of additional methane adsorbed or held in the interstitial permeability within the coal surface in the canister.

Using a spreadsheet and following the procedure outlined by Griffin (1978) the total desorbable methane (TDM) for each of the samples (canisters) can be calculated.

The rate of desorption experiment is initiated by slowly opening a valve on the sorption canister and measuring the reduction in canister weight over a set period of time. For this analysis the weight change was recorded at 1 min intervals for the first 10 min of the rate of desorption experiment. From 10 to 30 min the weight reduction was measured every 5 min. From 30 to 120 min, 15 min intervals were used, and finally a 60 min interval to the end of the rate of desorption experiment at 3 h.

All weight loss from the canister is due to the release of methane. Initially, a large proportion of the methane emitted is the free gas component held in the interstitial permeability pathways. Barker-Read (1984) noted that the free gas component was emitted within 4 min of the initiation of the experiment for coal of 200 μm size. Thereafter all gas emitted consists of that previously held in an adsorbed state on the molecular pore surface of the coal.

There are several factors other than fracture frequency which can affect the rate at which methane desorbs from coal. These include: (1) temperature; (2) methane pressure within the canister; (3) ash and moisture contents; (4) maceral composition; and (5) particle size distribution of grains.

To minimize the effect of these factors the experimental procedure was developed to maintain the sorption canisters at a constant temperature of 25°C, pressure was maintained to +/−0.05 bar (absolute) during the adsorption procedure and all moisture was removed from the coal using a high vacuum. Differences in ash and maceral composition were noted after the experiments and their effects on the rate of methane desorption detailed later. Grain size differences in the sample lead to variance in the rate of desorption following the equation (Airey 1968)

$$t_0 \propto x^2 \qquad (3)$$

where $x$ = particle size and $t_0$ = rate of desorption (63% TDM).

During the maceral analysis the particle size distribution of the coal tested was also investigated and was shown to be highly consistent. Variances in the emission rate caused by differences in particle size distribution between each sample were minimized. By minimizing the physical differences between each sample any variation noted between the rate of desorption from each of the nine coals tested can only be attributable to the presence or absence of fracture-induced permeability between the samples investigated.

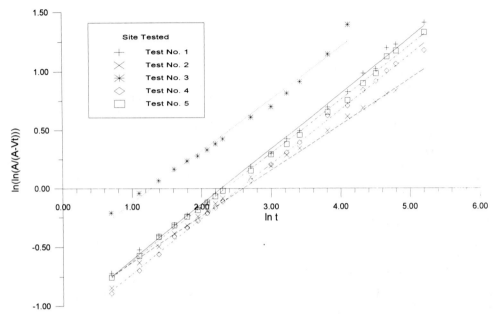

**Fig. 5.** Linear rate of desorption isotherms for test numbers 1–5. See text for explanation.

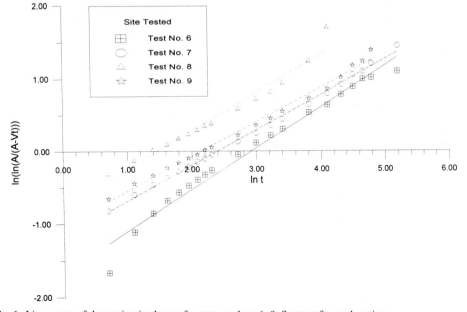

**Fig. 6.** Linear rate of desorption isotherms for test numbers 6–9. See text for explanation.

## Results

The desorption results are shown graphically in two ways. Firstly, the linear form of the desorption isotherm is used. This was developed by Griffin (1978) to determine more accurately Airey's emission constants $n$ and $t_o$ (Fig. 5). Airey's empirical emission equation is

$$V_t = A\{1 - \exp[-(t/t_0)^n]\}$$

Secondly, the cumulative percentage isotherm (Fig. 8), the shape of which shows the emission characteristics of the coal tested.

### Linear plot of desorption data

This form of graphical representation is based upon an equation by Airey (1968) Fig 6. This linear form utilizes the following transformations of the $x$ and $y$ axis data and was introduced by Fisher (1977)

$$x\text{-axis} = \text{time} = \ln t$$

$$y\text{-axis} = \text{methane emitted} = \ln\left[\ln\left(\frac{A}{A - V_t}\right)\right]$$

where $A$ = total desorbable methane (i.e. the total methane at 15 bar minus 1 bar) and $V_t$ = volume of gas emitted per unit mass up to time $t$.

Using a least-squares regression technique an equation can be determined for the line representing the data from which the gradient and intercept of the line can be determined.

Airey's constants are recorded from

$$n = \text{the curve gradient}$$

$$t_0 = -\exp\left(\frac{\text{intercept}}{\text{gradient}}\right)$$

where $n$ = Airey's crack index and $t_0$ = time taken to emit 63% of TDM.

### Cumulative percentage graphs

This technique approximates Airey's emission constants, but does not account for the non-linearity of emission rates. The percentage of total methane emitted during the desorption experiment is plotted on the $y$-axis and time is plotted on the $x$-axis. This graph is used to approximate the time taken to emit 63 and 75% of the TDM adsorbed onto the coal sample; also the shape of the isotherm is indicative of the emission characteristics (Fig. 8).

## Correlation of the hand drill penetrometer and the gravimetric sorption apparatus: rate of desorption technique

### Results

Table 2 shows the HDP rates recorded for each of the nine test sites investigated. Variations within each test are generally small, with the exception of sites 5 and 8 which, as discussed earlier, have fairly large standard deviations.

There is a wide range of values for the penetration rate recorded by this sampling suite. Test sites 1, 3 and 8 have very high penetration rates—generally very soft *in situ* coal which has been extensively tectonized and which has a greater proportion of fracture discontinuities. Sites 2, 6 and 7 record very low penetration rates, i.e. more competent coals which appear to have a lower frequency of fracture discontinuities.

If the presence of tectonically induced fractures in normally tight anthracite significantly affects the methane emission rate, test sites 1, 3 and 8 should have a much higher initial desorption rate than sites 2, 6 and 7 due to the reduced distance required for gas diffusion.

Table 3 shows the time taken to desorb 63 and 75% of TDM. Faster desorption rates are associated with increased permeability. For the determination of 63% TDM, site 3 is the fastest emitter, closely followed by sites 8 9 and 1; sites 6 2 and 4 are the slowest emitters. The same pattern is noted for the time taken to emit 75% of TDM from the coal.

Figures 5 and 6 show the linear form of desorption isotherm for sites 1 to 5 and 6 to 9, respectively. The lines which cut the $x$-axis closest to the origin are those which desorb and emit 63% of their methane more quickly. Sample 1 has very similar desorption and emission characteristics to

**Table 3.** *Mean HDP rate and time taken to emit 63 and 75% TDM*

| Site number | HDP (cm s$^{-1}$) | 63% (min) | 75% (min) |
|---|---|---|---|
| 1 | 0.97 | 9.77 | 21.2 |
| 2 | 0.22 | 13.5 | 27.75 |
| 3 | 1.51 | 3.59 | 8.03 |
| 4 | 0.41 | 12.84 | 26.32 |
| 5 | 0.59 | 10.43 | 22.28 |
| 6 | 0.17 | 18.50 | 31.4 |
| 7 | 0.32 | 10.94 | 25.57 |
| 8 | 1.00 | 4.31 | 8.63 |
| 9 | 0.67 | 8.53 | 17.5 |

both samples 5 and 9. Figure 7 shows the two fastest and two slowest emitters on one graph.

Figure 8 shows the cumulative percentage graph of the two fastest and slowest emitters. The fastest sites 3 and 8 have a much steeper gradient during the initial period of desorption compared with sites 2 and 6. Also of note is that both sites 3 and 8 had desorbed almost 100% of TDM within 75 min of the start of the desorption experiment. By contrast, sites 2 and 6 had only emitted approximately 90% of TDM by the end of the experiment at 180 min.

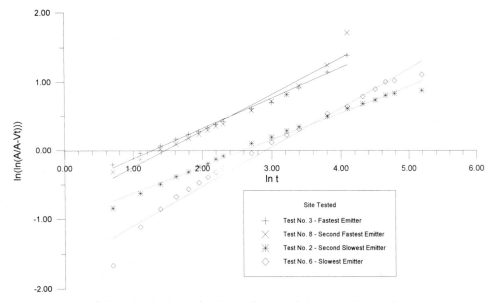

**Fig. 7.** Linear rate of desorption isotherms for the two fastest and slowest methane emitters.

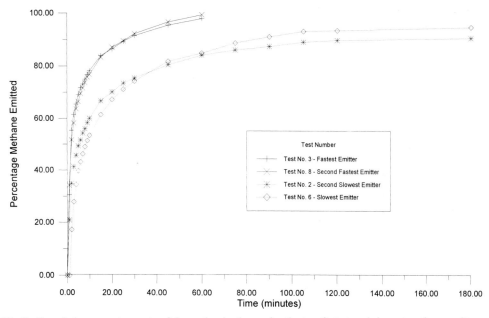

**Fig. 8.** Cumulative percentage rate of desorption isotherms for the two fastest and slowest methane emitters.

Site 6 initially desorbs less of its TDM than site 2; however, after 75% of TDM has been emitted at around 40 min into the experiment, the position changes and test site 6 emits more of its TDM for the remainder of the experiment.

From the data shown in Table 3, it appears that there is a relationship between the penetration rate and the desorption rate of a tectonically deformed coal seam. Close examination, however, shows that a perfect relationship does not exist, e.g. site 4 has the sixth slowest penetration rate, but the seventh slowest desorption and emission rates. It was therefore necessary to investigate the correlation coefficients of the

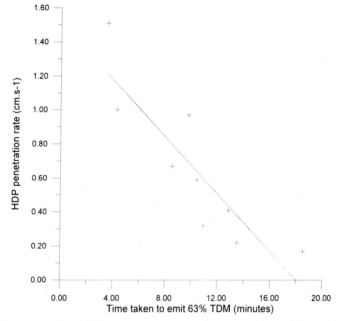

**Fig. 9.** Correlation between hand drill penetrometer penetration rate and rate of desorption of 63% total desorbable methane (TDM).

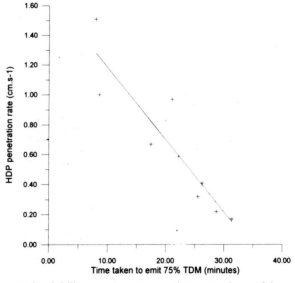

**Fig. 10.** Correlation between hand drill penetrometer penetration rate and rate of desorption of 75% total desorbable methane (TDM).

HDP and desorption rate. This was achieved using the least-squares regression method with analysis of the line conducted with Pearson's, correlation coefficient $r$. The following relates to comparisons between HDP rates and the time constants for 63 and 75% emission of TDM. Figure 9 shows the line of best fit correlating the time taken to emit 63% of TDM with HDP penetration rate. The correlation coefficient is +0.88.

Figure 10 shows the line of best fit correlating the time taken to emit 75% of TDM with HDP penetration rate through the two data sets. The correlation follows the trend noted in the previous comparison; the correlation coefficient is +0.91. Both are very satisfactory results considering the random distribution of coal friability within a coal seam.

The correlation between HDP penetration rate and methane emitted becomes more statistically acceptable with an increase in the amount of methane emitted. The correlation coefficient increasing from +0.88 to +0.91 for 63 and 75%, respectively. There is a definite relationship between the HDP rate and the time taken to emit 63 and 75% of TDM; more solid, less fractured coals take longer to emit methane from their internal surface than less solid coals.

## Factors affecting methane desorption and transmissibility

It was indicated earlier that differences between the ash, moisture, volatile and maceral composition could alter the rate of desorption of methane from a coal, as could variations of temperature and pressure at which the experiment was undertaken. In the experiment described earlier, the temperature, pressure and moisture content were kept constant and thus their effects have not been investigated. However, the samples had variable maceral composition and ash content, which may have influenced methane desorption in the experiments. To investigate these effects the composition of the coals tested was determined by a proximate analysis, giving the amount of ash (mineral matter) and volatile matter content. Petrological analyses were also carried out to establish the maceral composition and the vitrinite reflectance of the coals. Variations between these parameters are shown in Table 4 (for coal samples used in the previously described gas desorption experiment).

The vitrinite reflectance and the volatile matter content show significant variations between the nine samples, which is surprising as the samples were taken from an area of coal with a strike length of 20 m. The vitrinite reflectance varies by 17% whereas the volatile matter varies by 31%. The implied differences in coal rank are supported by a good correlation (+0.86) between the two rank parameters (Fig. 11), with the one exception of sample 1. It is not clear why this sample is anomalous, nor why the rank should vary over such a short distance. There is a possibility that it could relate to the presence of thrusting on the site, which is being currently studied by one of us (G.A.D.).

Table 5 shows the comparison of rate of desorption (to desorb 63% TDM) with the mean

**Table 4.** *Composite results for all analyses*

| Coal sample No. | 1 | 2 | 3 | 4 | 5 | 6 | 7 | 8 | 9 |
|---|---|---|---|---|---|---|---|---|---|
| Rank/maceral analysis | | | | | | | | | |
| $R_{mo}$ | 3.03 | 2.76 | 2.80 | 2.89 | 2.64 | 2.62 | 2.76 | 2.71 | 2.59 |
| $R_{max}$ | 3.63 | 3.30 | 3.44 | 3.44 | 3.15 | 3.16 | 3.57 | 3.50 | 3.12 |
| $R_{min}$ | 2.43 | 2.27 | 2.20 | 2.05 | 2.09 | 1.85 | 2.02 | 2.14 | 2.09 |
| %$V$ | 88.60 | 90.80 | 95.00 | 96.20 | 96.40 | 95.20 | 90.00 | 89.20 | 90.40 |
| %$I$ | 9.60 | 8.80 | 4.00 | 3.80 | 3.60 | 4.80 | 9.60 | 10.20 | 9.40 |
| %$I$ = semi-fusinite | 50 | >70 | 50 | 60 | 40 | 65 | 5 | 80 | 70 |
| %$L$ | 0.20 | 0.00 | 0.00 | 0.00 | 0.00 | 0.00 | 0.00 | 0.20 | 0.00 |
| Residual | 0.00 | 0.40 | 0.60 | 0.00 | 0.00 | 0.00 | 0.40 | 0.40 | 0.20 |
| *In situ* quantification | | | | | | | | | |
| HDP (cm s$^{-1}$) | 0.97 | 0.22 | 1.51 | 0.41 | 0.59 | 0.17 | 0.32 | 1.00 | 0.67 |
| Gas emission times (min) ($t_0$) | | | | | | | | | |
| 0.63 | 9.77 | 13.50 | 3.59 | 12.84 | 10.43 | 18.50 | 10.94 | 4.31 | 8.53 |
| 0.75 | 21.20 | 27.75 | 8.03 | 26.32 | 22.28 | 31.40 | 25.57 | 8.63 | 17.50 |
| Proximite analysis (%) | | | | | | | | | |
| Moisture | 2.10 | 1.80 | 1.29 | 1.62 | 0.98 | 0.93 | 0.72 | 0.68 | 0.57 |
| Ash | 7.70 | 6.80 | 1.33 | 4.40 | 4.20 | 7.49 | 2.30 | 3.50 | 5.30 |
| Vol. matter (DAF) | 8.00 | 7.19 | 7.59 | 6.58 | 8.61 | 7.88 | 7.19 | 7.31 | 8.25 |

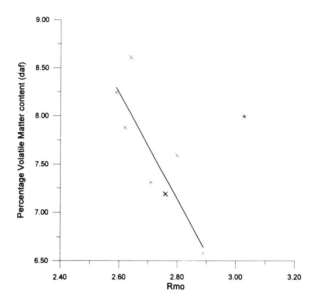

**Fig. 11.** Correlation between $R_{mo}$ and volatile matter content (daf).

**Table 5.** *Factors affecting methane desorption and transmissibility*

| Comparison | $r^2$ | $r$ | $n$ |
|---|---|---|---|
| $t_0$ and mean reflectance | 0.02 | −0.13 | 9 |
| $t_0$ and volatile matter content | 0.003 | −0.06 | 9 |
| $t_0$ and ash content | 0.44 | 0.66 | 9 |
| $t_0$ and inertinite content | 0.05 | 0.22 | 9 |

reflectance values for each site analysed. The goodness of fit of the data to the best fit line is 0.02 with a correlation coefficient of 0.13, suggesting a lack of correlation between these two parameters. Similarly, a comparison between rate of desorption (again 63% TDM) and volatile matter content shows a very poor goodness of fit of the data to the best fit line of 0.003, with a correlation coefficient of −0.06. There appears to be no correlation between rate of desorption and the rank variations of vitrinite reflectance and volatile matter content recorded for the anthracite tested.

Although there is little or no correlation between the rank indicators and rate of desorption, a comparison between the ash content and the desorption time to emit 63% of TDM records a correlation coefficient of 0.66.

Finally, comparisons between the percentage inertinite and rate of desorption showed a very poor goodness of fit of the data to the best fit line of 0.05. The correlation coefficient of 0.22 was similarly poor (Table 5). From the data it appears there is no relationship between the amount of inertinite in the samples tested and the emission characteristics of the coal.

## Conclusions

The HDP rates varied extensively between sites 1 to 9. Variations within each site tested were generally low, with the exception of sites 5 and 8, which gave greater than normal standard deviations.

Similarly, variations between the desorption and emission rates from each of the nine sites tested were considerable. For 63 and 75% TDM emitted, sites 3 and 8 were the fastest to desorb, whereas sites 2 and 6 were the slowest.

Correlation between those coals that were most easily penetrated by the HDP and the coals that emitted 63 and 75% of TDM was very good at 0.88 and 0.91, respectively.

Ash, moisture content, rank and maceral composition were investigated to determine their effects on the above correlations. Owing to the crushed and sized nature of the coal, the probable effects of maceral changes on gas emission were minimized, e.g. differences in the fusinite proportion can affect gas emission rates in solid coal due to its persistent bedding-parallel fractures. With the exception of the percentage of ash they have been shown not to affect the desorption and emission characteristics of the coals tested. The correlation between the percentage of ash and emission rate has previously been shown to affect only subtly the emission rate of methane from coal; the significance of increased fracture permeability has been shown to be far more important.

The present research on the South Wales anthracites has demonstrated the relationship between drill penetration rate and fracture frequency. The greater the extent of fracturing within the coal the less competent the coal becomes. Although the coals tested for this analysis were intact within the seam, this is not always so, as coals are sometimes completely particulate and granular—a result of intense tectonic layer-parallel movement. These coals record faster HDP ($>3\,\mathrm{cm\,s^{-1}}$) rates than those documented in this study and are commonly present within several operating sites within the South Wales coalfield. Thus it appears that the extent of this fast methane emitting coal style is widespread. It is likely that more highly particulate coals will emit methane more rapidly than those in this experiment.

The significance of the relationship documented here between coals with greater fracture frequency and those giving faster desorption rates is extremely important for the development of CBM projects within areas previously considered unsuitable because of the low emission rates of methane from anthracite. The distribution of structurally deformed coal within the northwest anthracite region is associated with areas of intense tectonically controlled shortening—a situation common in this area. Large aerial extents of this deformed coal, if present in a continuous nature, may preserve significant gas reservoirs. In addition to the possibility of creating large gas reservoirs, these tectonically induced fracture sets increase the inherent permeability. More importantly, the induced fracture sets are multi-oriented and overprint the earlier bedding and cleat fractures; this multi-orientation means that regardless of the current *in situ* stress conditions (especially at depth), numerous permeability pathways will be open; this will allow rapid gas migration (Harris 1995b).

The main disadvantage of these particulate coals is the problems associated with evacuating coal product with the water and methane from a CBM well. It would be imperative in these regions to plan adequately for an increase in particulate material.

We are grateful to British Coal Opencast for allowing access to data and sites and sponsoring I.H.H., who also received a EPSRC research studentship during the period of the study. In particular, we thank all on-site staff for their assistance in the opencast sites.

# References

AIREY, E. M. 1968. Gas emission from broken coal, an experimental and theoretical investigation. *International Journal of Rock Mechanics and Mining Sciences*, **5**, 475–494

BARKER-READ, G. R. 1980. *The geology and related aspects of coal and gas outbursts in the Gwendraeth valley.* MSc Thesis, University of Wales, Cardiff.

——1984. *A study of the gas dynamic behaviour of coal measure strata, with particular reference to West Wales outburst prone seams.* PhD Thesis, University of Wales, Cardiff.

CLOSE J. 1993. Natural fractures in coal. *In*: LAW, B. E. & RICE, D. D. (eds) *Hydrocarbons From Coal*, AAPG Studies in Geology, **38**, 119–133.

CREEDY, D. P. 1985. *The origin and distribution of firedamp in some British Coalfields.* PhD Thesis, University of Wales, Cardiff.

DECKER, A. D., SCHRAUNFNAGEL, R., GRAVES, S., BEAVERS, W. M., COOPER, J., LOGAN, T., HORNER, D. M. & SEXTON, T. 1986. *Petroleum Frontiers*, **3**, 1.

FISHER, R. J. 1977. *A study of variations in the physical characteristics of West Wales anthracite.* PhD Thesis, University of Wales, Cardiff.

FRODSHAM, K. 1990. *An investigation of geological structure within opencast coal sites in South Wales.* PhD Thesis, University of Wales, College Cardiff.

GAMSON, P. D., BEAMISH, B. B. & JOHNSON, D. P. 1993. Coal microstructure and micropermeability and their effects on natural gas recovery. *Fuel*, **72**, 87–104.

GAN, H., NANDI, S. P. & WALKER, P. L. 1972. Nature of the porosity in American coals. *Fuel*, **51**, 272–277.

GAYER, R. A. 1993. The effect of fluid over-pressuring on deformation, mineralization and gas migration in coal-bearing strata. *In*: PARNELL, J., RUFFELL, A. H. & MOLES, N. R. (eds) Contributions to an International conference on fluid evolution, migration and interaction in rocks. *Geofluids '93 extended abstracts, Torquay*, 186–189.

——, COLE, J., FRODSHAM, K., HARTLEY, A. J., HILLIER, B., MILIORIZOS, M. & WHITE, S. C. 1991. The role of fluids in the evolution of the South Wales coalfield foreland basin. *Proceedings of the Ussher Society*, **7**, 380–384.

GRAY, I., 1987. *Reservoir Engineering in Coal Seams: Part 1—The Physical Process of Gas Storage and Movement in Coal Seams.* SPE Reservoir Engineering, February 1987.

GRIFFIN, P. E. 1978. *Methane sorption studies with particular reference to the outburst prone anthracites of West Wales.* MSc Thesis, University of Wales, Cardiff.

HARRIS, I. H. 1995*a*. Newly developed techniques to determine proportions of undersized (friable) coal during prospective site investigations. *In*: WHATELY, M. K. G. & SPEARS, D. A. (eds) *European Coal Geology*, Geological Society, London, Special Publication, **82**, 99–114.

——1995*b*. *The remote and* in situ *quantification of friable anthracite in the South Wales Coalfield.* PhD Thesis, University of Wales, Cardiff.

JONES, J. A. 1991. A mountain front model for the variscan deformation of the South Wales coalfield. *Journal of the Geological Society, London*, **148**, 881–891.

RODDY, D. Y. & DAVIS, L. K. 1977. Shatter cones formed in large-scale experimental explosion craters. *In*: RODDY, D. J., PEPIN, R. D. & MERRILL, R. B. (eds) *Impact and Explosion Cratering*. Pergamon, New York, 715–749.

STACH, E. 1982. *Stachs textbook of Coal Geology*, 3rd edn. Gebrunder Brontraeg.

WHITE, S. C. 1992. *The tectono-thermal evolution of the South Wales coalfield.* PhD Thesis, University of Wales, Cardiff.

# Model study of the influence of matrix shrinkage on absolute permeability of coal bed reservoirs

## JEFFREY R. LEVINE

*Consulting Geologist, 2715 Seventh Street, Tuscaloosa, AL 35401, USA*

**Abstract:** The matrix volume of coal shrinks when occluded gases desorb from its structure. In coalbed gas reservoirs, matrix shrinkage could cause the fracture aperture width to increase, causing an increase in permeability. A computer model was developed, based on elastic rock mechanics principles, to evaluate the potential effect of matrix shrinkage on the absolute permeability of coalbed reservoirs as fluid pressure is drawn down during gas production. The model predicts that the fracture width can potentially increase, depending on the combined influence of a number of parameters, particularly Young's modulus of elasticity, Poisson's ratio, fracture spacing and matrix shrinkage parameters. Each of these parameters vary depending on coal composition, so each individual coal will behave differently.

A sensitivity study was conducted to evaluate the influence of each model parameter using a geologically reasonable range of input values. 'Base case', upper and lower limits were selected, based on published data. Gas production was simulated by reducing fluid pressure from 1290 PSI (8.89 MPa) to 100 PSI (0.70 MPa). The matrix shrinkage parameter $\varepsilon_{max}$ was found to produce the largest effect on permeability. Permeability changes as large as $+250\,mD$ are predicted for the upper case value of $\varepsilon_{max}$. If $\varepsilon_{max}$ is small, however, the predicted permeability change will be negligible. An increase in Young's modulus, Poisson's ratio and fracture spacing each cause a predicted increase in permeability. Results of this model study should be verified through additional modelling, laboratory and field-based studies.

Underground coal seams typically contain large amounts of methane sorbed within their microstructure. Sorbed methane can potentially be recovered economically in the form of natural gas (so-called coalbed methane or CBM), but only if the gas is able to migrate through the coal seam at a satisfactory rate. The matrix of the coal, where most of the methane is sorbed, typically has a very low permeability, but coalbeds are usually highly fractured and these fractures may potentially provide permeable conduits for reservoir drainage provided that they are sufficiently open to fluid flow. The principal fracture network in coal, termed *cleat*, is comprised of sets of closely spaced, more or less vertical fractures. In most instances, (at least) two sets of cleat are overprinted at any given locality, at roughly 90° to one another, termed the face cleat and butt cleat. These fractures divide the coal into discrete matrix blocks, bounded on the sides by fractures and on the top and bottom by bedding surfaces. Gas transport through coal is usually modeled as comprising two steps: (1) diffusion of gas from the coal matrix into the cleat network, followed by (2) Darcy flow through the cleat.

In most instances, the cleat network of coal *in situ* is initially saturated with water. In such instances, before gas production can commence, water must be pumped from the reservoir, thus reducing the fluid pressure, allowing occluded methane to diffuse from the coal matrix into the fractures. Gas bubbles may then coalesce and begin to flow, once a certain critical gas saturation is reached.

A distinctive feature of coalbed gas wells is the 'negative decline' in gas production rates commonly observed during production. In contrast with most conventional reservoirs, gas production rates from coal may increase over time while the reservoir pressure decreases or remains roughly constant. Negative decline is probably due largely to the increasing relative permeability of the gas phase (where the relative permeability is equal to the absolute permeabilty multiplied by the relative permeability coefficient, which is usually less than unity). It is possible, however, that the absolute permeability of coal may also increase over time, in response to volumetric shrinkage of the coal matrix due to methane desorption. Any such increase in absolute permeability would be important in that it could influence the gas production rates and, consequently, the overall economics of coal seam gas recovery. On the other hand, in the absence of matrix shrinkage, the fracture network in coalbeds could become 'self-sealing' due to pore volume compressibility as fluid pressure decreases (Briggs & Sinha 1933; Gray 1987). In either case, a thorough understanding of temporal changes in permeability is required for accurate reservoir modelling and production forecasting.

*From* Gayer, R. & Harris, I. (eds) 1996, *Coalbed Methane and Coal Geology*, Geological Society Special Publication No 109, pp 197–212.

Three approaches may be taken to evaluate changes in absolute permeability due to matrix shrinkage: (1) a field-test approach, where formation permeability is measured; (2) an experimental approach, where changes in absolute permeability are directly measured in a controlled laboratory setting; or (3) a theoretical modelling approach, where fracture width may be calculated according to principles of rock mechanics, as described by Gray (1987). Each approach is subject to its own advantages and limitations. The precision and accuracy of field measurements are probably insufficient to discern and verify subtle changes in absolute permeability. Large permeability changes, however, might potentially be measureable via field tests. The experimental approach is also difficult. *In situ* conditions are difficult (and perhaps impossible) to simulate in the laboratory. Harpalani & Zhao (1989) and Harpalani & Schraufnagel (1990a; 1990b) describe laboratory experiments where methane was flowed through core samples of coal confined in a triaxial load cell under incrementally decreasing fluid pressure. These tests were interpreted to indicate large increases in permeability with decreasing fluid pressure, which they interpreted to be caused by matrix shrinkage. Their conclusions, however, have been the subject of much controversy. Among the many problems inherent in the experimental approach are: (1) the high fluid pressure gradients required to produce measureable flow-rates, which produce a high longitudinal gradient of effective stress; (2) the long equilibration times required for the coal to adjust to changes in fluid pressure; (3) the difficulty in interpretation of the Klinkenberg effect; and (4) the difficulty in sample preparation. Thus although these laboratory experiments appear to document the influence of matrix shrinkage on permeability, the conclusions are ambiguous. In any case, it is not possible to quantitatively translate the results of these laboratory tests to a field setting.

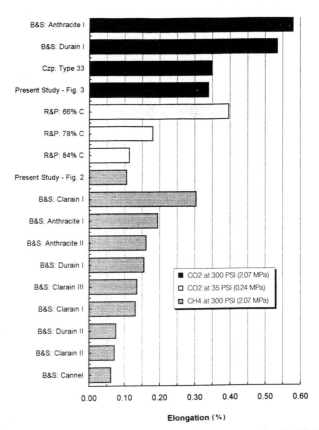

**Fig. 1.** Axial strains in coal due to gas sorption at various pressures (Czp, Czaplinski 1971; R&P, Reucroft & Patel 1986; and B&S, Briggs & Sinha, 1933). The data from the present study and from Czaplinski (1971) were measured over a range of pressures and were interpolated at 300 PSI (2.07 MPa).

In the present paper, a simple rock mechanics model is used to evaluate matrix shrinkage effects. Theoretical changes in permeability are calculated as a function of changes in reservoir fluid pressure, taking into account volumetric shrinkage of the coal matrix and pore volume compressibility of the fracture network.

## Measurements of coal matrix shrinkage

There are few reliable measurements available for strains caused by gas sorption/desorption in coal. The lack of data is due to the difficulty of making these measurements and to few such studies having been conducted. Much more data are needed to document this phenomenon. Nevertheless, the available data show a fairly consistent trend. For the range of pressures typically encountered in CBM reservoirs, (about 0–1600 PSIG, or 0–11 MPa) axial strains for methane sorption are of the order of a few tenths of 1% (Figs 1 and 2). In the most detailed work carried out to date on this topic, Briggs & Sinha (1933) used a mechanical micrometre to measure sorption-related strains of coal at 300 PSI (2.07 MPa) in methane and in carbon dioxide. Their samples were a few millimetres in diameter and a centimetre or so in length. Axial strains $(d//l$, represented here by the Greek letter $\varepsilon$) for methane ranged from 0.06 to 0.30% (Fig. 1). Briggs & Sinha (1933) found that coal rank, petrographic composition, mineral matter content, orientation with respect to bedding and sorbate composition all influence the results. Anthracites showed the greatest amount of linear strain among the samples studied, consistent with anthracites having generally high methane sorption capacities. Clarain (bright banded coal) showed a strain of 0.0030 (0.0015 per MPa) when measured perpendicular to bedding, but only 0.0013 (0.0006 per MPa) measured parallel to bedding. The smallest strains (0.0006, or 0.0003 per MPa) were exhibited by a sample of cannel (i.e. sporinite-rich) coal. This coal, however, had a high ash yield (41.5%), so it is unclear whether the low sorption strain was related to its unusual petrographic composition or to its high mineral matter content.

Axial strains for carbon dioxide sorption are higher than those for methane, ranging to more than 0.5% (Figs 1 and 3). This is consistent with the sorption capacity of the coal for carbon dioxide being nearly twice that of methane (Fig. 4). Reucroft & Patel (1986) indicate that carbon dioxide sorption swelling decreases with increasing coal rank through the lower half of the bituminous rank range (Fig. 1). This is consistent with the observation that carbon dioxide surface area decreases with increasing rank over this same rank interval (Thomas & Damberger 1976; Levine 1993).

The data in Fig. 1 show sorption strain at a single pressure value. To evaluate the response of coal to changes in pressure requires that the sorption strain is known over a range of pressures. Unfortunately, there are very few data available documenting this relationship. Czaplinski (1971) reports sorption strain for a Polish coal (type 33, or high-volatile bituminous) in carbon dioxide at pressures ranging up to 55.4 atm (814 PSI or 5.61 MPa), but the data are erratic. A study conducted for the US Department of Energy (Levine & Johnson 1992) provides more consistent results (Figs 2 and 3), but the data are few. Measurements were conducted under dry conditions on a high-volatile C bituminous coal from Illinois. The sample blocks used for the measurements were several centimetres in length and width and were polished on one face to provide a smooth surface for adhesion of the strain gauges. An area of homogeneous coal, absent of any visible fractures, was used to fix the strain gauge rosettes. It is assumed that the sorption strain measurements are representative of the coal 'matrix'. Simultaneous measurements taken on two replicate samples (A and B) demonstrate the consistency of the readings. The samples were fully immersed in an atmosphere of high-pressure gas, so the measured strains represent the net combined influences of sorption-related swelling, plus (any?) mechanical compression of the coal matrix.

Using 300 PSI (2.07 MPa) as a reference pressure, the methane sorption strain of this coal falls about midway in the range of data shown in Fig. 1. Axial strains were observed to range up to 0.18% at 750 PSI (5.17 MPa) for methane and up to 0.5% at 450 PSI (3.10 MPa) in carbon dioxide. Sorption strain for both methane and carbon dioxide is shown to be reversible on increasing and decreasing pressure, with no apparent hysteresis. Briggs & Sinha (1933) previously showed that sorption strain is recoverable on reduction of ambient pressure. Other workers, however, have reported a hysteresis in the strain curve (Moffat & Weale 1955; Czaplinski 1971). Strains were slightly greater perpendicular to bedding than parallel to bedding for both methane and carbon dioxide (Figs 2 and 3), a result observed previously in numerous other solvent swelling studies on coal. Volumetric strain, which may be calculated as two times the strain parallel to bedding plus strain perpendicular to bedding, was around

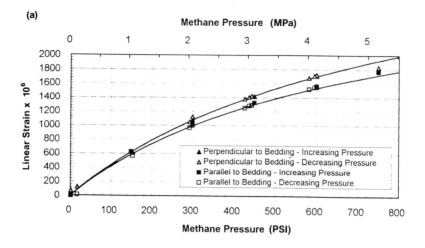

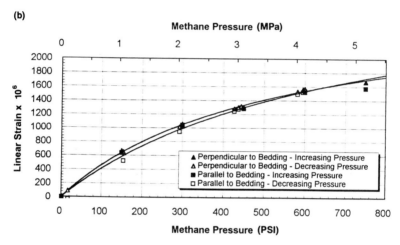

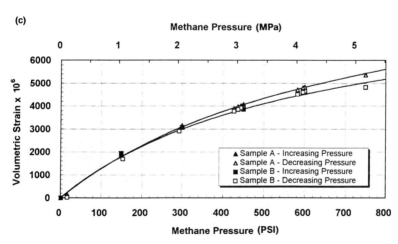

**Fig. 2.** Strain during sorption and desorption of methane, measured under dry conditions on a sample of Carboniferous, high-volatile bituminous coal from Illinois (UACP 162): (**a**) Sample A, axial strain; (**b**) sample B, axial strain; (**c**) samples A and B, volumetric strain.

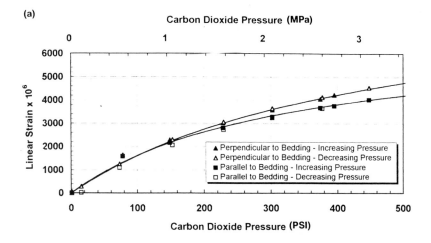

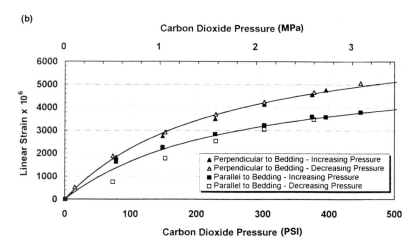

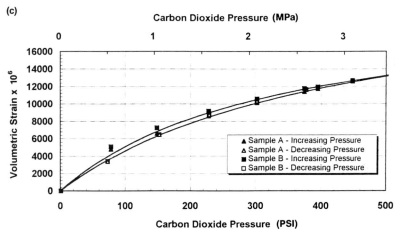

**Fig. 3.** Strain during sorption and desorption of carbon dioxide, measured under dry conditions on a sample of Carboniferous, high-volatile bituminous coal from Illinois (UACP 162).(**a**) Sample A, axial strain; (**b**) sample B, axial strain; (**c**) samples A and B, volumetric strain.

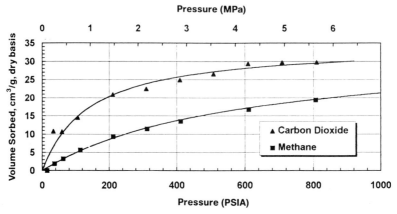

**Fig. 4.** Sorption isotherms for methane and carbon dioxide for Illinois coal, determined gravimetrically, under dry conditions, using an electromicrobalance (see Levine *et al.* 1993).

0.52% in methane at 750 PSI (5.17 MPa) and 1.25% in carbon dioxide at 450 PSI (3.10 MPa). In other words, the volume of the coal increases by over 1% via sorption of carbon dioxide at 450 PSI (3.10 MPa).

The data also show that sorption strain is not linear with pressure, but exhibits a curvilinear form that is steeper at low pressure, becoming flatter at higher pressure, resembling the shape of the sorption isotherm (Fig. 4). Over this pressure range, sorption strain appears to be proportional to the quantity of gas sorbed. (Moffat & Weale (1955) have shown that at much higher pressures, greater than 2200 PSI (15.2 MPa), the matrix volume may begin to decrease with pressure, owing to the mechanical compression of the coal structure, but these pressures are well beyond those typically encountered in coalbed reservoirs.)

In a previous evaluation of the present model, as a simplification, sorption strain was assumed to be linear with pressure, but this was found to produce an exaggerated influence matrix shrinkage effect, especially at high pressure. A curvilinear function which closely follows the actual data yields more accurate results. Methane and carbon dioxide sorption data can be modelled fairly accurately using the Langmuir isotherm model (cf. Yee *et al.* 1993). Therefore an equation having the same mathematical form as the Langmuir equation was used in the present study to 'fit' the sorption strain data depicted in Figs 2 and 3:

$$\varepsilon_s = \varepsilon_{max} \cdot \frac{P}{(P + P_{50})} \qquad (1)$$

where $\varepsilon_s$ is the linear sorption strain at pressure $P$. $\varepsilon_{max}$, which corresponds mathematically to

the 'Langmuir volume' in the Langmuir equation, represents the theoretical maximum strain, approached asymptotically at infinite pressure. $P_{50}$, which corresponds mathematically to the 'Langmuir pressure', represents the pressure at which the coal has attained 50% of its maximum strain. The lower the value of $P_{50}$, the steeper is the sorption strain curve at low pressures.

The slope of the function in Equation (1) at any pressure, provides a measure of the strain rate ($d\varepsilon/dp$)

$$d\varepsilon/dp = \frac{(\varepsilon_{max} \cdot P_{50})}{(P_{50} + P)^2} \qquad (2)$$

$\varepsilon_{max}$ and $P_{50}$ were estimated by best fitting a linearized form of Equation (1) where $\varepsilon/P$ is plotted versus $\varepsilon$. The slope of this linearized form is equal to $1/P_{50}$, and the $y$-intercept is equal to $\varepsilon_{max}/P_{50}$. The trend lines plotted in Figs 2 and 3 were calculated in this manner, and appear to fit the data fairly well. Values for the coefficients $\varepsilon_{max}$ and $P_{50}$ appear in Table 1. It should be emphasized, however, that (1) this is an entirely empirical correlation and (2) extrapolating the measured data to pressures beyond the measured range entails an increasing degree of uncertainty.

Measurements of sorption-related strain of coal are subject to analytical difficulties which may render the results unreliable. For example, if resistance-type strain gauges are used, they may not adhere properly to the coal, or may not deform homogeneously with the coal. Lengthy equilibration times can also pose problems, especially for samples larger than a few millimetres in dimension. For example, the samples used by Briggs & Sinha (1933) were only a few

**Table 1.** *Matrix shrinkage parameters for Langmuir-type model*

| Expt. No. | Sorbate | Specimen | Orientation | $P_{50}$ | $\varepsilon_{max}$ |
|---|---|---|---|---|---|
| 7 | $CH_4$ | A | $\perp$ Bedding | 833.3 | 4060.7 |
| 7 | $CH_4$ | A | // Bedding | 714.3 | 3357.5 |
| 7 | $CH_4$ | A | Volumetric | 769.2 | 11 001.5 |
| 7 | $CH_4$ | B | $\perp$ Bedding | 526.3 | 2887.4 |
| 7 | $CH_4$ | B | // Bedding | 714.3 | 3351.4 |
| 7 | $CH_4$ | B | Volumetric | 625.0 | 9225.0 |
| 8 | $CO_2$ | A | $\perp$ Bedding | 476.2 | 9261.0 |
| 8 | $CO_2$ | A | // Bedding | 333.3 | 6933.0 |
| 8 | $CO_2$ | A | Volumetric | 434.8 | 24 598.7 |
| 8 | $CO_2$ | B | $\perp$ Bedding | 243.9 | 7597.0 |
| 8 | $CO_2$ | B | // Bedding | 263.2 | 5972.4 |
| 8 | $CO_2$ | B | Volumetric | 333.3 | 22 051.7 |

millimetres in diameter and a centimetre or so in length, but required 30–100 h to equilibrate to changes in pressure. Reucroft & Patel (1986), using a similar specimen preparation and measurement procedure, report that their samples did not attain equilibrium even after exposure times as long as 200 h. Larger samples would probably require even longer equilbration times, owing to longer diffusion pathways. Any such measurement errors would yield strain measurements lower than reality. The high potential for mismeasurement must, therefore, be borne in mind when evaluating published data. The data depicted in Figs 2 and 3 collected using resistance-type strain gauges, appear to provide accurate results, but these results were attained only after considerable effort was spent perfecting measurement procedures.

The methane sorption capacity of coal is suppressed by the presence of water. Suppression of methane capacity by water ranges from a few per cent to over 50% of the capacity of dry coal, depending on coal rank (Joubert *et al.* 1974). Thus the coal sample depicted in Figs 2–4 would be expected to exhibit lower methane sorption capacity and correspondingly lower shrinkage if it had been measured on a moisture-containing basis. Nevertheless, the methane sorption capacity of this coal on a dry basis is comparable, or even lower than the typical methane sorption capacity of coals from the better producing areas of the Black Warrior Basin (Alabama) and San Juan Basin (New Mexico and Colorado). Thus the measured shrinkage values would appear to provide a valid approximation for modelling Black Warrior and San Juan Basin coals, assuming that the matrix shrinkage coefficients are proportional to the amount of gas sorbed.

## Model approach

Coalbeds are highly complex and heterogeneous at many different scales. Accordingly, modelling of coalbed reservoirs necessarily entails a great deal of simplification. In the present model (Fig. 5), gas transport is represented as occurring by laminar flow through a closely spaced set of parallel cracks (cleat). In nature, cleat fractures are irregular, discontinuous, non-uniform in width and non-uniformly distributed through the coal seam. Accordingly, gas flow paths in nature are much more tortuous. Moreover, larger scale fracture networks, which are not part of the cleat system *per se*, may significantly influence production rates from coalbed gas wells, so modelling the permeability of the cleat network may only partly account for reservoir behaviour. Nevertheless, the present model, even if not correct in detail, provides a reasonable approach to understanding the fundamental influences of matrix shrinkage on reservoir permeability.

For laminar fluid flow through regularly spaced fracture, permeability, $k$ is related to fracture width, $b$, and fracture spacing, $s$ according to the following relationship (Gray 1987, after Snow 1968)

$$k = \frac{(1.013 \times 10^9) \cdot b^3}{12 \cdot s} \tag{3}$$

where $b$ is fracture width in millimetres and $s$ is fracture spacing in millimetres.

Fracture width, $b$, has a very strong influence on permeability, as it is raised to the third power in Equation (3). Accordingly, a small increase in fracture width could potentially produce a significant increase in flow-rates (as shown later). Fracture width is dependent on the

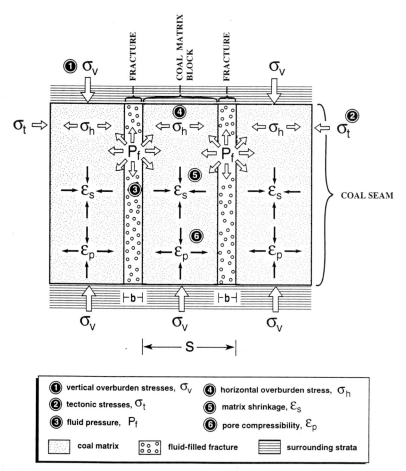

**Fig. 5.** Schematic representation (not to scale) of the mechanical model of coal bed reservoir used in the present study. Actual fracture widths, *b*, *in situ* are probably of the order of several micrometres to tens of micrometres. Fracture spacing, *s*, is typically of the order of millimetres to tens of millimetres. Hollow arrows represent stress ($\sigma$) vectors, whereas solid arrows represent strain ($\varepsilon$) vectors. $\sigma_v$ and $\sigma_h$ represent stresses due to the weight of overlying strata. $\sigma_t$ represents horizontal tectonic stresses, which are set at zero in the present model. $P_f$ represents reservoir fluid pressure. $\varepsilon_s$ represents matrix shrinkage due to desorption of occluded gases from the coal matrix and $\varepsilon_p$ the mechanical compressibility due to fluid pressure decrease.

combined influences of (at least) five factors as shown in Table 2.

It is assumed here that coal behaves as an elastic material, but this is not certain. Under some stress conditions coal may deform plastically (e.g. Newman 1955), or perhaps viscoelastically; however, the non-elastic mechanical behaviour of coal is not well documented and most mechanical tests, especially under conditions of low strain, have documented elastic behaviour (Berkowitz 1979). Nevertheless, the presupposition of elastic behaviour is one of the simplifying assumptions made in the present model.

**Table 2.** *Factors influencing openness of fractures in coalbed reservoirs*

(1) Horizontal compressive stress ($\sigma_h$) due to the weight of the overlying strata, which tends to squeeze the fractures shut
(2) *In situ* 'tectonic stress', $\sigma_t$, which may be either compressive or tensile across the fracture plane
(3) Hydraulic fluid pressures, $P_f$, within the coal, which tend to hold the fractures open
(4) The mechanical strength of the coal, which represents its ability to resist stresses
(5) Volumetric shrinkage of the coal matrix due to desorption of occluded substances, notably methane

If elastic behaviour is assumed, then the strength of the coal (Table 2, No. 4) would be reflected in its two elastic constants, Young's modulus ($E$) and Poisson's ratio ($\nu$). Young's modulus ($E$) provides a measure of the amount of stress ($\sigma$) required to produce a certain amount of elastic strain ($\varepsilon$)

$$E = \frac{\sigma}{\varepsilon} \qquad (4)$$

Stress ($\sigma$), which represents force per unit area, is reported in many different measurement units—pounds per square inch (PSI) being the most common in petroleum engineering publications. Strain represents the ratio of change in length per unit length ($dl/l$). If the same units of length are used in both numerator and denominator, then this ratio is dimensionless. Therefore, Young's modulus is represented in the same measurement units as stress (e.g. PSI). The smaller the strain for any given stress, the larger is Young's modulus. Thus mechanically strong rocks, such as sandstone, have a much larger Young's modulus than weak rocks, such as coal.

In the present study, tectonic stresses (Table 2, No. 2) are assumed to be zero, so overburden stresses (Table 2, No. 1) provide the sole driving force for compression of the fractures. Any rock, when vertically loaded by overburden stresses, will shorten parallel to the vertical load and elongate to some degree perpendicular to the load. Poisson's ratio ($\nu$) represents the negative ratio of (vertical) axial shortening to (horizontal) transverse elongation. Horizontal elongation produces a horizontal stress, $\sigma_h$ (Table 2, No. 1). In rocks subject to loading from overlying strata, Poisson's ratio can be used to calculate the translation of vertical gravitational compressive stress ($\sigma_v$) to horizontal gravitational compressive stress ($\sigma_h$)

$$\sigma_h = \frac{\nu}{(1 - \nu)} \cdot \sigma_v \qquad (5)$$

For an ideally incompressible fluid, Poisson's ratio is 0.5, indicating that horizontal stresses and vertical stresses are equal. A mechanically strong rock, such as sandstone, typically has a low value for Poisson's ratio, thus producing a smaller horizontal stress due to gravity. Coal falls somewhere between these two.

In virgin coal, all stresses are in equilibrium. The compressive stresses due to the weight of overlying strata are resisted by two forces: (1) the elastic mechanical strength of the coal (Table 2, No. 4) and (2) the pressure of fluids residing within the coal ($P_f$, Table 2, No. 3). During production of a coal gas reservoir, fluid pressures progressively decrease ($\Delta P_f = P_2 - P_1$), whereas the overburden stresses remain essentially constant. Therefore, the 'effective stress' (mechanical stress minus the fluid pressure) progressively increases. As the effective stresses increase, vertical fractures will close. Fracture closure strain due to pressure change, $\varepsilon_p$, is calculated according to standard elastic rock mechanics relationships as

$$\varepsilon_p = \frac{1}{E} \cdot (1 - 2\nu) \cdot \Delta P_f \qquad (6)$$

Decreasing the reservoir fluid pressure has the additional effect of allowing occluded gases, particularly methane, to desorb from the coal structure, causing a decrease in bulk volume of the coal matrix, which tends to cause the fractures to open (Table 2, No. 5). Matrix shrinkage strain, $\varepsilon_s$, is a function of the matrix shrinkage coefficient, $M_s$, and the pressure change, $\Delta P_f$

$$\varepsilon_s = M_s \cdot \Delta P_f \qquad (7)$$

where

$$M_s = \frac{(dl/l)_{SORPTION}}{dp}$$

$M_s$, which represents the slope of a curve relating shrinkage strain to change in pressure, is a function of the two matrix shrinkage parameters $\varepsilon_{max}$ and $P_{50}$ [Equation (2)].

Following a decrease in reservoir fluid pressure, the new fracture width is equal to the sum of three terms:

new fracture width ($b_2$)

= previous fracture width ($b_1$)

+ closure due to fracture compressibility

+ opening due to matrix shrinkage.

As the strain terms in Equations (6) and (7) are calculated as ratios, they must be multiplied by the spacing between the fractures, $s$, to yield the total amount of closure and/or opening of each individual fracture

$$b_2 = b_1 + \varepsilon_p \cdot s + \varepsilon_s \cdot s \qquad (8)$$

Equations (2), (3), (6), (7) and (8) provide the basis for the computer model used in the present study.

## Parameter sensitivity

To solve Equations (2), (3), (6), (7) and (8), values for six variables must be specified: initial fracture width ($b_1$), Young's modulus ($E$), Poisson's ratio ($\nu$), fracture spacing ($S$), and the two matrix shrinkage parameters ($\varepsilon_{max}$ and

$P_{50}$). Each of these will vary significantly from seam to seam and between different lithotypes within a single coal seam. A sensitivity study was conducted to evaluate the potential impact of these six parameters on the model. For each parameter, a 'base case' value was selected, along with geologically reasonable upper and lower limits (Table 3). In selecting values for the study, consideration was given to representing the potential variability among all types of coals, while at the same time approximating the conditions in the San Juan Basin, the world's most prolific basin for coalbed gas production.

*Parameter values*

*Cleat spacing.* Cleat development in coal is related both to geological stresses and to the loss of volatile by-products during coalification (Close 1993; Levine 1993). Cleat spacing varies from a few millimetres to a few centimetres, depending on rank, petrographic composition, mineral matter content and stress/strain history. Given the complexity of these interrelationships, it is impossible (at present) to predict cleat development reliably. Cleat may be directly observed, however, via core analysis or downhole camera, or inferred via geophysical logs. At the COAL research test site in the northwestern part of the San Juan Basin, Mavor *et al.* (1991) report cleat spacings on the order of 2.5 mm to 1 cm. Cleat is much less well developed, however, in the lower rank, lower permeability coals in the south-central part of the basin. For the parameter sensitivity study, a base case value of 10 mm was selected, with 20 and 5 mm representing upper and lower limits (Table 3).

*Initial permeability.* Absolute permeabilities of coals vary over several orders of magnitude, ranging from microdarcies to several tens of millidarcies (and possibly higher). As economic production rates depend on a combination of many factors, there is no absolute lower limit for the permeability of coal bed wells; however, as a

rule of thumb, permeabilities in the range of a few tens of millidarcies are considered excellent, and permeabilities less than 1 mD are considered poor. Permeabilities in the northern part of the San Juan Basin are generally high. At the COAL site, 20 mD or higher has been estimated (Mavor *et al.* 1991). Nevertheless, as the present study was intended, in part, to evaluate the effect of matrix shrinkage on low permeability coals, a baseline value of 1 mD was selected, with 100 $\mu$D and 10 mD representing the lower and upper values, respectively (Table 3).

*Elastic parameters.* There are few published data on the elastic parameters of coal, owing largely to the difficulty of making accurate measurements. The existing data show a considerable range in values (e.g. Fig. 6). Berkowitz (1979, his Table 4.5.2) compiled data from a number of sources, showing Young's modulus ranging from less than $1 \times 10^{10}$ dynes/cm$^2$ to greater than $14 \times 10^{10}$ dynes/cm$^2$ (equivalent to 145 000 to 2 030 000 PSI, or 1 to 14 GPa). According to these data Young's modulus is higher for dull coal bands than for bright bands. It is not certain, however, to what extent this relationship is influenced by maceral composition as opposed to mineral matter content. Qualitatively, high ash coals are stronger than 'clean' (low ash) coals. Young's modulus appears to be higher parallel to bedding than perpendicular to bedding, but this may be partly a consequence of stress release when the sample is removed from the ground.

Berkowitz's (1979) data show that Young's modulus varies with coal rank, but the rank effects are not clear. Not unexpectedly, anthracites exhibit higher values of Young's modulus than bituminous coals; however, variation within the bituminous rank range is not well documented. In accordance with other 'strength' indicators, Young's modulus might be expected to decrease through the bituminous rank series, passing through a minimum for medium to low volatile bituminous coals, but there are insufficient data to document such a trend.

**Table 3.** *Values selected for parameter sensitivity study*

| Parameter | Symbol | Lower value | Base case | Upper value |
|---|---|---|---|---|
| Cleat spacing | $s$ | 5 mm | 10 mm | 20 mm |
| Initial permeability | $k$ | 100 $\mu$D | 1 mD | 10 mD |
| Young's modulus | $E$ | 145 000 PSI | 493 000 PSI | 725 000 PSI |
| Poisson's ratio | $\nu$ | 0.22 | 0.32 | 0.42 |
| Shrinkage: $\varepsilon_{max}$ | $\varepsilon_{max}$ | 1995 microstrains | 3414 microstrains | 6647 microstrains |
| Shrinkage: $P_{50}$ | $P_{50}$ | 212.1 PSI | 697.5 PSI | 1407.2 PSI |

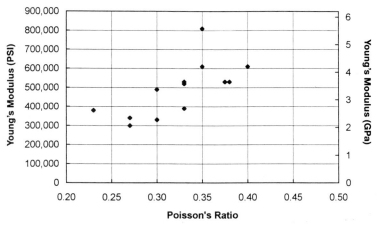

**Fig. 6.** Elastic mechanical properties of coal (after Bell & Jones 1989).

An additional problem in measuring coal strength parameters pertains to scale-dependent variability. Larger samples, which encompass many fractures and inhomogeneities, typically indicate different strength measurements from small samples (Schatz & Olszewski 1993). The strength parameters used in the present model are intended to pertain to the coal 'matrix', i.e. the coal material falling between the cracks, but it is uncertain whether or not the published values are actually measuring this.

A Young's modulus value of 521 000 PSI was measured for a sample of Fruitland coal from the COAL site in the San Juan Basin (Mavor *et al.* 1991); however, these coals tend to be fairly high in ash, which would give them a particularly high Young's modulus. Therefore, based on the range of data in Fig. 6 a slightly lower value of 493 000 PSI (3.4 GPa) was used as the base case value, with 145 000 PSI (1.0 MPa) for the lower limit and 725 000 PSI (5.0 MPa) as the upper limit.

Similar to Young's modulus, estimates of Poisson's ratio for coal are fairly variable, spanning a range from around 0.2 to 0.4 (Fig. 6). Mavor *et al.* (1991) report a value of 0.21 for a Fruitland coal from the 'COAL' site, but this unusually low value (if correct) may also be related to the high mineral matter content. (An additional problem in this regard would entail sample selection. A coherent section of core, which would tend to be selected for mechanical testing, may have remained coherent because of compositional differences, hence would not be representative of the entire seam.)

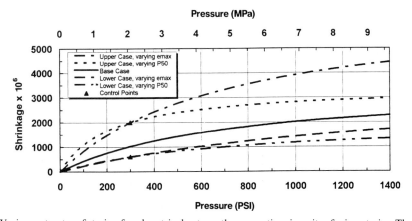

**Fig. 7.** Various estmates of strain of coal matrix due to methane sorption, in units of microstrains. The 'base case' (solid line) represents an average of the curves shown in Figs 2a, 2b, 3a and 3b. The upper and lower cases were calculated to pass through the upper and lower control points, indicated by the solid triangles.

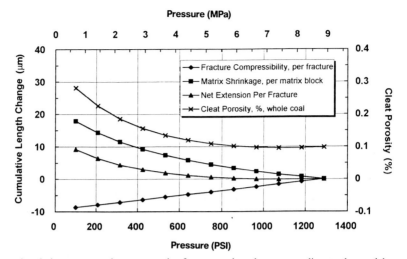

**Fig. 8.** Dimensional changes occurring as a result of pressure drawdown, according to the model.

Following Gray (1987), and in accordance with the range of values in Fig. 6, 0.32 was used as the base case for Poisson's ratio, with 0.22 and 0.42 used as the lower and upper limits.

*Matrix shrinkage parameters.* There is little basis for estimating the value of matrix shrinkage parameters for unknown coals as so few data are presently available. This notwithstanding, and subject to the limitations discussed previously, the data shown in Figs 1–3 provide some indication of reasonable range of values. For the base case (represented by the solid line in Fig. 7), an arithmetic mean was calculated for

the four sorption–strain curves shown in Figs 2 and 3. Using Fig. 1 as a guide, a value of 0.2% (at 300 PSI) was used as an upper limit and 0.06% (at 300 PSI) as a lower limit. Even if these two points are constrained, however, there is still a range of possible sorption–strain curves which could pass through these points (Fig. 7). Therefore, two sets of curves were derived. For one set of limits, $P_{50}$ was held at the base case level and $\varepsilon_{max}$ was adjusted so that the sorption–strain curves would pass through the upper and lower control points (represented by the dashed–dotted curves in Fig. 7). For the other set of limits, $\varepsilon_{max}$ was held at the base case level, and

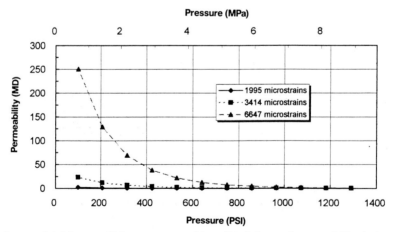

**Fig. 9.** Influence of shrinkage coefficient $\varepsilon_{max}$ on model-predicted changes in permeability during pressure drawdown. End-member examples represent the dashed–dotted curves in Fig. 7.

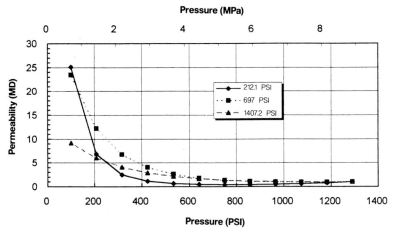

**Fig. 10.** Influence of shrinkage coefficient $P_{50}$ on model predicted changes in permeability during pressure drawdown. End-member examples represent the broken curves in Fig. 7.

$P_{50}$ was adjusted so that the sorption–strain curves would pass through the control points (represented by the broken curves in Fig. 7). The parameter values selected appear in Table 3.

### Model results

In all model runs, the coal was assumed to be initially normally pressured (freshwater gradient of 0.43 PSI/ft) at 3000 ft (914 m) depth (equivalent to 1290 PSI, or 8.90 MPa). The model results represent changes as fluid pressure decreases in 11 incremental steps to a final pressure of 100 PSI (0.69 MPa).

Using the base case values, the fractures are interpreted to increase in width by almost 10 μm

per fracture (Fig. 8). This represents the combined effect of 18 μm widening per fracture due to matrix shrinkage, plus almost 9 μm closure per fracture due to compressibility effect. Calculated in terms of porosity, this represents an increase from 0.1% initial fracture porosity up to nearly 0.3% at the end of the run.

Results of parameter sensitivity runs are depicted in Figs 9–14. The three curves on each diagram represent the base case and the two end-member cases. The permeability coefficient $\varepsilon_{max}$ is the parameter having the strongest influence on final permeability (Fig. 9). The upper case value predicts an increase in permeability from 1 to 250 mD. The base case predicts a more modest increase of only 25 mD, whereas the

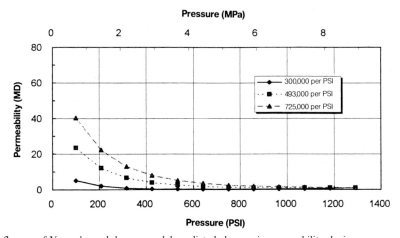

**Fig. 11.** Influence of Young's modulus on model predicted changes in permeability during pressure drawdown.

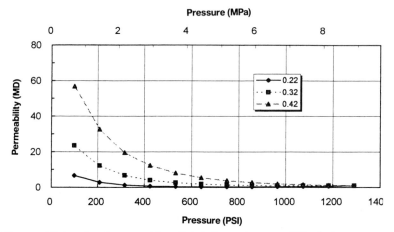

**Fig. 12.** Influence of Poisson's ratio on model predicted changes in permeability during pressure drawdown.

lower level case predicts essentially no change (i.e. matrix shrinkage is exactly offset by fracture compressibility). According to the model, coals with high matrix shrinkages have the potential of exhibiting significant increases in reservoir permeability. Notably, however, the greatest increases are observed at lower pressure, owing to the steeper slope of the sorption–shrinkage curve at low pressure.

The matrix shrinkage parameter $P_{50}$ has a more subtle effect on the model, as it affects the shape of the curve more than the total magnitude of the change (Fig. 10). The lower case value (212.1 PSI) produces little change at high pressures where the sorption–strain curve is nearly flat, but accounts for rapid change at lower pressures where the curve becomes much

steeper (cf. Fig. 7). Thus the lower case curve actually crosses the middle and upper case curves at pressures below 300 PSI (2.07 MPa).

Parameter sensitivity to Young's modulus shows that, other factors being equal, 'stronger' coals exhibit a greater increase in permeability than weaker coals (Fig. 11). This is because stiffer coals will exhibit less pore volume compression as the reservoir pressure drops. Figure 12 shows that matrix shrinkage increases with increasing Poisson's ratio. Even coals with the lower case value for Poisson's ratio (0.22) show a significant increase in predicted permeability.

Figure 13 shows that, other factors being equal, the greater the initial fracture spacing, the greater will be the predicted increase in absolute

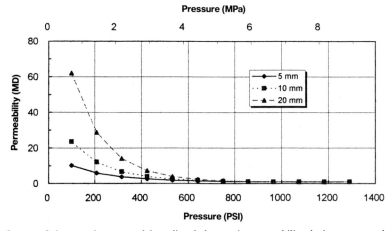

**Fig. 13.** Influence of cleat spacing on model predicted changes in permeability during pressure drawdown.

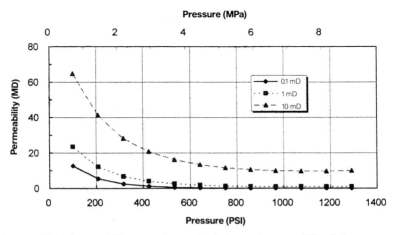

**Fig. 14.** Influence of initial permeability on model predicted changes in permeability during pressure drawdown.

permeability. This is a consequence of the exponential term in Equation (3). For any given amount of sorption strain, the amount of extension of each individual fracture will be greater if the fractures are more widely spaced. As the permeability is proportional to the cube of fracture width, it might be advantageous, from the standpoint of permeabililty change, if the strain is concentrated into a smaller number of more widely spaced fractures.

Finally, Fig. 14 depicts the influence of differences in initial permeability. In this set of curves, the initial fracture width was allowed to vary to produce [using Equation (3)] desired initial permeabilities of $100\,\mu$D, 1 mD and 10 mD. The results show that the larger the initial permeability, the greater the absolute increase in permeability. Low permeability coals, however, show a larger proportional increase than high permeability coals. The lower case curve shows a 10-fold increase in permeability, whereas the upper case curve shows only a six-fold increase.

## Discussion

This study shows that significant changes in permeability could occur during gas production due to matrix shrinkage but that rank, petrographic composition, mineral matter content and sorbate composition are each important factors controlling the magnitude of the effect. The model results show that the predicted change in permeability is strongly dependent on the matrix shrinkage coefficient and the elastic moduli. These factors are highly variable from coal to coal and so must be directly

measured to assess their impact. Thus each coal reservoir will behave differently and some effort should be made to derive estimates for each parameter as accurately as possible.

Clearly, within the constraints of the model many factors acting in combination with one another bring about the predicted changes in absolute permeability. In the parameter sensitivity study only one variable at a time was changed, but, in nature, all of these factors are interrelated and will vary simultaneously with one another. For example, Fig. 11 indicates that the permeability should increase more for coals with a higher Young's modulus; however, coals with a higher Young's modulus will tend to have a correspondingly lower matrix shrinkage coefficient as well and would probably actually exhibit a smaller increase in permeability. The results of the parameter sensitivity study suggests that there may be some optimum combination of factors which would produce the greatest net increase in permeability.

It has been generally assumed that closer cleat spacing is better for reservoir drainage, but Fig. 13 implies this may not necessarily be so. Coals with a wider fracture spacing could exhibit a greater increase in permeability during production. On the other hand, coals with a wider cleat spacing probably have lower shrinkage coefficients, so the interrelationships are complex.

This modelling approach, although subject to some limitations of its own, is potentially useful in that it may be readily incorporated into reservoir simulator design. Some effort should be made, however, to verify the validity of these theoretical calculations through field testing and, possibly, through additional laboratory tests.

The author is grateful to M. McCollum, who assisted in sorption strain experiments, and to P. M. Johnson of the University of Alabama for advice and input. Support for this work came in part through a contract from the US Department of Energy (DE-FG22-89PC89764).

# References

BELL, G. J. & JONES, A. H. 1989. Variation in mechanical strength with rank of gassy coals. *In*: *Proceedings of the 1989 Coalbed Methane Symposium*. University of Alabama, Tuscaloosa, 65–74.

BERKOWITZ, N. 1979. *An Introduction to Coal Technology*. Academic, New York

BRIGGS, H. & SINHA, R. P. 1933. Expansion and contraction of coal caused respectively by the sorption and discharge of gas. *Proceedings of the Royal Society of Edinburgh*, **53**, 48–53.

CLOSE, J. C. 1993. Natural fractures in coal. *In*: LAW, B. E. & RICE, D. D. (eds) *Hydrocarbons from Coal*. American Association of Petroleum Geologists, Studies in Geology, **38**, 119–132.

CZAPLINSKI, A. 1971. Simultaneous investigations concerning the kinetics of expansion and the kinetics of sorption in the system coal-carbon dioxide. *Archiwum Gornictwa*, **16**, 227–234.

GRAY, I. 1987. Reservoir engineering in coal seams. Part 1—the physical process of gas storage and movement in coal seams. *Reservoir Engineering (SPE)*, 28–40.

HARPALANI, S. & SHRAUFNAGEL, R. A. 1990a. Shrinkage of coal matrix with release of gas and its impact on permeability of coal. *Fuel*, **69**, 551–556.

—— & ——1990b. Influence of matrix shrinkage and compressibility on gas production from coalbed methane reservoirs. *SPE Paper 20729, Proceedings of 65th Annual Meeting, New Orleans, September 23–26 1990*. 171–179.

—— & ZHAO, X. 1989. The unusual response of coal permeability to varying gas pressure and effective stress. *In*: KHAIR, A. W. (ed.) *Rock Mechanics as a Guide for Efficient Utilization of Natural Resources. Proceedings of the 30th US Symposium*. Balkema, Rotterdam, 65–72.

JOUBERT, J. I., GREIN, C. T. & BIENSTOCK, D. 1974. Effect of moisture on the methane capacity of American coals. *Fuel*, **53**, 186–191.

LEVINE, J. R. 1993. Coalification: the evolution of coal as source rock and reservoir rock for oil and gas. *In*: LAW, B. E. & RICE, D. D. (eds) *Hydrocarbons from Coal*. American Association of Petroleum Geologists, Studies in Geology, **38**, 39–77.

—— & JOHNSON, P. W. 1992. *Permeability Changes in Coal Resulting from Gas Desorption*. Final Report, Contract No. DE-FG22–89PC89764, US Department of Energy, Pittsburgh Energy Technology Center.

MAVOR, M. J., CLOSE, J. C. & PRATT, T. J. 1991. *Summary of the Completion Optimization and Assessment Laboratory (COAL) site*. Report GRI-91/0377, Gas Research Institute, Chicago.

MOFFAT, D. H. & WEALE, K. E. 1955. Sorption by coal of methane at high pressures. *Fuel*, **34**, 449–462.

NEWMAN, P. C. 1955. Plastic flow in coal. *British Journal of Applied Physics*, **6**, 348–349.

REUCROFT, P. J. & PATEL, H. 1986. Gas-induced swelling in coal. *Fuel*, **65**, 816–820.

SCHATZ, J. F. & OLSZEWSKI, A. J. 1993. Size-scale dependence of mechanical properties: application to coalbed methane wells. *In*: *Proceedings of the 1993 International Coalbed Methane Symposium*. University of Alabama, Tuscaloosa, 667–674.

SNOW, D. T. 1968. Rock fracture spacings, openings, and porosities. *Journal of the Soil Mechanics and Foundations Division, Proceedings of the American Society of Civil Engineers*, **94**, 73–91.

THOMAS, J. JR. & DAMBERGER, H. H. 1976. *Internal Surface Area, Moisture Content, and Porosity of Illinois Coals—Variations with Rank*. Illinois State Geological Survey Circular, **493**.

YEE, D., SEIDLE, J. P. & HANSON, W. B. 1993. Gas sorption on coal and measurement of gas content. *In*: LAW, B. E. & RICE, D. D. (eds) *Hydrocarbons from Coal*. American Association of Petroleum Geologists, Studies in Geology, **38**, 203–218.

# Characterization of anthracite

M. I. DAVIDSON, R. BRYANT & D. J. A. WILLIAMS

*Department of Chemical Engineering, University of Wales Swansea,*
*Singleton Park, Swansea SA2 8PP, UK*

**Abstract:** Examination of the surface properties of anthracite provides information of assistance in the interpretation of *in situ* behaviour and use of the material as a micronized fuel and as an inexpensive support substrate for growth of micro-organisms. The characteristics of anthracite particles ($<10 \mu$m) produced by crushing and grinding larger material (obtained from Cynheidre Colliery, South Wales UK) are reported. Mineral particles, liberated from this anthracite matrix, subsequent to radiofrequency-induced plasma (RFIP) oxidation, were found to constitute 2 wt%, 0.6 vol% of that matrix. X-Ray fluorescence analysis and analytical transmission electron microscopy of these particles indicated silicon, aluminium and iron to be the predominant electropositive elements present, with calcium the only other element detected at a level greater than 5 wt% of the mineral component. X-Ray diffractometry (XRD) of this component indicated that hematite, pyrite, siderite and a variety of aluminosilicate minerals were present in the original matrix.

Mineral particles liberated by RFIP oxidation of the planar (cut) surface of anthracite were initially captured in plastic replica films and thence transferred to carbon films. Electron microscopic examination of these films indicated that mineral particles present in the anthracite matrix possessed diverse chemistries, were submicron in size and suggested their distribution throughout the carbonaceous matrix to be homogeneous.

Estimates of the surface area of $<10 \mu$m anthracite particles ($107 \pm 15 \, \text{m}^2 \, \text{g}^{-1}$) made using nitrogen adsorption/desorption (BET) analysis, were significantly greater than those predicted for non-porous particles of this size. Equilibration of the surface with nitrogen was reached $\approx 1.2 \times 10^4$ s after the imposition of changes in pressure. Hysteresis of the adsorption/desorption isotherms was pronounced, suggesting that the material contained 9% $v/v$ pores, of which 75% were estimated to be $\approx 2$ nm diameter.

Potentiometric (aqueous, acid/base, batchwise) titration of the material, at a variety of ionic strengths ($I$), indicated long equilibration times between $H^+/OH^-$ and the surface ($\approx 3 \times 10^5$ s). The dependence of the analytical surface charge density ($\sigma_a$) on pH and $I$ were found to be significant though small, with points of zero titration (pH$_{pzt}$) in the range $4 < \text{pH}_{pzt} < 7$ which increased with $I$. At pH $<$ pH $< 9$, $\sigma_a$ was virtually independent of $I$, suggesting a point of zero salt effect to be located in this region or that solute ions were unable to penetrate the surface. The range of $\sigma_a$ (based on the surface area determined by BET) was found to be small ($-7 < \sigma_a < 6 \, \mu\text{C cm}^{-2}$) and its nature consistent with that of surfaces of minerals shown and identified by electron microscopy and XRD analysis to be present at low concentration within (and exposed at the surface of) the carbonaceous matrix.

In recent years a significant interest has developed in the use of pulverized coal, a few micrometres in size, suspended in water or oils, for use as fuels. The surface properties of these coals will have a significant effect on suspension stability and rheology. These factors are likely to affect the combustion efficiency of such fuels and the surface adsorption/desorption behaviour of the matrix *in situ*. However, the surface properties of coal have received only limited attention and problems associated with the measurement of surface charge have been identified (Barrett-Gultepe *et al.* 1989). Some evidence is available which suggests that the surface potential of coals is determined by $H^+$ and $OH^-$ ions (Latiff-Ayub *et al.* 1985; Sun & Campbell 1966).

Of the various organic functional groups that could be responsible for the surface charge on coals, it has been suggested (Blom *et al.* 1957) that contributions of both carboxyl and methoxy groups to surface chemistry are of little significance in coals $>80$ wt% carbon. Phenolic and carbonyl groups appear to be more prevalent in such high-rank coals (Blom *et al.* 1957; Abdel-Baset *et al.* 1978; Hayatsu *et al.* 1983). Despite possessing a partially graphitized structure, Arnett (1986) observed the acid–base properties of anthracite to be better represented by those of bituminous coals than by those of graphite. Perhaps this suggests that the carbonaceous component of coal exerts little influence on surface electrical properties.

Increases in the concentration of metal carboxylates with increased surface oxidation have been observed for a bituminous coal whose original surface possessed phenolic acid groups

*From* Gayer, R. & Harris, I. (eds) 1996, *Coalbed Methane and Coal Geology,*
Geological Society Special Publication No 109, pp 213–225.

(Teo *et al.* 1982). The formation of carbonium ions at heterocyclic rings has been postulated as a source of sites of positive surface charge in coals (Garten & Weiss 1967).

Surface oxidative processes have been attributed to the observed increase in negative charge on coals (Celik & Somasundaran 1980). Anthracite electrodes with freshly cut surfaces have been found to possess a positive charge in water which becomes negative after prolonged exposure in air (Laskowski & Parfitt 1988). It is almost inevi-table that during the pulverization of coals they will be exposed to oxidative environments, but the magnitude of the effect that this has on the surface electrical properties of coal particles remains uncertain.

Examination of the porous nature of coals has previously been made using transmission electron microscopy (TEM) (Harris & Yust 1981) and from the examination of replicas of the surface (Pooley *et al.* 1966). Small angle X-ray scattering, despite some limiting assumptions, has indicated coal specimens to contain pores from 3 to 22 nm in size (Lin *et al.* 1978). The surface charge of many materials, e.g. silica (Michael & Williams 1984), appears to result from significant contributions made from functional groups located within a porous surface layer.

The exposure of mineral inclusions at the surfaces of coals could provide numerous small areas with a high density of functional groups (typical of electrostatically bonded materials) capable of developing a surface charge. The interface between mineral inclusions and the carbon matrix in anthracite has been examined using high resolution electron microscopy and found to consist of numerous ledges and microledges which serve to enhance exposure of the mineral surface relative to the contribution that it makes to the mass of the matrix (Little & Gronsky, 1985).

Herein the characteristics of the inorganic mineral particulate component of a Welsh anthracite and the surface area, porosity and surface charge density of the matrix in which it occurs are presented and the origin of the surface charge on anthracite particles <10 $\mu$m) discussed in relation to *in situ* behaviour.

## Experimental

### Apparatus

All glassware was cleaned in an aqueous solution of Neutracon for $8 \times 10^5$ s and exposed to ultrasound before rinsing with water and steaming for $8 \times 10^4$ s.

Conductivity measurements were made using an Autobalance Universal Bridge (Wayne Kerr Model B642) and pH measurements were made with WDA a CD330 pH meter.

### Chemicals

All reagents were of analytical-reagent grade. Water ($2 \times H_2O$) was twice distilled (to a specific conductance of <2 $\mu$S); the final distillation was made in the presence of potassium permanganate.

### Sample details

A sample of anthracite, a single cube (0.11 m³, 2 kg), was obtained, by hand, from the centre of the working face of the Big Vein seam (at a depth of $\approx$880 m) from Cynheidre Colliery, South Wales, UK, courtesy of K. P. Williams, University College Cardiff. The sample was taken remotely from intrusions of strata of other material (visble to the naked eye) from a region of $\approx$89 wt% carbon, 8 wt% volatile matter and $\approx$1 wt% moisture. The vein is surrounded by mudstones and shales.

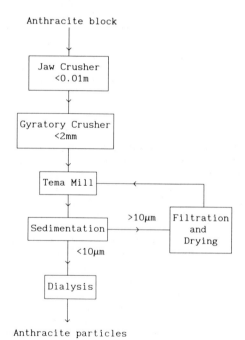

**Fig. 1.** Outline of the procedure for the comminution of anthracite.

The sample was comminuted (see Fig. 1) to <1 cm using a jaw crusher and a few pieces of material were retained. The remainder was reduced in a gyratory mill (<2 mm). The material was then processed in a Tema mill for 30 s and the product suspended in water (2 vol%). The pH of the suspension was adjusted to $\approx 9$ by the addition of NaOH (1 M) to improve suspension stability and allowed to equilibrate $(4 \times 10^5 \text{ s})$. The suspension was then thoroughly shaken and allowed to settle. A particle fraction was collected by repeated decantation and resettling.

The (>10 $\mu$m) fraction was filtered, dried and repulverized. Particles <10 $\mu$m were placed in Visking tubing, dialysed against $2 \times H_2O$ and equilibrated for $8 \times 10^5$ s. The water was then replaced and the process repeated until the conductivity of the aqueous phase remained below <5 $\mu$S at the end of the period of equilibration.

The material was then dried and stored under nitrogen. Pycnometric analysis of anthracite/ethanol suspensions indicated the density of the material to be 1500 kg m$^3$.

## Particle size analysis

Particle size measurements were made using electronic particle counting (Coulter Counter Model TAII) using 30 and 50 $\mu$m orifices. Particles were suspended in aqueous sodium chloride (1% w/v) which had been pre-filtered through capillary pore membranes (0.22 $\mu$m pore diameter). These suspensions were treated with ultrasound for $6 \times 10^2$ s before analysis.

## Mineral material

Mineral material, sufficient for elemental analysis, was liberated by oxidation from samples of the <10 $\mu$m particles using a radiofrequency-induced oxygen plasma (RFIP) at various temperatures. A small piece of anthracite (1 cm) was cut with a disc grinder to produce a flat surface which was ultrasonically cleaned and then partially oxidized, as described earlier to liberate mineral particles embedded within its immediate surface. Small areas (10 mm$^2$) were then coated with an aqueous solution of polymer which was allowed to dry and then removed (Pooley et al. 1966 and references cited therein). The resulting plastic films were carbon-coated and then immersed in water, whereupon the plastic dissolved releasing particle-bearing carbon films which were floated onto gold electron microscope grids.

## X-ray diffractometry

Samples of mineral particles were suspended in water and pipetted onto glass slides. Some preferential orientation of mineral particles probably resulted. The resulting specimens were examined using a Phillips X-ray diffractometer operated using $Co K\alpha$ radiation with a goniometer scanning at $0.25^\circ 2\theta \text{ min}^{-1}$.

## X-ray fluorescence

Specimens were prepared by mixing and compressing known masses of mineral matter with a cellulose binder. They were analysed using a Phillips X-ray fluorescence unit (Model 1400) fitted with a wavelength dispersive analyser. For the purposes of quantification, samples were compared with a variety of standard materials and mixtures thereof.

## Analytical electron microscopy

Specimens were prepared by placing a few drops of an aqueous mineral suspension on glass slides. These were allowed to dry and then carbon-coated. Sections of the carbon film were then floated on to gold transmission electron microscopy (TEM) grids (3 mm).

Mineral particles were examined using a Phillips EM400T transmission electron microscope fitted with a Kevex energy dispersive X-ray analyser and an EDAX multichannel spectrum analyser. An on-line DEC minicomputer was used for data analysis. Elements of atomic number <11 were not detected. X-Ray counts were processed to provide the oxide composition of the irradiated material using the method described by Pooley (1977). Estimates of the average composition of the specimen were made by the irradiation of large areas of the specimens which contained many particles.

The chemistries of individual particles were examined by selective irradiation using a focused electron beam.

## Scanning electron microscopy.

Particles of anthracite (<10 $\mu$m) were suspended in ethanol and aliquots thereof placed directly on polished brass stubs and allowed to dry in a dust-free atmosphere. These were coated with gold and examined using a JEOL 120C scanning electron microscope.

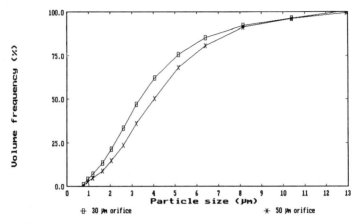

**Fig. 2.** Particle size distributions of comminuted anthracite obtained by electronic particle counting.

### Surface area and porosity measurement

Particulate surface areas were measured by nitrogen adsorption/desorption and the BET method (Brunauer *et al.* 1938) using a standard apparatus (British Standard 1969). Helium was used to evaluate the dead space. Porosity and pore volume distributions were obtained from adsorption/desorption isotherms using the method of Cranston & Inkley (1957).

### Potentiometric titration

Potentiometric titrations were performed using a modified version of the procedure described by Uehara & Gillman (1981). Batches of particles ($<10\,\mu$m 450 mg) were suspended in electrolyte (15-ml) and treated with ultrasound for 1 h and left to equilibrate for two days. Acid or base was then added using a micrometer syringe. Suspensions were purged with carbon dioxide-free nitrogen and left to equilibrate for four days at 25°C after which the pH of the supernantant liquid was measured. Particle-free (blank) solutions were prepared and treated in an identical manner to the corresponding suspensions. Data were processed in the manner described by James & Parks (1980) to produce profiles of the dependency of analytical or titratable surface charge ($\sigma_a$) on pH.

### Results and discussion

#### Particle size

Particle size (volume) distributions, obtained by electronic particle counting (Fig. 2) and electron microscopy, possess mean diameters of 4.5–5.0

and 3.1 $\mu$m, respectively. It is probable that the latter value reflects non-representative sampling of material for SEM analysis. In view of the relatively large number of particles sampled during electronic particle counting, the former

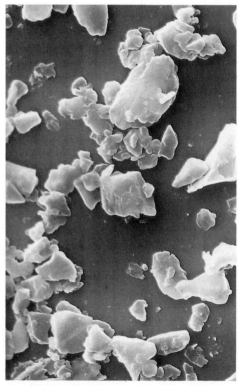

**Fig. 3.** Scanning electron micrograph of comminuted anthracite particles.

value is the more reliable. Particles in this range are typical of those present in micronized fuels. The particles are approximately isometric and angular, with smooth surfaces devoid of macroscopic surface features (see Fig. 3). Assuming the particles to be smooth and spherical, the maximum specific (geometric) surface area is $\approx 2\,m^2\,g^{-1}$.

## X-Ray diffraction

The qualitative features of the X-ray diffraction patterns for the mineral component of this anthracite, obtained by ashing specimens at 150 and 450°C (Table 1), indicate the possible presence of a variety of mineral oxides and sulphides. The absence of a peak at 16.8° $2\theta$ in the data for the ash obtained at 450°C is indicative of the decomposition of lepidocrocite, which is present in ash obtained at 150°C. The mineral responsible for the peak at 11.2° $2\theta$ (450°C ash) has not been unequivocally identified. Peaks indicative of the presence of hematite are present in data for the 450°C ash, but absent from those of the 150°C ash. Data for this low

temperature ash are indicative of the presence of siderite and pyrite. There is no evidence to suggest that these minerals were present in the high temperature ash. It appears that these minerals decompose between 150 and 450°C, but are constituents of this particular anthracite. Many of these minerals have been previously identified as being common constituents of a variety of coals (Renton 1982).

## Elemental analysis

The chemical composition of the mineral particles produced by X-ray fluorescence (XRF) analysis of ash (Table 2) indicate silicon, aluminium and iron to be the dominant electropositive constituent elements—with calcium and sodium present in low concentrations and no significant levels of other elements (<0.1 wt%). Data obtained using analytical electron microscopic simultaneous irradiation of many mineral particles (<1 μm diameter) indicate a composition broadly similar to that obtained by XRF, but suggesting that the major elements may be distributed unevenly across the size range.

Table 1. Summary of features of X-ray diffraction patterns of mineral ash of anthracite

| Angle (degrees) $2\theta$ | d-spacings (Å) 150°C | 450°C | Possible minerals |
|---|---|---|---|
| 11.20 | | 7.89 | |
| 13.45 | 7.64 | | Gypsum |
| 14.35 | 7.17 | | Kaolinite |
| 16.8 | 6.13 | | Lepidocrocite |
| 20.50 | | 4.33 | Gypsum, quartz |
| 23.9 | | 3.72 | Hematite |
| 24.15 | 4.28 | | Gypsum, quartz |
| 25.3 | | 3.52 | Kaolinite, siderite |
| 25.9 | 3.99 | | Diaspore |
| 26.4 | | 3.37 | Lepidocrocite, quartz |
| 27.15 | 3.81 | | |
| 28.9 | 3.59 | | Kaolinite, quartz |
| 30.6 | | 2.92 | Ankerite, anhydrite |
| 30.95 | 3.35 | | Lepidocrocite, quartz |
| 33.00 | | 2.71 | Pyrite, hematite |
| 33.90 | 3.07 | | Gypsum |
| 35.4 | | 2.53 | Hematite |
| 35.95 | 2.90 | | Ankerite |
| 37.4 | 2.79 | | Siderite |
| 38.6 | 2.71 | | Pyrite |
| 43.3 | 2.43 | | Lepidocrocite, pyrite |
| 44.75 | 2.35 | | Kaolinite, diaspore, siderite |
| 47.65 | 2.23 | | Pyrite |
| 49.7 | 2.13 | | Diaspore, siderite |
| 55.7 | 1.92 | | Quartz, pyrite, siderite |

Table 2. Elemental oxide composition of anthracite ash

| Element | Oxide wt%: Method | |
|---|---|---|
| | XRF | EDAX |
| Si | 36.84 | 25.83 |
| Al | 25.15 | 31.03 |
| Fe | 29.33 | 34.99 |
| Mg | 1.97 | 1.71 |
| Ca | 6.20 | 1.10 |
| Na | 1.65 | 1.30 |
| K | 0.29 | 0.73 |
| Ti | 0.92 | 2.03 |
| P | 0.09 | |
| Mn | 0.09 | 0.00 |
| Ba | 0.52 | |
| Rb | 0.00 | |
| Sr | 0.01 | |
| Y | 0.00 | |
| Nb | 0.00 | |
| Zr | 0.01 | |
| Cr | 0.05 | |
| V | 0.02 | |
| As | 0.06 | |
| Pb | 0.04 | |
| Zn | 0.07 | |
| Cu | 0.26 | 1.22 |
| Co | 0.02 | |
| Ni | 0.06 | 0.00 |
| Nd | 0.05 | |

## Chemistries of individual particles

Systematic analysis of mineral particles by TEM provided X-ray spectra that may be broadly described in 12 categories (Table 3). Despite the absence of information for elements of atomic number <11, many of these spectra indicate chemistries typically produced by the alumino-silicate and iron minerals identified as being present in the material using X-ray diffraction (XRD). These minerals are typical of those previously found in other coals from a variety of locations (Hsieh & Wert 1981).

The most frequently encountered particles were those with chemistries typical of alumina, kaolinite, pyrite and siderite. These were assumed to be the major constituent mineral. The composition of ash (as elemental oxides)

**Table 3.** *Individual analysis of mineral particles*

| Relative abundance of elements | Frequency of observation (%) | Probable minerals |
|---|---|---|
| | 25 | Alumina |
| | 22 | Pyrite |
| | 6 | Mullite |
| | 1 | |
| | 9 | |
| | 11 | Kaolinite |
| | 8 | Na-kaolinite |
| | 6 | Mica |
| | 8 | Siderite/lepidocrocite |
| | 1 | Rutile |
| | 1 | Quartz |
| | 2 | Copper mineral |

Na Mg Al Si S K Ti Cu Fe

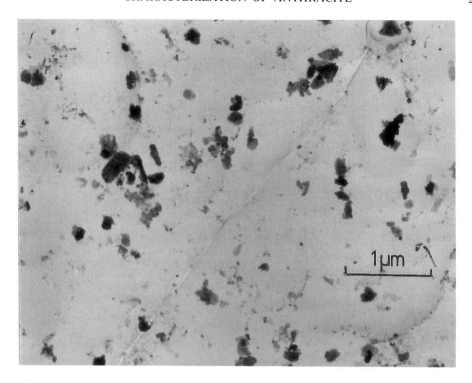

**Fig. 4.** Transmission electron micrograph of mineral particles liberated on oxidation of anthracite particles.

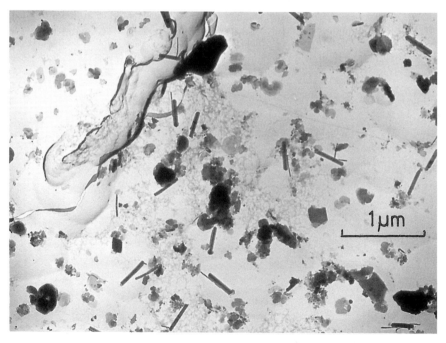

**Fig. 5.** Transmission electron micrograph of mineral particles recovered from the replica of an oxidized anthracite surface.

and those of these major mineral constituents were used to provide an estimate of specimen mineralogy (Pooley 1977), from which the average specific volume of the mineral material in anthracite was obtained.

Mineralogical analysis suggests the mineral material to be approximately 50% kaolinite, 29% siderite/lepidocrocite, 16% pyrite and 5% alumina, with an average specific gravity of 3.3. The contribution that this makes to the volume of the anthracite is $\approx 0.64\%$.

## Mineral particle morphology

Electron micrographs suggest the presence of particles both fibrous and platy in habit (Fig. 4), with dimensions in the sub-micrometre range. Images obtained from replicas of the oxidized surface of anthracite (Fig. 5) suggest that (on the scale of several micrometres) mineral particles are uniformly distributed throughout the carbonaceous matrix.

## Surface area and porosity measurement

Analysis of the surface area of this anthracite was a protracted affair ($>24$ h). Equilibrium was reached relatively slowly ($\approx 3$ h) after each change in the pressure of nitrogen during adsorption/desorption, suggesting the presence of pores of a size capable of retarding the transport of nitrogen molecules. The surface area was found to be $106 \pm 15 \, \text{m}^2 \, \text{g}^{-1}$, which is

**Table 4.** *Specific surface areas of anthracites obtained using the BET method*

| Particle size ($\mu$m) | Adsorbate molecule | Temperature (K) | Surface area ($\text{m}^2 \, \text{g}^{-1}$) | Source |
|---|---|---|---|---|
| <246 | $CO_2$ | 195 | 175 | * |
| <246 | $N_2$ | | 11 | * |
| 208–1168 | Methanol | 293 | 122 | † |
| 208–1168 | $CO_2$ | 195 | 165 | † |
| 208–1168 | $CO_2$ | 298 | 234 | † |
| 210–420 | $N_2$ | 77 | 6 | ‡ |
| 210–420 | $CO_2$ | 298 | 426 | ‡ |
| 210–420 | $N_2$ | 77 | 7 | ‡ |
| 210–420 | $CO_2$ | 298 | 408 | ‡ |
| 210–420 | $N_2$ | 77 | <1 | ‡ |
| 210–420 | $CO_2$ | 298 | 231 | ‡ |
| 38–53 | $CO_2$ | 195 | 170 | § |
| <10 | $N_2$ | 77 | 107 | ¶ |

\* Walker & Geller (1956).
† Spitzer & Ulicky (1976).
‡ Gan *et al.* (1972).
§ Ramsey (1965).
¶ Present work.

nearly two orders of magnitude larger than the geometrical surface area and much larger than previous measurements made on anthracites using $N_2$ (Table 4). This specific surface area is, however, of a similar order to measurements obtained from samples composed of much larger particles using carbon dioxide or methanol as adsorbates (Table 4). This suggests that any

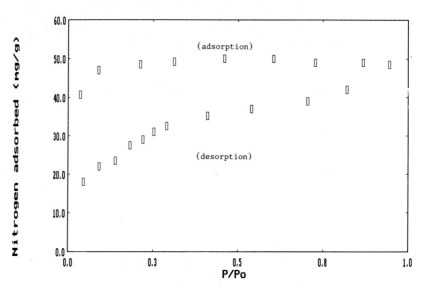

**Fig. 6.** Adsorption/desorption isotherm of nitrogen on the surface of anthracite ($<10 \, \mu$m).

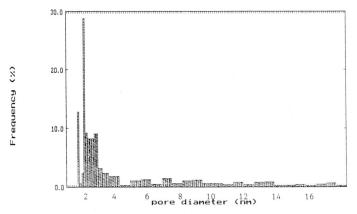

**Fig. 7.** Frequency distributions of pore volume as a function of pore diameter for anthracite particles ($<10\,\mu$m).

system of pores which provides for such an estimate of surface area in coals may be inherent in the material rather than simply an artifact of comminution, but that these pores may be more readily probed by nitrogen after comminution. Nitrogen does not appear to be a suitably labile adsorbate for coarse materials, where hindered transport leads to prolonged equilibration.

A typical isotherm for $<10\,\mu$m anthracite particles possess noticeable hysteresis (Fig. 6), further suggesting the surface to be highly porous. The isotherm can be classified as type 1 with type H4 hysteresis (Gregg & Sing 1982). The volume concentration of micropores estimated from the uptake of nitrogen at a relative pressure of 0.95 was found to be 8.7% and is of a similar order to that of King & Wilkins (1944).

Frequency distributions of pore volume and surface area (as a function of pore diameter), obtained from discretization of the hysteresis loop, after the method of Cranston & Inkley (1957) (Figs 7 and 8), possess modal values of $\approx$2 nm and suggest that 75% of the pore volume is contributed by pores with diameters less than this and that the surface area is dominated by that contributed by pores $<3$ nm in diameter. Surface features of these dimensions (2–20 nm) have been observed on specimens of vitrinite (Harris & Yust 1981) using TEM and Pooley et al. 1966) have suggested anthracite to possess a micellular structure composed of units 30–80 nm in diameter. Small-angle X-ray scattering indicates coals to possess average micropore diameters of 3 nm and those of mesopores to be 22 nm. Many American coals appear to possess sharp unimodal pore size distributions with a similar volume contributed by pores of $\approx$3 nm (Gan et al. 1972). The diameter of a spherical particle of equivalent density to this Welsh anthracite that would provide the same

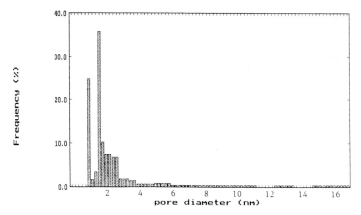

**Fig. 8.** Frequency distributions of surface area as a function of pore diameter for anthracite particles ($<10\,\mu$m).

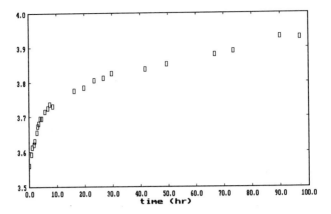

**Fig. 9.** Response of anthracite to a change of pH of supporting electrolyte.

surface area is ≈40 nm, well within the scale of the micellular structure observed by Pooley *et al.* (1966). The porous properties of anthracite could therefore be intimately associated with the structural fabric of a matrix which possesses inherent weaknesses at micellular boundaries, which may suffer disruption by expulsion of gas and provide pathways for (albeit restricted) mass transfer of gas to regions of lower chemical potential or pressure.

### Surface charge

The response of suspensions of anthracite to addition of acid and base was found to be slow and equilibrium was effectively obtained after 100 h (Fig. 9), suggesting that the transport and reaction of protons and hydroxyl ions was

occurring in a highly resistive porous medium in which they are consumed. Titration curves for anthracite in various concentrations ($I$) of NaCl and KCl (Figs 10 and 11) suggest that the relationship between analytical surface charge density ($\sigma_a$), $I$ and pH is complex.

The titrations curves obtained at various values of $I$ lie parallel at pH < 7, with a tendency to converge at pH < 7. At low pH the curves for both electrolytes diverge and those of NaCl (but not KCl) also diverge at high pH. The overall range in $\sigma_a$ over the pH range 3–9 is of the order of −7 to +6 $\mu$C cm$^{-2}$ and is reasonably low in comparison with oxide minerals such as silica and alumina and mixture thereof which exhibit charge densities of −50 to +50 $\mu$C cm$^{-2}$ (Schwarz *et al.* 1984). However, it is of a same order as a non-porous illitic clay of similar specific surface area (Beene *et al.* 1991). Protons and hydroxyl ions, by virtue of

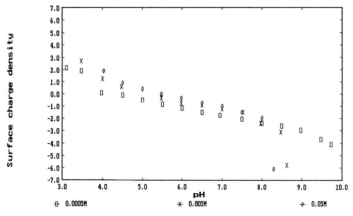

**Fig. 10.** Analytical surface charge of anthracite as a function of pH in various solutions of NaCl.

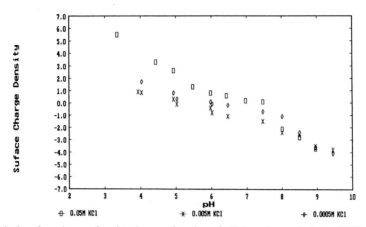

**Fig. 11.** Analytical surface charge of anthracite as a function of pH in various solutions of KCl.

their small size, may gain access, albeit slowly, to surfaces that cannot be probed by molecules such as nitrogen. The surface charge densities reported herein are therefore probably overestimates. Barrett-Gultepe *et al.* (1989) have estimated a surface charge density for an unspecified coal of $800\,\mu C\,cm^{-2}$ (in $0.2\,M\,KBr$ at pH 8), which is an order of magnitude above that capable of developing on a hexagonally close-packed oxide surface and is indicative of the use of an highly inappropriate estimate of specific surface area.

The surface charge on anthracite may arise from interactions between (i) polar functional groups, (ii) sites on the hydrocarbon skeleton and (iii) inorganic mineral impurities with solution cations and anions. The increase in $\sigma_a$ at low pH ($<5$) could be due to protonation of phenolic groups or of amphoteric metal oxide/hydroxide groups. Mild oxidation of the anthracite surface has been shown to increase the number of acidic sites available and hence reduce the positive surface charge (Laskowski & Parfitt 1988) and consequent acid shifts of isoelectric points have been observed (Harris & Yust 1981).

In the neutral to alkaline region (pH < 7) the convergence of the titration curves obtained using various concentrations of electrolyte suggests that the surface is relatively insensitive to the concentration of electrolyte. This could be due to either a point of zero charge in this region or to exclusion of such ions from pores on the basis of some effective physicochemical size. The noticeable divergence of the titration curves, obtained using NaCl at pH 8.5, which indicates an increase in $-\sigma_a$ with $(I)_{pH}$, suggests that (compared with $K^+$) $Na^+$ ions are more able to screen charged surface sites that deprotonate near this pH.

The profiles of $\sigma_a$ with pH are similar to those produced by mixtures of acidic and basic oxides (such as silica and alumina) previously studied by Schwartz *et al.* (1984), except that the range of $\sigma_a$ for their mixtures is $\approx 6$ times larger. The only other oxides likely to be present in anthracite either before or after exposure to oxidizing environments are those of iron. Such oxides do not, in general, possess significant net negative surface charge at pH < 5 and their presence would therefore tend to neutralize the acid produced by the deprotonation of more acidic oxides. It is therefore possible that the various mixtures of Fe, Si and Al oxides present in anthracite contribute to a significant proportion of its electrically active sites. If the maximum site density of $80\,\mu C\,cm^{-2}$ is assumed for the mineral components and these contribute to the matrix surface in proportion to their volume fraction, then the carbonaceous matrix of anthracite needs to contribute surface charge of low density (consistent with the hydrocarbon nature of the matrix) to obtain the observed magnitudes of $\sigma_a$.

An assessment of the influence of the inorganic mineral component on the surface charge of anthracite is complicated by both the chemical complexity of that component and a precise knowledge of its surface distribution. Although the distribution of mineral material, in this anthracite, appears to be homogeneous at the scale of $\approx 1\,\mu m$, this scale is large compared with the estimated size of the pores. The formation of pores from a fracturing of the surface could be assisted by a local mechanical weakness caused by the presence of mineral particles. If this were so then it is probable that the contribution to the surface area made by mineral material would be disproportionately large compared with that estimated from its volume concentration. Whether or not the mineral component exerts an influence on matrix permeability therefore remains an open question. Comparison of the densities of graphite $(2350 \, kg \, m^{-3})$ and this anthracite $(1504 \, kg \, m^{-3})$ suggests that the carbon in the latter could accommodate an extensive pores network, of which only a fraction $(\approx 8\%)$ is detected by $N_2$, and therefore for many purposes can be considered to be impermeable. Unfortunately, no information is available as to the nature and amount of any gas released from the sample during comminution.

Materials such as anthracite may exert a significant influence on the distribution of solutes present in waters percolating through coalbeds. These interactions will be rather slow as access to the buffering capacity of electrically active sites in such a porous medium is likely to be diffusive in nature. The electrical circuit analogue of this material may be described as a large capacitor in series with a high resistance. If the amphoteric properties of anthracite, exhibited in the laboratory, are operative *in situ*, then the material will tend to bring the pH of percolating waters into the neutral range. The exclusion of solutes from pores may lead to increases in their relative concentration in regions close to pore entrances. Surface electrical forces in macroscopic pores and flowpaths which originate locally or extend from fine pores into these regions may also influence the release and capture of colloidal particles which may modify the geometry of local flow-paths and consequently their resistance to mass transfer processes.

This present work suggests that appropriate methodology be used in the characterization of anthracite to provide data within physically realistic bounds and that encompass equilibrium conditions. The relatively slow equilibrium attained for adsorption of both nitrogen and $H^+/OH^-$ ions suggests a surface whose response to changes in the environment is likely to be protracted. Comminution might not significantly alter the particulate surface, but might simply serve to facilitate easier access to its fine structure.

## Conclusions

Estimates of the specific surface area (with respect to $N_2$) of this anthracite are $\approx 100 \, m^2 \, g^{-1}$ and larger than those reported for samples composed of larger particles.

This paper reports the first estimates of $\sigma_a$ for anthracite, which indicate that the density of surface sites is much lower than those on mineral oxide particles, but is physically realistic in relation to the surface chemistry and composition of the material. Exposed mineral surfaces could provide a highly significant contribution to the overall surface charge of anthracite.

## References

ABDEL-BASET, Z., GIVEN, P. H. & YARZAB, R. F. 1978. Re-examination of the phenolic hydroxy contents of coals, *Fuel*, **57**, 95–99.

ARNETT, E. M., 1986. *Acid–base Properties of Coals and Other Solids*. Duke University, Durham, DOE/PC/80521-T4.

BEENE, G. M., BRYANT, R. & WILLIAMS, D. J. A. 1991. Electrochemical properties of illites. *Journal of Colloids and Interfacial Science*, **147**, 358–369.

BLOM, L., EDELHAUSEN, L. & VAN KREVELEN, D. W. 1957. Chemical structure and properties of coal XVIII—Oxygen groups in coal and related properties. *Fuel*, **36**, 135–153.

BARRETT-GULTEPE, M. A., GULTEPE, M. E. McCARTHY, J. L. & YEAGER, E. B. 1989. A study of steric stability of coal–water dispersions by ultrasonic absorption and velocity measurement. *Journal of Colloids and Interfacial Science*, **132**, 144–160.

BRITISH STANDARD 1969. *B. S. 4369*. British Standards Institute, London.

BRUNAUER, S., EMMETT, P. H. & TELLER, E. 1938. Adsorption of gases in multimolecular layers. *Journal of the American Chemical Society*, **60**, 309–319.

CELIK, M. S. & SOMASUNDARAN, P. 1980. Effect of pretreatments on flotation and electrokinetic properties of coal. *Colloids and Surfaces*, **1**, 121–125.

CRANSTON, R. W. & INKLEY, F. A. 1957. The determination of pore structures from nitrogen adsorption isotherms. *Advances in Catalysis*, **9**, 143–154.

GAN, H., NANDI S. P. & WALKER, P. L. JR 1972. Nature of the porosity in american coals. *Fuel*, **51**, 272–277.

GARTEN, V. A. & WEISS, D. E. 1957. A new interpretation of the acidic and basic structures in carbons II. The chromene–carbonium ion couple in carbon, *Australian Journal of Chemistry*, **10**, 309–328.

GREGG, S. J. & SING, K. S. W. 1982. *Adsorption, Surfaces Area and Porosity*, 2nd edn. Academic, London.

HARRIS, L. A. & YUST, C. S. 1981. The ultrafine structure of coal determined by electron microscopy. *In*: GORBATY, M. L. & OUCHI, K. (eds) *Coal Structure*. American Chemical Society, Advances in Chem. Series, **192**.

HAYATSU, R., SCOTT, R. G., WINANS, R. E., MCBETH, R. L. & BOTTO, R. E. 1983. Organic structures in coals: a new oxidative depolymerization technique. *In: Proceedings, International Conference on Coal Science*, Pittsburgh, Pennsylvania, 322–325.

HSIEH, K. C. & WERT, C. A. 1981. Sulphide crystals in coal. *Materials Science and Engineering*, **50**, 117–125.

KING, J. G. & WILKINS, E. T. 1944. The internal structure of coal. *In*: *Proceedings, Conference on Ultrafine Coals and Cokes*. British Coal Utilisation Reserach Association, 46–56.

JAMES, R. O. &. PARKS, G. A. 1980. Characterisation of aqueous colloids by their electrical double layers and intrinsic surface chemical properties. *In*: MATIJEVIC, E. (ed.) *Surface and Colloid Science*, Vol. 12. Plenum, New York, 119–216.

LASKOWSKI, J. S. & PARFITT, G. D. 1988. Electrokinetics of coal–water suspensions. *In*: BOTSARIS, G. D. & GLAZMAN, Y. M. (eds) *Interfacial Phenomena in Coal Technology*, Marcel Dekker, New York.

LATIFF-AYUB, A., ANAKAWA, K., AL TAWEEL, A. M. & KWAK, J. C. T. 1985. Surface properties of coal fines in water. II. Isotherms, electrokinetics and chain length dependence for cationic surfactant adsorption. *Coal Preparation*, **2**, 1–17.

LIN, J. S., HENDRICKS, R. W., HARRIS, L. A. & YUST, C. S. 1978 Microporosity and micromineralogy of vitrinite in a bituminous coal. *Journal of Applied Crystallography*, **11**, 621–625.

LITTLE, J. A. & GRONSKY, R. 1985. The microstructure of high-rank coals at lattice resolution. *Journal of Microscopy*, **138**, 79–89.

MICHAEL, H. Ll. & WILLIAMS, D. J. A. 1984. Electrochemical properties of quartz. *Journal of Electroanalytical Chemistry*, **179**, 131–139.

POOLEY, F. D., 1977 The use of an analytical electron microscope in the analysis of mineral dusts. *Philosophical Transactions of the Royal Society of London*, **286**, 625–638.

——, PLATT, J. & HENDERSON, W. J. 1966. Surface microstructute of some vitrains of British coals. *Nature (London)*, **210**, 179–181.

RAMSEY, J. W., 1965. Calculation of surface area of anthracite from carbon dioxide adsorption data. *Fuel*, **44**, 277–284.

RENTON, J. J. 1982. Mineral matter in coal. *In*: MEYERS, R. A. (ed) *Coal Structure*. Academic Press, New York.

SPITZER, Z. & ULICKY, L. 1976. Specific surfaces of coals determined by small-angle X-ray scattering and by adsorption of methanol. *Fuel*, **55**, 21–24.

SUN, S. C. & CAMPBELL, J. A. L. 1966. Anthracite lithology and electrokinetic behaviour. *Advances in Chemistry Series*, **55**, 363–375.

SCHWARZ, J. A., DRISCOLL, C. T. & BHANOT, A. K. 1984. The zero point of charge of silica–alumina oxide suspensions. *Journal of Colloids and Interfacial Science*, **97**, 55–61.

TEO, K. C., FINORA, S. & LEJA, J. 1982. Oxidation states in surface and buried coal from the Fording River deposit. *Fuel*, **61**, 71–76.

UEHARA, G. & GILLMAN, G. 1981. *The Mineralogy, Chemistry and Physics of Tropical Soils with Variable Charge Clays*. Series No. 4, Westview Tropical Agriculture.

WALKER, P. L. JR & GELLER, I. 1956. Change in surface area of anthracite with heat treatment. *Nature*, **178**, 1001.

# Measurement of gas permeability of coal and clastic sedimentary rocks under triaxial stress conditions

PAVEL KONEČNÝ & ALENA KOŽUŠNÍKOVÁ

*Czech Academy of Sciences, Institute of Geonics, Studentská 1768, 708 00 Ostrava-Poruba, Czech Republic*

**Abstract:** A method of gas permeability for coal and sedimentary rocks under triaxial stress conditions is outlined. The transmissibility of compressed air through these rocks is shown to vary markedly, depending on both the rock lithology and containment pressure.

The Laboratory of Petrology and Rock Properties within the Institute of Geonics, Ostrava-Poruba deals with the measurement and evaluation of the physico-mechanical properties of rocks. The laboratory has been investigating the potential exploitation of coalbed methane (CBM) from coal seams of the Upper Silesian Coal Basin.

To understand the variations of gas permeability within different rock types, it is necessary to understand the physical properties of the material tested and how these influence gas permeability. Influencing factors include: texture; bedding; jointing; porosity; temperature; and compression (Hatala & Trancík 1983; Duracan & Edwards 1986). Conventional laboratory measurements of gas permeability are carried out at atmospheric pressure and temperature; however, at the depth where sedimentary rocks of the coal-bearing Carboniferous occur the stress is often tens of megapascals (for rocks with an average weight of $2500 \, \text{kg m}^{-3}$, the stress at 1000 m is 25 MPa).

## Measurement method

The laboratory is equipped to measure gas permeability under triaxial stress conditions. The triaxial (Karman's) cell is made by UNIPRESS of Poland and allows compressed gases to be emitted to the test sample. The equipment can also measure the rate of gas migration through the sample (Fig. 1).

The confining pressure of the triaxial cell is supplied by hydraulically pressured oil, whereas the axial pressure is created by a ZWICK 1494 computer-controlled mechanical press (Konečný & Poláček 1989). The equipment is able to measure the following ranges: sample diameter, 41–43 mm; sample height, 80–90 mm; confining pressure in triaxial cell, 0–50 MPa; axial force, 0–500 kN; and gas pressure, 1–1.5 Mpa.

The experiment was conducted on medium-grained sandstones and low-volatile bituminous coal from the Czech part of the Upper Silesian Coal Basin. The bedding planes were oriented perpendicular to the direction of gas flow. The sample moisture contents were calculated to be approximately 0.5%. The gas pressure (in this experiment compressed air) was maintained at a constant 1.5 MPa throughout the experiment. At the start of the experiment the axial stress was 2.0 MPa; however, this increased throughout the experiment in parallel with increases in the confining pressure. The confining pressure varied from 3 to 50 MPa in 3 or 5 MPa steps (according to the required measurement frequency). The resultant loading of the test body was essentially hydrostatic. Gas which had been transmitted through the test body (i.e. the rate of flow) was measured volumetrically.

The calculation of gas permeability is based on several theories which describe gas flow (Somerton *et al.* 1975; Harpalani & McPherson 1985; 1988; Majewska & Marcak 1989). With reference to the work of Šatěrba *et al.* (1989), laminar flow is assumed for these experiments, and thus for the calculation of permeability coefficient Darcy's law can be applied

$$k = \frac{2Q_0 \mu L p_0}{S(p_1^2 - p_2^2)}$$

where $k$ is the coefficient of gas permeability of the rock ($\text{m}^2$); $Q_0$ is the volumetric rate of flow at reference pressure $p_0$ ($\text{m}^3 \text{s}^{-1}$); $\mu$ is the fluid viscosity (Pa s); $L$ is the height of sample (m); $p_0$ is the reference pressure – in this instance it is equal to $p_2$ (Pa); $S$ is the cross-sectional flow area ($\text{m}^2$); and $p_1$, $p_2$ is the upstream and downstream pressures of gas (Pa).

## Results and conclusions

Gas permeability decreases with and increase in confining pressure. The development of the

*From* Gayer, R. & Harris, I. (eds) 1996, *Coalbed Methane and Coal Geology*, Geological Society Special Publication No 109, pp 227–229.

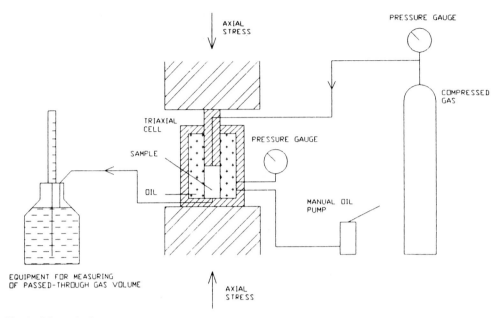

**Fig. 1.** Schematic diagram of permeability apparatus.

dependence of gas permeability on confining pressure is illustrated in Figs 2 and 3.

When comparing the Fig. 2 with Fig. 3, it can be seen that the gas permeability of coal at low confining pressure is several times greater than

the gas permeability of sandstone. With increasing confining pressure, however, the gas permeability of coal decreases more rapidly than that of sandstone, such that at pressure levels above 25 MPa the gas permeability of coal is already

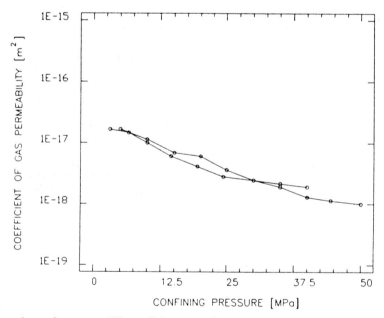

**Fig. 2.** Dependence of gas permeability coefficient on confining pressure (samples of sandstone).

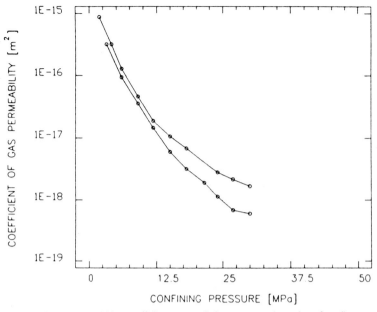

**Fig. 3.** Dependence of gas permeability coefficient on confining pressure (samples of coal).

lower than that of sandstone. This phenomenon is the subject of additional detailed research. The importance of this research, especially for CBM, is that the effective permeability of coal decreases markedly with increased confining pressures.

This paper has been supported financially by the Grant Agency of Czech Republic (reg. grant No. 105/93/2409) and the Czech–American Scientific and Technical Program (Project No. 930 65).

# References

DURACAN, S. & EDWARDS, J.S. 1986. The effects of stress and fracturing on permeability of coal. *Mining Science and Technology*, **3**, 205–216.

HARPALANI, S. & McPHERSON, M. J. 1985. Effects of gas pressure on permeability of coal. *In: 2nd US Mine Ventilation Symposium, Reno, NV*, 369–375.

——1988. An experimental investigation to evaluate gas flow characteristics of coal. *In: Proceedings of the Fourth International Ventilation Congress, Brisbane, Queensland*, 175–182.

HATALA, J. & TRANČÍK, P. 1983. *Mechanika Hornin a Masivu (Mechanics of Rocks and Rock Massif)*. Scripta VST BF, Kosice.

KONEČNÝ, P. & POLÁČEK, J. 1989. *Triaxialni Mereni Pevnostnich a Pretvarnycg Vlastnosti Hornin (Triaxial Measurement of Stress–Strain Properties of Rocks)*. Sbornik Stabilita svahu, VUHU Most.

MAJEWSKA, Z. & MARCAK, H. 1989. The relationship between acoustic emission and permeability of rock under stress. *Mining Science and Technology*, **9**, 169–179.

SOMERTON, W. H., SOYLEMEZOGLU, I. M. & DUDLEY, R. C. 1975. Effect of stress on permeability of coal. *International Journal of Rock Mechanics, Mining Science & Geomechanical Abstracts*, **12**, 129–145.

ŠATEŘBA, L. et al. 1989. *Vyzkum Specialnich Vlastnosti Hornin k Prohloubeni Analyzy (Detailed Research of Special Properties of Rocks)*. PHP pro Dul Slany. Technicka zprava uk. c.633 800. VUGI Brno.

# Relationship between the hydrogen content of coal and the lithological characteristics of rocks overlying the coal seam

## A. KOŽUŠNÍKOVÁ

*Academy of Sciences of the Czech Republic, Institute of Geonics, Studentská 1768, 708 00 Ostrava–Poruba, Czech Republic*

**Abstract:** The Prokop and 38a seams were analysed during exploration of the Frenštát-East and Frenštát-West coal fields (the southern part of the Ostrava–Karviná Coalfield in the Czech Republic). The analyses were designed to show possible relationships between hydrogen content $H$(daf) of coal and other parameters, especially the character of the strata overlying the coal seams. Where originally very porous sediments, such as those of fluvial channel facies, form the roof of seams, a pronounced influence on the elementary composition of coal may be demonstrated. In particular, a definite reduction in hydrogen content may occur. In contrast, where strata of low porosity, such as the siltstones and mudstones of swamp or oxbow lake facies, form the roof, the decrease in hydrogen in the coal seams is not so pronounced. It is argued that the retention of coalbed methane in a coal seam will be influenced by the porosity of the roof rock.

The Ostrava–Karviná Coalfield represents the southern part of the Upper Silesian Coal Basin and is located in the northeast of the Czech Republic. In the Ostrava–Karviná coalfield the coal-bearing strata can be separated into the Ostrava and Karviná formations (Dopita & Kumpera 1993). The paralic Ostrava Formation (Namurian A) is up to 3000 m thick and

it contains more than 170 seams. The continental Karviná Formation (Namurian B, C and Westphalian A), deposited after tectonic inversion, is up to 1000 m thick. It contains up to 90 coal seams. The formation is subdivided into the Saddle, Suchá and Doubrava members. The coal seams mainly comprise bituminous coal. At the lower boundary of the limnic Karviná

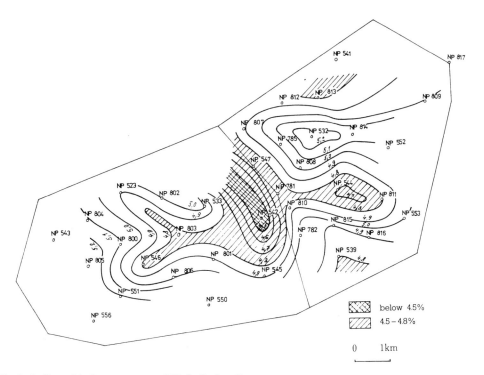

**Fig. 1.** Isolines of hydrogen content, $H$(daf), Prokop Seam.

*From* Gayer, R. & Harris, I. (eds) 1996, *Coalbed Methane and Coal Geology*, Geological Society Special Publication No 109, pp 231–236.

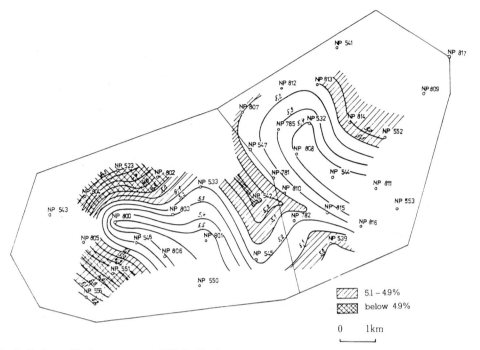

**Fig. 2.** Isolines of hydrogen content, $H$(daf), 38a Seam.

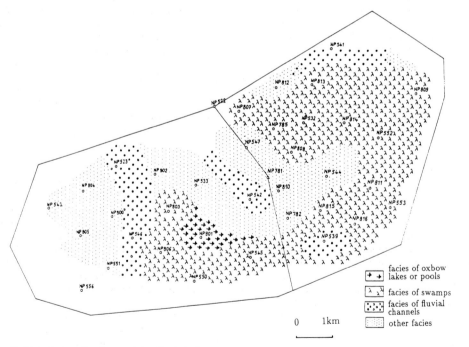

**Fig. 3.** Lithofacies of rocks overlying Prokop Seam.

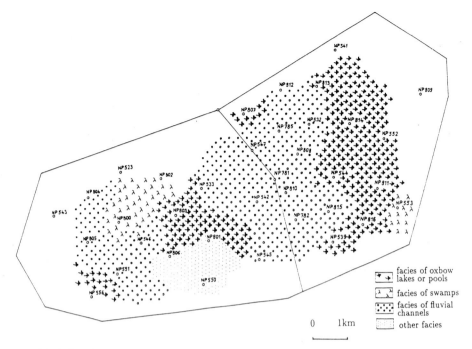

**Fig. 4.** Lithofacies of rocks overlying 38a Seam.

Formation, Saddle Member there is an abrupt increase in the abudance of inertinite.

Analyses have been carried out on the Prokop seam (Saddle Member, No. 504) and the 38a seam (Saddle Member, No. 522) in the exploration areas of Frenštát-East and Frenštát-West in the southern part of the Ostrava–Karviná Coalfield. These analyses were intended primarily to determine the hydrogen content, $H$(daf), of the coal (see Figs 1 and 2). The analysis results were related to other parameters, especially to the character of rocks overlying the coal seam (see Figs 3 and 4).

## Results and discussion

Owing to the thickness of weathered carboniferous rock mantle (Mikytová 1984), hydrogen values affected by post-Carboniferous processes were excluded from the correlations. Furthermore, values of hydrogen content obtained from boreholes NP 553, NP 815 and NP 816 drilled into seam 38a were also excluded. In these locations a diorite–porphyrite dyke penetrates the seam transforming it into natural coke with a greatly reduced hydrogen content (Blumenthal et al. 1981).

A comparison of the isolines of $H$(daf) and volatile matter contentV.M. (daf), clearly shows

that they follow different courses and that no extreme value occurs (see Figs 1, 2, 5 and 6).

Maceral analyses shows that the average inertinite contents of the Prokop and 38a seams in the Czech part of the Upper Silesian Coal Basin are 50% (Havlena 1964) and 47.3% (Weiss et al. 1975), respectively. Notably higher inertinite levels were noted in the Prokop seam in boreholes NP 803, NP 816 and NP 811 (Fig. 7) and in the 38a seam around borehole NP 814 (Fig. 8). As the inertinite group macerals have a lower hydrogen content than those of other groups, high inertinite coals have inherently low hydrogen contents.

The Prokop seam shows a pronounced reduction in hydrogen content in the vicinity of boreholes NP 539, NP 542 and NP 541, where the seam roof consists of very coarse sandstones and poorly sorted conglomerates. These roof measures are of fluvial channel origin and have a high original porosity. Roof strata of fluvial channel facies can also be seen in Fig. 3 to pass in a north-south direction between boreholes NP 523 and NP 546. The hydrogen content of the coal in these boreholes was reduced, but the hydrogen isolines in this area were also influenced by the high inertinite content recorded in borehole NP 803.

Reduced hydrogen levels were recorded in the coal of the 38a seam in the vicinity of boreholes

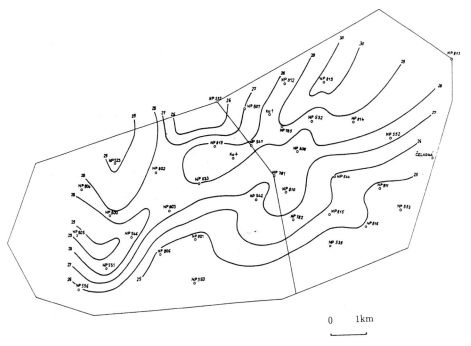

**Fig. 5.** Isolines of volatile matter, Prokop Seam.

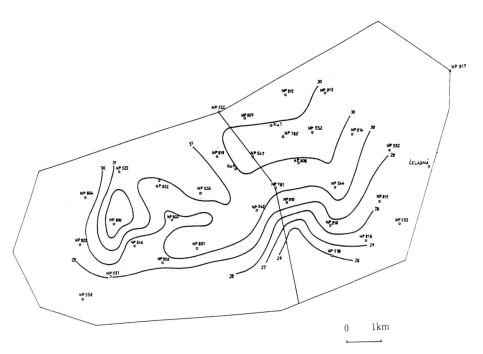

**Fig. 6.** Isolines of volatile matter, 38a Seam.

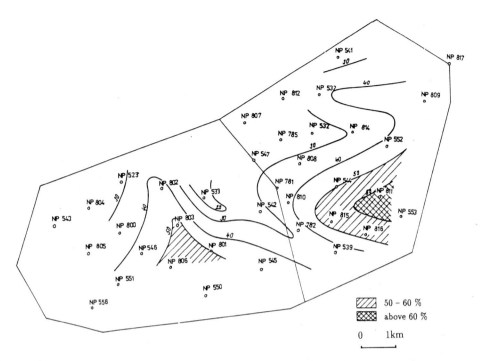

**Fig. 7.** Isolines of content of inertinite macerals, Prokop Seam.

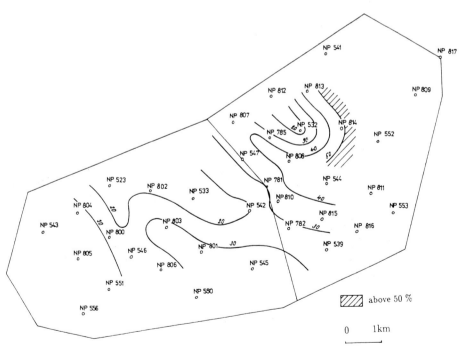

**Fig. 8.** Isolines of content of inertinite macerals, 38a Seam.

NP 523, NP 804, NP 551 and NP 802. In all these instances the immediate seam roof was recorded as very coarse sediment of fluvial channel facies. Overlying strata of a similar nature were also recorded in a zone defined roughly by boreholes NP 785, NP 532, NP 542, NP 533, NP 545, NP 815, NP 810 and NP 782. The hydrogen content of the coal is reduced in this area, but not so distinctly as in the west of the exploration area. Only in borehole NP 542 did the hydrogen value decrease to less than 5%.

## Significance for coalbed methane

The results indicate that hydrogen, possibly associated with methane and other alkanes generated in the coals, has migrated into porous roof rock sandstones, but has been retained in the coal where the roof rock is impermeable (siltstones and mudstones). Although the timing of this migration is not indicated by the study, it can be predicted that the CBM content of coals would be lowered where porous roof rocks overlie the seam. Conversely, where the roof rock is impermeable it is likely that a significant amount of the generated methane will be retained in the coal.

## Conclusions

A significant correlation can be shown between the lithological character of overlying rocks and the hydrogen content of the coal seams studied. This relationship can be demonstrated by comparing the isolines of hydrogen content with maps of the facies of overlying strata (excluding thoselow values which could have been influenced by the factors mentioned earlier).

Where sediments of fluvial channel facies with consistently high original porosity were deposited immediately above the coal seam, a pronounced decrease in hydrogen content of the coal was often observed. In contrast, where siltstone or mudstone of swamp or oxbow lake facies, with a low original porosity, formed the roof of a seam, reduced hydrogen levels in the coal were not so pronounced. Roof sediments of alluvial plain facies, with great variability in original porosity due to variable grain size and clay content, produced no consistent effect on the hydrogen content of the underlying coal.

This study emphasizes that the physical and chemical properties of coal, including the retention of CBM, can be influenced not only by primary factors connected with the materials and environment of formation of the seam, but also by factors such as the character of directly overlying strata which come into effect at a later stage of coal formation.

## References

BLUMENTHAL, J. et al. 1981. Výpočet zásob průzkumného pole Frenštát–východ (The Coal Resources in the Exploration Area of Frenštát-East). Archiv GPO, Ostrava.

DOPITA, M. & KUMPERA, O. 1993. Geology of the Ostrava–Karviná coalfield, Upper Silesian Basin, Czech Republic, and its influence on mining. International Journal of Coal Geology, 23, 291–321

HAVLENA, V. 1964. Geologie uhelných ložisek (Geology of Coal Deposits) 2. Nakladatelství SAV, Prague.

MIKYTOVÁ, M. 1984. Studie zvětralinového pláště karbonu Freštát–Trojanovice (Studies of the Weathered Carboniferous Rock Mantle). Archiv GPO, Ostrava.

WEISS, G. et al. 1975. Surovinová studie čsl. části hornoslezské pánve (Raw Materials Studies of the Czechoslovak Part of the Upper Silesian Coal Basin). Archiv GPO, Ostrava.

# Analysis of the problems associated with the use of image analysis for microlithotype analysis on solid coal mounts

EDWARD LESTER[1], MARTIN ALLEN[2], MICHAEL CLOKE[1] &
BRIAN ATKIN[2]

[1] *Coal Technology Research Group, Department of Chemical Engineering,
University of Nottingham, University Park, Nottingham NG7 2RD, UK*
[2] *Department of Mineral Resources Engineering, University of Nottingham, University Park,
Nottingham NG7 2RD, UK*

**Abstract:** An automated system has been developed by the Coal Technology Research Group at Nottingham University which is capable of producing microlithotype and maceral data on solid coal mounts, compliant with International Standard requirements for particulate coal analysis. The maceral composition of a $50 \times 50\,\mu m$ area (or 'point') analysed can be used to classify the microlithotype. However, some problems in classification can occur due to the increased precision of the automated analysis over manual analysis and, because of this, up to 2% of the points cannot be classified when using the ICCP classification scheme. Experimental investigation has revealed that the repeatability of maceral and microlithotype results depend strongly on the scanning pattern used. Using a grid sampling pattern, 100 points produce microlithotype results with barely acceptable levels of repeatability, but the use of 500 points produces results with a level of repeatability compliant with the International Standard.

Petrographic analysis of coal has long been recognized as an important tool in the description and investigation of the genesis, vertical and lateral variations and continuity of coal. With ever increasing importance being placed on optimizing fuel performance, driven by environmental as well as economic forces, coal petrography has become a vital tool in the experimental and commercial utilization of coal (Kojima 1976; Thomas *et al.* 1991; Lester *et al.* 1994*a*).

Petrographic analysis of coals can be used to provide information about rank, maceral and microlithotype compositions. However, the petrographic characterization of coals is usually limited to rank and/or maceral analysis. This limitation is primarily due to the time required to perform such analyses. Classical manual maceral analysis requires a skilled operator and analyses can take a significant time, depending on the complexity of the coal being analysed. For example, rank and maceral analyses may each take between 20 min and 1 h to perform, depending on the petrographic variability within the coal. Microlithotype analysis, however, may take substantially longer, with analysis times varying from 2 to 4 h.

Coupled with these points is the fact that maceral and microlithotype results can have relatively low levels of reproducibility due to high levels of subjectivity of the operator performing the analyses. The acceptable limits (British Standard 1981) of reproducibility for a maceral with an abundance of 50.0% is for 95%

of the results to fall within a 12.67% range centred on the actual abundance. Therefore, for 19 out of 20 analyses, the measured abundance could lie anywhere between 43.7 and 56.3% and one out of 20 analyses will be outside this range. With a coal which is particularly difficult to analyse, these limits of reproducibility may be difficult to achieve due to the subjectivity of individual operators. This problem is significantly worse with microlithotype analysis.

To overcome these problems, much effort has been placed on automating petrographic analyses, principally using image analysis systems as a substitute for the manual operator. The principal problem in using image analysis to identify the individual macerals has been in distinguishing liptinite from the resin binder used in the preparation of the polished particulate coal mounts (Riepe & Steller 1984). This can been overcome by a number of different strategies, including the use of morphological manipulation of individual images (Lester *et al.* 1994*b*), fluorescence microscopy (Cloke *et al.* 1995) and full colour image processing (Prado unpublished data). A number of image analysis based systems are now routinely used to produce non-subjective, highly reproducible, precise and fast, rank and maceral analysis results (Lester 1994). Despite this, little work has been undertaken into the automation of microlithotype analysis of polished coal mounts. Microlithotype analysis differs from normal maceral analysis in that each microlithotype 'point' is a summary of

*From* Gayer, R. & Harris, I. (eds) 1996, *Coalbed Methane and Coal Geology*,
Geological Society Special Publication No 109, pp 237–248.

10–20 individual maceral identifications. These 20 points are spread over a $50 \times 50 \mu m$ area in a grid pattern defined by a Kötter graticule (see Fig. 1). For this reason microlithotype analysis represents a more significant challenge to an image analyser than just division into main maceral groups. Microlithotype analyses are important in that they provide key information about the distribution of the individual macerals within the coal. Maceral analysis, by itself, only provides information about the relative abundance of the individual macerals, but the association of these macerals within the coal structure can be vital when explaining the behaviour of the coal in subsequent coal comminution and utilization processes. For example, consider two extreme solid coals with identical proportions of individual macerals, one composed entirely of liptite, vitrite and inertite, the other entirely composed of trimacerite. After comminution and flotation to remove mineral matter (and excess inertinite) the maceral composition of the two coals may be radically different (Kizgut *et al.* 1994) and subsequent behaviour during utilization will be equally incomparable. It is therefore useful to have an understanding of the association of macerals, as well as their relative abundances.

To investigate the automation of microlithotype analyses using image analysis, two experiments were planned. The first experiment was devised to investigate the repeatability of automated maceral analyses performed on a single coal. The second experiment involved the manual and automated analysis of three test coals to allow comparison of the results and to assess the repeatability of the automated analyses. Both of these experiments were performed on solid coal blocks rather than particulate blocks for the following two reasons.

(1)  Although the detection and measurement of liptinite is possible through various means when performing maceral analysis (Prado unpublished data; Cloke *et al.* 1995), microlithotype analysis presents a different set of problems as regards identification. According to petrographic standards (British Standard 1990), a microlithotype 'point' refers to a discrete particle. The detection of single particles is a complex procedure when attempting to use image analysis. The detection of the maceral proportions within each particle requires the separation of any touching particle, the establishment of a definite particle boundary, followed by an assessment of the abundance of each maceral

within the particle walls. This requires greater computer processing power than was available to us for this study.

(2)  When we consider coalbed methane production, diffusion and emission, petrographic characterization is perhaps best performed on solid coal material, as breaking the lithotype structure of the seam would lead to the loss of valuable information.

## Experimental

### Preparation of the polished coal mounts

Three 30 mm polished coal mounts were prepared, each containing a single fragment of coal of about 20 mm square. The coal blocks were placed in the centre of a 30 mm mould and then surrounded by Simplex Rapid, a powdered dental resin. The mould was then pressed at 150°C and 2 bar pressure for 20 min, followed by 3 min of cooling time, in a Presi Mecapress C. The blocks, once removed from the moulds, were then polished using a Struers Pedemat Rotapol polisher to produce a high quality polished solid coal mount.

The three test coals selected were from Colombia, Poland and Russia, and were chosen because of their different maceral compositions. The Colombian and Russian coals were Cretaceous coals and the Polish coal was Carboniferous. The Polish and Russian blocks contained similar maceral species, i.e. similar ranked vitrinite and high reflectance fusinite. The proportions of each maceral, however, were distinctly different. The liptinite content of the Polish coal was considerably higher and the inertinite content of the Russian coal was greater than in the Polish coal. The Colombian coal was different to the other two because it contained more semi-fusinite and much less liptinite.

### Manual maceral and microlithotype analyses

Maceral analysis (British Standard 1981) was carried out using a Swift point counter, measuring 500 separate maceral points, with graduated movements of 1/3 mm. The Leitz Ortholux II POL-BK microscope was fitted with a 32× oil immersion objective and a 10× eyepiece.

Microlithotype analysis (British Standard 1990) was performed using a 20 point Kötter graticule (shown in Fig. 1), which measured $50 \mu m$ across, with the 32× objective. A total of 500 counts was made on each block, across the bedding plane, to minimize the possibility of

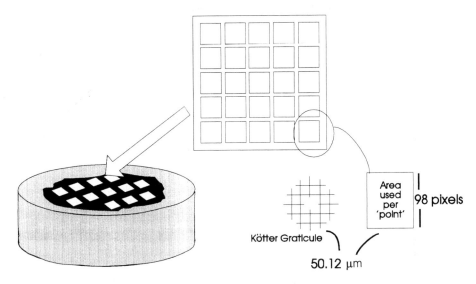

**Fig. 1.** Schematic diagram of procedure for scanning a solid coal particle, including Kötter graticle used for manual analysis.

following a single layer, which would generate unrepresentative results. Tables 1 and 2 show the microlithotype classifications used in the tests.

## Automated maceral and microlithotype analyses

The image analysis was performed using the same microscope connected to an IBAS 2000 Image Processing System via a Hamamatsu C2400-07 (Newvicon) video camera.

Each image was digitized by the image processing system into a $512 \times 512$ pixel bitmap using 256 grey levels. With a $32\times$ objective, the

image scaling was measured as $0.5115 \mu m$ per pixel, giving an image size of $261.89 \times 261.89 \mu m$. Each image was then subdivided into a $5 \times 5$ array of image blocks of $98 \times 98$ pixels, which is equivalent to the area covered by the graticule used for manual microlithotype analyses ($50.13 \times 50.13 \mu m$).

Each of the three coals was first subjected to a grey scale analysis to obtain a cumulative histogram for 50 images. These cumulative histograms were then used to determine sensible threshold levels for the liptinite/vitrinite and vitrinite/inertinite grey level boundaries. These thresholds were then used to separate the individual macerals. The boundaries were established arbitrarily, based on the position of the minimum between phases, as seen on the histogram plot. The thresholding between

**Table 1.** *Microlithotype nomenclature. Classification of microlithotypes based on B.S.6127: 4 (1990)*

| Liptinite (%) | Vitrinite (%) | Inertinite (%) | Microlithotype classification |
|---|---|---|---|
| ≥95.00 | <5.00 | <5.00 | Liptite |
| <5.00 | ≥95.00 | <5.00 | Vitrite |
| <5.00 | < 5.00 | ≥95.00 | Inertite |
| ≥5.00 | ≥5.00 | <5.00 | Clarite |
| ≥5.00 | <5.00 | ≥5.00 | Durite |
| <5.00 | ≥5.00 | ≥5.00 | Vitrinertite |
| ≥5.00 | ≥5.00 | ≥5.00 | Trimacerite |

All three logical conditions have to be met before the point is clasified: >, greater than; <, less than; ≥, greater than or equal to; and ≤, less than or equal to.

**Table 2.** *Microlithotype nomenclature. How certain points cannot be classified*

| Liptinite = 94% | Vitrinite = 3.00% | Inertinite = 3.00% | Microlithotype classification |
|---|---|---|---|
| No | Yes | Yes | Liptite |
| No | No | Yes | Vitrite |
| No | Yes | No | Inertite |
| Yes | No | Yes | Clarite |
| Yes | Yes | No | Durite |
| No | No | No | Vitrinertite |
| Yes | No | No | Trimacerite |

Three positive responses are required to classify a point.

phases could also be performed based on a mathematical Gaussian distribution of vitrinite. This would be a useful tool, particularly when attempting totally to automate petrography, although with these experiments this was not possible due to the processing limitations of the image analysis system.

The image analysis system then performed a maceral analysis based on grey levels of each of these blocks, producing 25 results per image. The polished coal mounts were scanned using a $10 \times 10$ array of images, each position in the array being orthogonally separated from its neighbours by 1080 $\mu$m, approximately four image widths. Thus the scan dimensions of $10.80 \times 10.80$ mm allowed a substantial part of the polished surface of each coal fragment to be covered. Figure 1 shows the schematic scanning process on the surface of the coal block. This scanning pattern, in combination with the image subdivision, allowed maceral data to be collected for a total of 2500 'points' and each of these 'points' represented a standard microlithotype field of $50 \times 50$ $\mu$m. The entire scan took 1 h 42 min, or about 61 s per image, or about 2.5 s per 'point'. The scanning patterns used are described in the next section.

## Results

### Repeatability of automated maceral analyses

To assess the repeatability of the automated maceral analyses the entire $50 \times 50$ array of points produced by the first experimental scan on the Russian coal were subsampled using a number of sampling patterns. These included: (1) 50 rows of 50 points; (2) 50 columns of 50 points; (3) 49 $(7 \times 7)$ blocks of 49 $(7 \times 7)$ points; (4) 49 grids of 49 points (each point being orthogonally separated by six points); and (5) 50 random samples of 50 points (taken from the entire $50 \times 50$ array of points).

Figure 2 shows a schematic representation of each scanning sequence. With the exception of the random sampling, all sampling took place without replacement. In each instance the abundance of vitrinite was totalled, averaged and the standard deviation of all the subsamples was calculated.

Vitrinite was selected because, of the three macerals present, its abundance is closest to 50.00% and the greatest theoretical standard deviation between results occurs for phases with abundances of 50.00%.

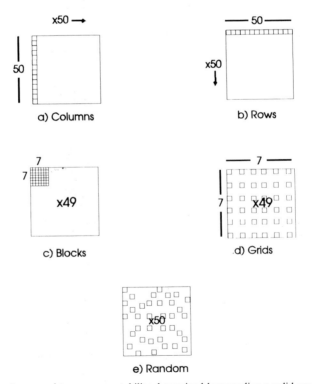

**Fig. 2.** Scanning patterns used to assess repatability determined by sampling a solid coal black.

**Table 3.** *Observed standard deviations for a number of scanning patterns*

| Sampling pattern | Observed standard deviation (%) |
|---|---|
| 50 rows of 50 points | 6.80 |
| 50 columns of 50 points | 6.77 |
| 49 blocks of 49 points | 7.34 |
| 49 grids of 49 points | 2.47 |
| 50 random sampls of 50 points | 2.81 |

The observed standard deviations are those for the abundance of vitrinite in each point. The mean abundance of vitrinite 57.76%.

Table 3 presents the observed standard deviations calculated for the individual scanning patterns. From these results it can be observed that the greatest standard deviation of 7.34% occurs when subsampling blocks. Slightly smaller standard deviations are achieved when subsampling either rows or columns. However, the smallest standard deviations are achieved by subsampling either grids or random patterns. The observed standard deviations in each instance of less than 3% are comparable with the theoretical standard deviations expected for a manual determinations based on 500 points. Therefore, maceral analyses based on using just 50 points sampled in a grid, or sampled randomly, meet the requirements for repeatability and reproducibility set out in the British Standard.

## Repeatability of automated microlithotype analyses

The maceral proportions of each 'point' were converted into microlithotypes using the logic tree given in Fig. 3. It can be seen that a series of logical steps can filter the maceral data into the correct categories. Tables 4 and 5 present the results of the manual and automated maceral

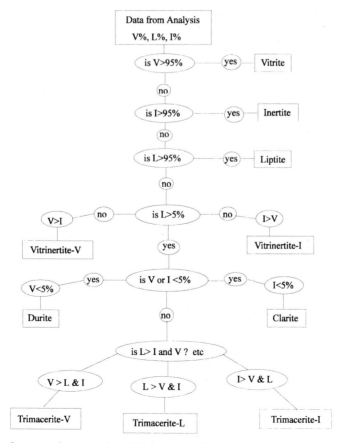

**Fig. 3.** Logic tree for converting maceral analysis into microlithotype 'points'.

**Table 4.** *Comparison of manual and automated maceral results*

| Maceral | Russian coal | | | Colombian coal | | | Polish coal | | |
|---|---|---|---|---|---|---|---|---|---|
| | Manual (500)* | Automated | | Manual (500)* | Automated | | Manual (500)* | Automated | |
| | | (2500)* | SD† | | (2500)* | SD† | | (2500)* | SD† |
| Liptinite | 6.6 | 8.1 | 1.64 | 3.4 | 3.7 | 0.92 | 13.4 | 15.5 | 2.36 |
| Vitrinite | 50.0 | 53.0 | 2.80 | 62.4 | 55.4 | 5.83 | 70.8 | 82.0 | 1.76 |
| Inertinite | 38.6 | 38.9 | 3.97 | 34.2 | 40.9 | 6.02 | 15.0 | 2.5 | 0.83 |

\* Number of data points.
† Standard deviation.

**Table 5.** *Comparison of manual and automated microlithotype results*

| Maceral | Russian coal | | | Colombian coal | | | Polish coal | | |
|---|---|---|---|---|---|---|---|---|---|
| | Manual (500)* | Automated | | Manual (500)* | Automated | | Manual (500)* | Automated | |
| | | (2500)* | SD† | | (2500)* | SD† | | (2500)* | SD† |
| Vitrite | 9.80 | 5.36 | 2.25 | 7.80 | 8.60 | 3.18 | 36.10 | 31.00 | 4.19 |
| Liptite | 0.00 | 0.04 | 0.20 | 0.00 | 0.00 | 0.00 | 1.33 | 0.48 | 0.65 |
| Inertite | 9.00 | 1.20 | 0.91 | 15.70 | 6.40 | 1.98 | 4.00 | 0.00 | 0.00 |
| Clarite | 1.20 | 4.72 | 2.51 | 0.90 | 0.76 | 0.93 | 28.07 | 57.40 | 6.46 |
| Durite | 1.20 | 0.28 | 0.05 | 1.10 | 0.40 | 0.50 | 0.00 | 0.00 | 0.00 |
| Vitrinertite | 47.00 | 45.16 | 9.29 | 64.00 | 67.20 | 5.21 | 0.00 | 2.80 | 1.50 |
| Trimacerite | 28.00 | 41.44 | 8.94 | 10.50 | 14.72 | 3.51 | 30.50 | 7.12 | 3.24 |
| Carbominerite | 3.80 | 0.00 | 1.15 | 0.00 | 0.00 | 1.47 | 0.00 | 0.00 | 1.00 |
| Unclassified | 0.00 | 1.80 | 0.00 | 0.00 | 1.92 | 0.00 | 0.00 | 1.20 | 0.00 |

\* Number of data points.
† Standard deviation.

and microlithotype analyses, respectively, for the three coals used in this study. The tables also include the standard deviation results for each sample.

To assess the repeatability of the automated microlithotype analyses, the sampling pattern of $10 \times 10$ images, each composed of 25 images, was broken down into 25 grids of $10 \times 10$ points by sampling the first point in each image, then the second and so on.

The standard deviations for the maceral abundances for the Russian and Polish coals are in agreement with the standard deviations encountered earlier in the repeatability study. The standard deviations for the Colombian coal are significantly larger (6.02 for inertinite) and can be explained by the larger scale of petrographic features in this coal. However, even these standard deviations are within the limits of repeatability required by the various standards for maceral analysis.

The standard deviations for the microlithotype abundances include some significant figures, the largest standard deviations (9.29 for vitrinertite and 8.94 for trimacerite) being associated with the Russian coal, not the Colombian coal as might be expected from the maceral repeatabilities. Again, this emphasizes the independence of maceral and microlithotype results from each other.

Some of the standard deviations for the subsampled microlithotype analyses are close to the acceptable limits of repeatability. However, these are for subsamples of only 100 points. The standard deviations for 500 points were significantly lower and the standard deviations for the full 2500 points are inevitably lower still.

## Comparison of manual and automated results

From Table 4 it can be seen that the manual and automated maceral results for the Russian and Colombian coals agree fairly closely. The results for the Polish coal, however, are clearly different. The difficulty has arisen with the distinction between vitrinite and inertinite. A 10% difference exists between the two. This has probably been caused by an incorrect grey scale threshold separating vitrinite and inertinite. The manual and automated results for the liptinite species agree fairly closely, which indicates that the scanning pattern used was sufficient to gain a representative result, but did not threshold the vitrinite and inertinite phases correctly.

The best agreement between microlithotype results came from the Colombian coal, and this can be seen in Table 5. The discrepancy between the inertinite fraction of this coal is made up with a higher trimacerite percentage, rather than vitrinertite. The Russian and Polish coals have some distinct differences between the manual and automated results, which are worth discussing.

The automated results for the Russian coal are lower in both vitrinertite and inertite. However, as the manual and automated maceral analyses (in Table 4) are similar, it is possible that the division of many points into microlithotypes was close to the 5% boundary mark. The difference between the automated and manual results could be a result of the enhanced accuracy of the automated system. This could explain why 10% more trimacerite was detected in the Russian sample. It is possible for a maceral to exist in proportions of >5% without lying under a Kötter graticule point. Conversely, it is possible for a maceral to exist in proportions <5% but still affect the manual microlithotype identification by falling under the graticule. At this stage, its inclusion would be at the discretion of the operator. A similar situation occurs during routine maceral analysis. Depending on the purpose of the study, i.e. geological or utility, an operator may chose to label the internal cells of inertinite as inertinite itself. With an automated study, this discretion is not possible.

The automated results for Polish coal are no doubt affected by this enhanced accuracy, causing a certain amount of discrepancy between the automated and manual results. Increased accuracy may well tend to cause more trimacerite. This is not so with the Polish coal, but as discussed for the maceral results, the separation of inertinite from the vitrinite was not good. As a result of this, the percentage of clarite was significantly higher than the manual results and the amount of trimacerite was much lower. Had the inertinite threshold been set correctly, the results would have probably been very different. The trimacerite percentage would have been much higher as a result of the increased percentage of inertinite identified in each image. Its inclusion in these results serves as an example of the GIGO (garbage in, garbage out) principle. Image analysis systems stand or fall on the accurate determination of thresholds, hence, as in this

instance, inaccurate threshold values seriously affect the results. Automated thresholding, either per image or per scan, based on statistical principles, would reduce the possibility of poor results being produced.

## Graphical representation of the maceral distribution within each microlithotype 'point'

The image analyser stores the maceral information for each discrete microlithotype point, hence each point can be plotted on a ternary diagram. Figures 4–6 represent the 2500 individual results for each of the three coals.

Major clustering around the vitrinite corner can be seen with the Polish coal in Fig. 4. This is a false result due to the incorrect choice of grey level differentiating between the inertinite and vitrinite. The Colombian coal (Fig. 5), containing the lowest amount of liptinite, tended to produce points along the line between the vitrinite and inertinite. The Russian coal, which contained more liptinite than the Colombian coal, tended to produce points which fell near the inertinite–vitrinite line, but sufficiently away towards the liptinite end-member to produce trimacerite points (Fig. 6).

The value of the automated method of microlithotype analysis is that it shows, for each field, the proportions of the component macerals, whereas the manual method merely assigns each field to a microlithotype based on arbitrarily defined limits. An enhanced 'visualization' of a coal's maceral structure could potentially be useful for workers interested in the migration and generation of methane within a coalbed. Understanding the layering of the macerals, as well as their reflectance levels (also useful from a utilization perspective), could well justify the incorporation of microlithotype analysis into the characterization of borehole samples for the oil and gas industry. The development of an automated system could only promote the cause and accessibility of the use of petrography.

Overlapping of the individual points on the graph is the main drawback of this type of figure. A reduction in the number of points on the graph would obviously reduce this problem. The way to overcome this problem is to produce three-dimensional ternary plots, showing population as well as distribution. In Figs 7–9 it is possible to see the same results as in Figs 4–6 but with the relative proportions of each point represented by a three-dimensional bar. There are some distinct differences between each coal type.

Inertinite

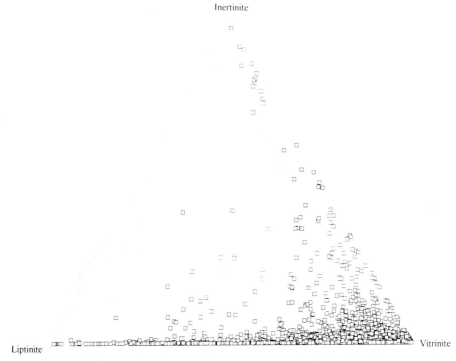

Liptinite                                                                 Vitrinite

**Fig. 4.** Two-dimensional plots of the maceral disribution of the  results for Polish coal.

Inertinite

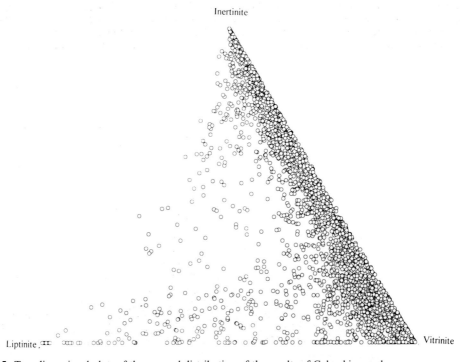

Liptinite                                                                 Vitrinite

**Fig. 5.** Two-dimensional plots of the maceral distribution of the results of Colombian coal.

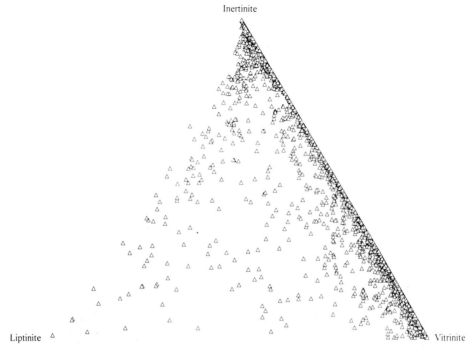

**Fig. 6.** Two-dimensional plots of the maceral distribution of the results of Russian coal.

The Polish coal, in Fig. 7 contained large amounts of vitrinite (as discussed earlier), hence large groups of peaks occur near the vitrinite corner. Most of the other columns occur along the vitrinite–liptinite line, corresponding to the clarite present in the microlithotype analyses. The Colombian coal (Fig. 8) has the opposite features, with most of columns sitting between vitrinite and inertinite, with an increase in frequency towards the vitrinite corner. The

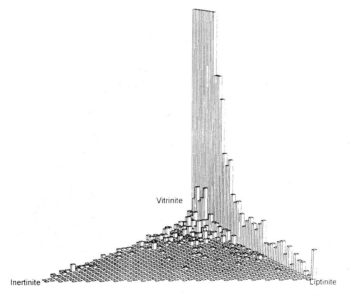

**Fig. 7.** Maceral distribution within a Polish coal as represented by a three-dimensional ternary diagram.

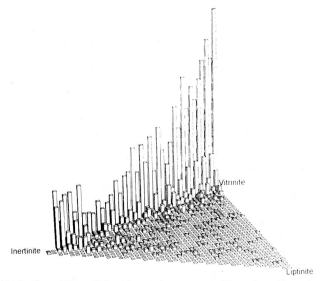

**Fig. 8.** Maceral distribution within a Colombian coal as repreented by a three-dimensional ternary diagram.

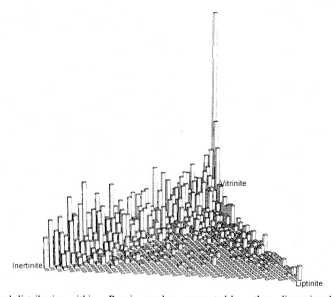

**Fig. 9.** Maceral distribution within a Russian coal as repreented by a three-dimensional ternary diagram.

Russian coal (Fig. 9) has the greatest spread of results, mainly focused around the vitrinite corner.

## Classification of microlithotypes used for automated analyses

As mentioned earlier, the classification of microlithotypes used for the automated ana-lyses is identical to that used for manual analyses (British Standard 1990) and is listed in Table 1. However, classification problems do arise using this scheme. For manual microlithotype analyses the abundances of the three maceral groups can only be an integer multiple of 5%. The abundances of the three macerals in automated maceral analysis results can be expressed with a precision dependent on the image size on which the analysis is based.

For example, in the 98 × 98 pixel array used in the analyses in this paper, the smallest abundance, $A_{min}$, that can be measured is

$$A_{min} = 100/(98 \times 98) = 0.0104\%$$

Problems arise when macerals with abundances less than 5% occur. For example, consider a point with abundances of liptinite, vitrinite and inertinite of 94.00, 3.00 and 3.00%, respectively. This point can almost be classified as liptite, but if the classification is adhered to strictly it cannot. The next nearest classification is clarite, but again it is not an accurate classification. In fact, a point containing these maceral abundances cannot be classified at all using this scheme (Table 2). Figure 10a shows the area on a ternary diagram where there is a problem with microlithotype classification. Figure 10b shows how the area of null classification can be eliminated. This measure is not necessary during manual analysis, as any point of the Kötter graticule represents ≥5% abundance. Hence if any maceral species lie under a cross-hair, then it is present in significant proportions to be considered in the microlithotype identification.

## Conclusions

Automated petrographic analysis of coals based on image analysis provides a non-subjective

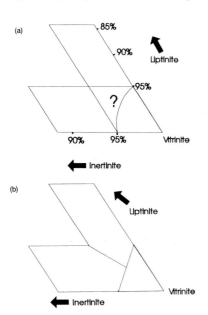

**Fig. 10. (a)** Corner section of a ternary diagram according to B. S. 6127. **(b)** Suggested alternative to the standard ternary corner to incorporate results from image analysis systems.

technique for both maceral and microlithotype analysis. The scanning pattern used can reduce the standard deviation of the results, depending on the size of the bedding planes. However, large-scale microlithotype bandings would be difficult to analyse with an acceptable degree of repeatability, with or without image analysis.

The use of three-dimensional ternary plots is perhaps the clearest way of representing microlithotype data, as once a maceral has >5% abundance in an image its presence is significant. This abundance could be 6 or 60% with manual microlithotype identification procedures. Only more detailed analysis techniques reveal more detailed information.

## Future work

This study represents an initial phase of the investigation. Future automation of microlithotype analyses will be extended to particulate coal and to investigations of the liberation and association of coal macerals on comminution.

Such a system would involve more complex image processing than the system described. The system would have to be capable of identifying individual particles, as well as identifying and separating two touching particles. Image processing strategies have been developed at Nottingham that are capable of overcoming these problems, and such a system, once developed, would allow results to be produced quickly to BSI standards.

An automated microlithotype analysis system capable of analysing particulate coal mounts could become part of future analysis systems which could analyse the composition of entire particles and produce data on both particle composition and particle size. Such a system would be an invaluable aid in future coal studies. The use of three-dimensional ternary diagrams at present could be seen as qualitative rather than quantitative, although in future it could be possible to assess statistically the distribution of points and, with the inclusion of the reflectance information of each maceral species, provide a useful correlation between a coal's petrography and its utilization behaviour.

The support of the Science and Engineering Research Council, UK is acknowledged in the funding of this work. The authors also acknowledge the help of F. Wigley, from Imperial College, for explaining how to draw three-dimensional diagrams in Excel, and D. Clift for assistance with petrographic work.

# References

BRITISH STANDARD 1981. Glossary of terms relating to the petrographic analysis of bituminous coal and anthracite, *B.S.6127 Part 1*. British Standards Institute, Milton Keynes, 8pp.

——1990. Method of determining microlithotypes, carbominerites and minerite composition, *B.S.6127 Part 4*. British Standards Institute, Milton Keynes, 12pp.

CLOKE, M., LESTER, E., ALLEN, M. & MILES, N. J. 1995. Automated maceral analysis using fluorescence microscopy and image analysis. *Fuel*, **74**, 659–669.

ISO STANDARD 1988. Methods for the petrographic analysis of bituminous coal and anthracite—Microlithotype Analysis. *ISO 7404:4*.

KIZGUT, S., RHODES, D., MILES, N. J. & CLOKE, M. 1994. Flotation response of bituminous coal macerals. *In: 5th International Mineral Processing Symposium, Cappadocia, Turkey, 6–8th September 1994*, Balkema, Rotterdam 313–319.

KOJIMA, K. 1976. Automatic system for evaluating coking coals. *Iron and Steel International*, **46**, 435–436.

LESTER, E. H. 1994. *Characterisation of coals for combustion*. PhD Thesis, Nottingham University.

——, ALLEN, M., CLOKE, M. & MILES, N. J. 1994a. An automated image-analysis system for major maceral group analysis in coals. *Fuel*, **73**, 1731–1734.

——, ——, —— & ——1994b. Maceral analysis by automated image analysis techniques. *In: 12th International Coal Preparation Congress 1994, Poland*. 531–537.

RIEPE, W. & STELLER, M. 1984. Characterisation of coal and coal blends by automatic image analysis. *Fuel*, **63**, 313–317.

THOMAS, C. G., SHIBAOKA, M., PHONG-ANANT, D., GAWRONSKI, E. & GOSNELL, M. E. 1991. Determination of percentage reactives under PF combustion conditions. *In: International Conference on Coal Science, IEA 1991*. Butterworth–Heinemann, Oxford, 48–51.

# Petrological and spectroscopic structural characteristics of Bohemian and Moravian coals and their possible relation to gas proneness

I. SÝKOROVÁ[1], M. NOVOTNÁ[2], H. PAVLÍKOVÁ[2] & V. MACHOVIČ[2]

[1] Institute of Rock Structure and Mechanics, Academy of Sciences of the Czech Republic, V Holešovičkách 41, 182 09 Prague 8, Czech Republic
[2] Institute of Chemical Technology, Technická 5, 166 28 Prague 6, Czech Republic

**Abstract:** Optical and infra-red microscopy, diffusion reflectance infra-red spectrometry and solid-state nuclear magnetic resonance spectrometry ($^{13}C$ CP/MAS NMR) were applied to the determination of the degree of coalification and structural characteristics of lignite and bituminous coal from deposits in the Czech Republic. Structural parameters such as the aliphatic and aromatic contents and carbonyl contents of coal and coal macerals could be valuable tools for determination of gas proneness, e.g. the $CH_2/CH_3$ ratio derived from infra-red spectroscopy. Aromaticity, $f_a$, determined by $^{13}C$ CP/MAS NMR, is shown to be a valuable parameter reflecting the degree of coalification, particularly for low and middle rank coal, thus supplementing data derived from the vitrinite reflectance $R_o$ and $(H/C)_{at}$ ratio data. The maceral group of huminite/vitrinite mirrored similar changes with increasing coalification degree as the parent coal. The structure of two macerals of the liptinite group, sporinite and cutinite, was investigated for bituminous coal, and the aromatic and oxygen group contents were higher in cutinite than in sporinite.

Coalbed gas generation is strongly related to the chemical and physical structure of coal. To obtain further information on gas proneness the petrological and chemical characteristics are of high value, especially maceral composition, vitrinite reflectance, aromaticity and the amount and types of functional groups.

Petrology is conventionally used for the classification and evaluation of coal in geological surveys and industrial processing. In this manner, information is obtained supplementing the basic technological parameters such as moisture, ash, sulphur and volatile matter, calorific value and elemental analysis (Lemos de Sousa et al. 1992).

Optical microscopy measures the light reflectance, mirroring the degree of coalification of vitrinite (in bituminous coal) or huminite (in lignite) and the maceral or microlithotype composition, which reflects the coal composition (Teichmüller & Teichmüller 1982).

Coal is heterogeneous and its chemical composition and maceral structure can be studied by physicochemical methods, particularly $^{13}C$ solid-state nuclear magnetic resonance spectrometry (CP/MAS NMR) (Retcofsky & VanderHart 1978; Botto et al. 1987; Choi et al. 1989; Kalkreuth et al. 1991) and Fourier transform infra-red spectrometry (FTIR) (Kuehn et al. 1982; Dyrkacz et al. 1984, 1991; Kister et al. 1986, Snowdon et al. 1986; Michaelian & Friesen 1990; Vassalo et al. 1991; McFarlane et al. 1993). Infra-red spectrometry is one of the most widespread techniques. Curve resolving methods have to be applied to derive quantitative data from the infra-red spectra. Appreciable differences in the spectra of macerals can be identified in this manner. Information about the aliphatic C—H bond contents can be obtained by integration over the wavenumber region of $3000–2800 \, cm^{-1}$, whereas information on the carbonyl groups present can be derived by band separation in the $1700 \, cm^{-1}$ range (Painter et al. 1981). The highest aliphatic contents have been found in liptinite, followed by vitrinite and inertinite, whereas the oxygen functional group content increases in the reverse order (Dyrkacz et al. 1984; Kister et al. 1986; Vassalo et al.1991).

Infrared microscopy begins to play an increasingly important part in the investigation of the chemical composition of organic sediments and coal macerals, thus replacing the tedious preparation of maceral concentrates (Landais & Rochdi 1990; Lin & Ritz 1993; Mastalerz & Bustin 1993; Mastalerz et al. 1993; Stasiuk et al. 1993). This technique allows chemical structure to be examined for samples whose spectra can be measured on an area of at least $20 \times 20 \, \mu m$. Coal samples are prepared in the form of thin slices or polished sections. In this manner, valuable structural parameters have been obtained, such as the aliphatic C—H bond and aromatic and oxygen structure contents (Lin & Ritz 1993).

The assets of the instrumental analytical methods include the fact that information is obtained rapidly. The methods find wide application, for instance, in geological facies

From Gayer, R. & Harris, I. (eds) 1996, *Coalbed Methane and Coal Geology*, Geological Society Special Publication No 109, pp 249–260.

studies (Diessel 1992), in characterizing palaeoenvironments (Stasiuk *et al.* 1993) and in evaluating the chemical structure of organic sediments with respect to oil and gas formation (Ganz & Kalkreuth 1990; Lin & Ritz 1993; Rozkošný *et al.* 1994). Technological coal processing constitutes another area of application (Senftle *et al.* 1984; Böhlmann *et al.* 1992; McFarlane *et al.* 1993).

This study aims to relate the degree of coalification of Bohemian and Moravian lignite and hard coal to the reflectance of vitrinite/huminite, maceral composition and aromatic, aliphatic and oxygen group contents, using methods of $^{13}C$ CP/MAS NMR and FTIR spectrometry. The paper also investigates the structure of individual macerals using FTIR microscopy.

## Experimental procedure

Maceral analyses were performed on a UMSP 30 petrological microscope–microphotometer (Opton-Zeiss) in incident light with the immersion objective on the leaf section. The light reflectance was measured on homogeneous surfaces of vitrinite or huminite at 542 nm with the immersion objective (magnification 45×) in an oil immersion ($n = 1.518$). Fluorescence analyses of liptinite macerals were made with the same microscope using the HBO discharge lamp and a Fl 09 reflector (Opton-Zeiss).

$^{13}C$ CP/MAS NMR spectra were measured using a Bruker 200 spectrometer operating at 50.32 MHz. The spectra were obtained with the following parameters: 1 ms contact time; sweep width of 29 kHz; 75 Hz line broadening; and 5 kHz spinning speed. The number of accumulations was from 1800 to 3600.

Aromaticity, $f_a$, was estimated as the ratio of the integrated area of aromatic carbons (about 90–165 ppm) to the integrated area of all carbons in the spectrum. The integrated intensities of the aromatic spinning sidebands were added to those of their parent peak and the aliphatic spinning sidebands were added to the sum of all carbons.

Diffuse reflectance infra-red Fourier transform spectrometry (DRIFT) was carried out on a Nicolet 740 FTIR spectrometer equipped with a TGS detector. Approximately 50 mg of the sample in neat form were measured with a diffuse reflectance accessory (Spectra Tech). Each sample spectrum, averaging 512 scans per sample, required 3 min of total measuring time. All spectra were recorded at a resolution of 2 cm$^{-1}$. Spectral manipulations (blank, smoothing, baseline correction) were performed by using Omnic Nicolet software.

The microscopic infra-red spectra were measured with an IR-Plan infra-red microscope (Spectra Tech) equipped with an MCT detector, coupled with a Nicolet Magna 700 FTIR spectrometer. A total of 1024 scans was co-added at a resolution of 4 cm$^{-1}$. The coal samples were moulded in an epoxy resin binder and polished to produce a smooth surface for both visual examination and FTIR measurements. The sample area was $20 \times 20 \, \mu m$. The spectra were recovered in the reflectance mode. A background spectrum was measured using the reference gold mirror. The reflectance infra-red spectra were corrected by using a Kramers–Kronig transformation algorithm (Omnic Nicolet software) to obtain the absorption infra-red spectra.

The Jandel Scientific PeakFit curve-fitting software was used for curve resolution of the aliphatic and aromatic C—H stretching modes between 3100 and 2700 cm$^{-1}$ and of carbonyl and skeletal vibrations between 1800 and 1500 cm$^{-1}$ (Lin & Ritz 1993). Voigt profiles were used in the curve fitting. The relative integrated peak intensities of the aromatic C=C (1600 cm$^{-1}$), aliphatic —CH$_3$ (2960 cm$^{-1}$), —CH$_2$ (2923 cm$^{-1}$), —CH (2891 cm$^{-1}$) and carbonyl C=O (1700 cm$^{-1}$) were normalized to 100%.

## Results and discussion

In the Czech Republic, brown coal is mined in the Podkrušnohoří region in Bohemia. Bituminous coal is mined in the Ostrava-Karviná Basin in Moravia. Coal deposits of minor importance occur in southern Moravia and in central and western Bohemia (Fig. 1).

The origin, genetic types and age of the samples studied are given in Table 1. Elemental and technical analysis data (see Table 2) indicate that the carbon content and calorific value increase with increasing degree of coalification, whereas the reverse is true of volatile matter. A higher mineralization (ash content $A^d$) of the lignite L1 and L4, bituminous coal BC2 and anthracite A is reflected in lower $C^{daf}$ and $Q_s^{daf}$ values. The chemico-technological parameters reflect not only the coalification and mineralization, but also the petrographic composition, which is given in Tables 3–5.

Photomicrographs of coal macerals are given in Fig. 2. The lowest mean reflectance of huminite was observed in the sample L1 of lignite from Mikulčice. This sample is xylite,

**Fig. 1.** Locations of Bohemian and Moravian coal deposits.

**Table 1.** *Sample information*

| Sample | Locality | Origin of sample | Age |
|--------|----------|------------------|-----|
| L1 | Mikulčice | South Moravia | Tertiary |
| L2 | Chabařovice | North Bohemia | Tertiary |
| L3 | Bílina | North Bohemia | Tertiary |
| L4 | Most | North Bohemia | Tertiary |
| L5 | Most | North Bohemia | Tertiary |
| L6 | Osek | North Bohemia | Tertiary |
| L7 | Osek | North Bohemia | Tertiary |
| BC1 | Plzeň | West Bohemia | Carboniferous |
| BC2 | Slaný | Middle Bohemia | Carboniferous |
| BC3 | Karviná | North Moravia | Carboniferous |
| BC4 | Rosice-Oslavany | South Moravia | Carboniferous |
| A | Jastrzebie | Poland | Carboniferous |

**Table 2.** *Proximate and ultimate analysis of coals*

| Sample | $A_d$ (%) | $S_s^d$ (%) | $V^{daf}$ (%) | $Q_s^{daf}$ (MJ/kg) | $C^{daf}$ (%) | $H^{daf}$ (%) | $N^{daf}$ (%) | $S_o^{daf}$ (%) | $O_d^{daf}$ (%) |
|--------|------|------|-------|---------|-------|------|------|------|-------|
| L1 | 13.62 | 1.68 | 55.77 | 25.83 | 65.32 | 5.18 | 1.09 | 1.24 | 27.17 |
| L2 | 5.65 | 0.30 | 48.26 | 28.79 | 70.71 | 5.33 | 1.05 | 0.30 | 22.64 |
| L3 | 3.37 | 1.35 | 47.91 | 29.77 | 72.14 | 5.37 | 0.95 | 0.67 | 20.87 |
| L4 | 16.16 | 1.58 | 52.42 | 29.77 | 70.62 | 6.60 | 1.18 | 1.26 | 20.34 |
| L5 | 3.77 | 1.79 | 45.19 | 28.79 | 71.43 | 3.58 | 1.62 | 1.82 | 21.54 |
| L6 | 8.59 | 1.45 | 48.23 | 29.85 | 71.65 | 5.55 | 1.19 | 0.97 | 20.64 |
| L7 | 4.13 | 0.50 | 50.32 | 30.11 | 72.46 | 5.62 | 1.03 | 0.45 | 20.44 |
| BC1 | 9.71 | 0.59 | 39.87 | 34.43 | 82.13 | 5.59 | 1.94 | 0.46 | 9.88 |
| BC2 | 11.66 | 8.50 | 36.88 | 32.05 | 75.21 | 5.03 | 1.38 | 0.53 | 17.85 |
| BC3 | 4.26 | 0.65 | 28.78 | 36.16 | 88.49 | 5.23 | 1.35 | 0.58 | 4.35 |
| BC4 | 7.16 | 3.85 | 26.12 | 35.68 | 86.15 | 4.82 | 1.66 | 1.20 | 6.17 |
| A | 10.86 | 0.70 | 8.74 | 34.57 | 90.04 | 3.49 | 1.23 | 0.55 | 4.69 |

**Table 3.** *Petrographic characterization*

| Sample | $R_o$ (%) | Huminite/vitrinite (%) | Liptinite (%) | Inertinite (%) |
|---|---|---|---|---|
| L1 | 0.28 | 97 | 2 | 1 |
| L2 | 0.37 | 91 | 7 | 2 |
| L3 | 0.35 | 92 | 6 | 2 |
| L4 | 0.36 | 85 | 13 | 2 |
| L5 | 0.40 | 100 | 0 | 0 |
| L6 | 0.38 | 90 | 8 | 2 |
| L7 | 0.39 | 90 | 9 | 1 |
| BC1 | 0.66 | 68 | 22 | 10 |
| BC2 | 0.70 | 90 | 8 | 2 |
| BC3 | 1.01 | 46 | 6 | 48 |
| BC4 | 1.24 | 91 | 0 | 9 |
| A | 2.13 | 86 | 0 | 14 |

**Table 4.** *Huminite macerals of lignites*

| Sample | Atrinite | Densinite | Textinite | Ulminite | Gelinite | Corpohuminite |
|---|---|---|---|---|---|---|
| L1 | 14 | 25 | 19 | 33 | 4 | 2 |
| L2 | 7 | 20 | 3 | 52 | 8 | 1 |
| L3 | 8 | 19 | 2 | 56 | 4 | 3 |
| L4 | 8 | 19 | 2 | 56 | 4 | 3 |
| L5 | 0 | 0 | 0 | 0 | 100 | 0 |
| L6 | 2 | 35 | 1 | 43 | 8 | 1 |
| L7 | 6 | 25 | 0 | 48 | 10 | 1 |

**Table 5.** *Liptinite macerals of coals*

| Sample | Sporinite | Liptodetrinite | Resinite | Cutunite | Others |
|---|---|---|---|---|---|
| L1 | 1 | 0 | + | 0 | 1 |
| L2 | 2 | 2 | 1 | + | 2 |
| L3 | 1 | 2 | 1 | 1 | 1 |
| L4 | 2 | 7 | 0 | 1 | 2 |
| L5 | 0 | 0 | 1 | 1 | 0 |
| L6 | 1 | 2 | 1 | 0 | 2 |
| L7 | 2 | 4 | 1 | 1 | 1 |
| BC1 | 20 | 0 | 1 | 1 | 0 |
| BC2 | 7 | 0 | 0 | 1 | 0 |
| BC3 | 6 | 0 | 0 | 0 | 0 |

showing a well-preserved wood structure with low gelification and an increased textinite content.

Lignite samples from the North Bohemian basin are ortho-lignites with reflectances of $R_o = 0.35–0.40\%$ and with appreciable differences in their petrographic composition. Sample L5 is nearly pure gelinite (Hurník 1991). The remaining samples from this region are gelinitized material with dominant ulminite and sparse textinite. Of the detritic component, densinite predominates, whereas the attrinite content is low (below 10%), as is the gelinite content. Fine liptodetrinite is dispersed in detrite in samples L4 and L7.

Bituminous coal BC1 from Plzeň contains numerous microscopic sites of vitrite and clarite with sporinite, including megaspores. Owing to

**Fig. 2.** Photomicrographs of coal macerals. A, L1 textinite; B, L3 texto-ulminite; C, L4 densinite; D, L4 liptodetrinite and resinite (fluorescent mode); E, BC1 vitrinite and cutinite; F, BC1 cutinite (fluorescent mode); G, BC1 sporinite (fluorescent mode); and H, BC3 inertinite and vitrinite. Actual length of field of view 0.22 mm.

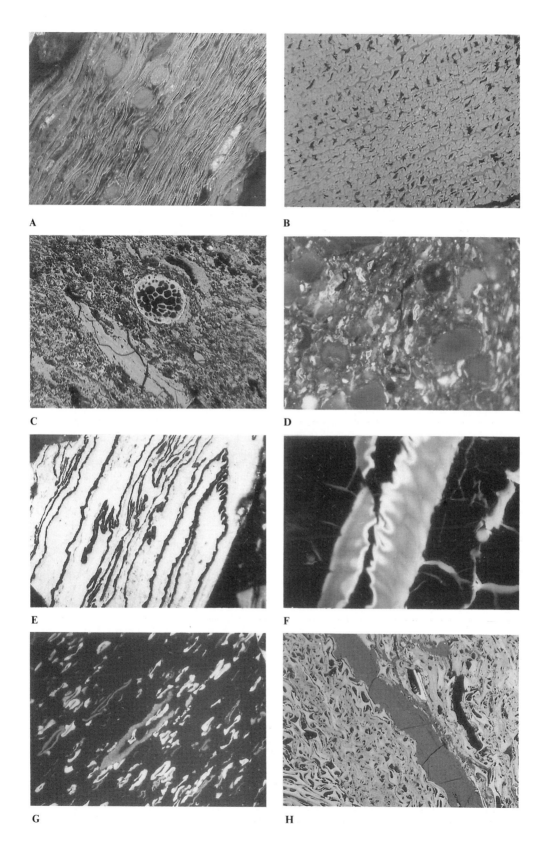

A

B

C

D

E

F

G

H

the lower inertinite content, trimacerit and durite are less frequent. Sample CB2 is vitrinite coal with dispersed sporinite and fragments of

inertinite. The sample contains many microscopic grains of both massive and framboidal pyrite. BC3 bituminous coal is a typical example of inertinite coal from the Saddle Member of the Upper Silesian basin (Dopita & Kumpera 1993). BC4 is representative of the Rosice-Oslavany Basin. The coalification series is complemented with anthracite A from the Polish part of the Upper Silesian Basin.

The coal samples were characterized by solid-state $^{13}C$ NMR (Fig. 3). The spectra contain two main peaks, one in the aliphatic (0–90 ppm) and the other in the aromatic region (90–165 ppm). The main peak in the aliphatic region has a maximum around 29 ppm and can be assigned to the carbon branched or long aliphatic chains and/or alicyclic carbons. A shoulder on this peak, seen in the higher rank coals, corresponds with carbon combined in short, straight alkyl groups. The chemical shift of the principal peak in the aromatic region is around 126 ppm and is assigned to unsubstituted aromatic carbons. The shoulder at 137 ppm, seen in the lower rank coals, corresponds with alkyl-substituted aromatic carbons. Oxygen-bearing carbons are found in three major areas: 50–90 ppm (aliphatic ethers, alcohols), 148–165 ppm (aromatic ethers, phenols) and from 165 ppm downfield carbons in carboxyls, carbonyls and aldehydes). The $^{13}C$ CP/MAS technique enabled the spectra to be rapidly obtained and quantitatively evaluated. The data,

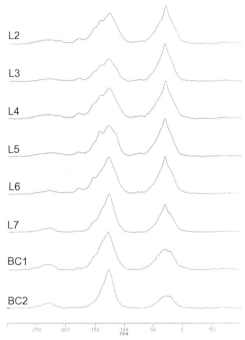

**Fig. 3.** $^{13}C$ CP/MAS NMR spectra of coals; see text for explanation.

**Table 6.** *Spectroscopic characterization of coals*

| Sample | $f_a$ | A CH$_3$ (%) | A CH$_2$ (%) | A CH (%) | A C=Car (%) | A C=O (%) | A CH$_2$/A CH$_3$ |
|---|---|---|---|---|---|---|---|
| **DRIFTS** | | | | | | | |
| L1 | nd | 8.10 | 14.25 | 4.13 | 34.51 | 39.01 | 1.76 |
| L2 | 0.58 | 5.85 | 13.73 | 0.73 | 45.53 | 34.16 | 2.35 |
| L3 | 0.55 | 7.46 | 14.68 | 3.03 | 38.50 | 36.33 | 1.97 |
| L4 | 0.48 | 7.94 | 18.38 | 3.87 | 23.91 | 45.90 | 2.31 |
| L5 | 0.58 | 7.52 | 10.19 | 3.08 | 30.27 | 48.95 | 1.35 |
| L6 | 0.54 | 9.15 | 16.02 | 4.52 | 32.90 | 37.41 | 1.75 |
| L7 | 0.50 | nd | nd | nd | nd | nd | nd |
| BC1 | 0.66 | 17.64 | 20.78 | 6.95 | 30.36 | 24.28 | 1.18 |
| BC2 | 0.73 | 6.97 | 35.41 | 1.88 | 27.84 | 27.90 | 5.08 |
| BC3 | 0.78 | 15.47 | 20.13 | 6.03 | 36.38 | 22.00 | 1.30 |
| BC4 | nd | 13.30 | 42.58 | 1.32 | 31.50 | 11.30 | 3.20 |
| A | nd | 4.75 | 19.29 | 7.21 | 52.49 | 16.25 | 4.06 |
| FTIR microscopy | | | | | | | |
| L1 huminite | nd | 4.61 | 6.59 | 0.65 | 41.97 | 46.18 | 1.43 |
| L2 huminite | nd | 5.28 | 7.29 | 0.95 | 32.93 | 53.35 | 1.38 |
| BC1 vitrinite | nd | 2.22 | 13.89 | 1.90 | 59.59 | 22.40 | 6.26 |
| A vitrinite | nd | 6.42 | 6.39 | 3.59 | 62.28 | 21.32 | 1.00 |
| BC1 sporinite | nd | 1.65 | 54.19 | 6.93 | 25.23 | 12.00 | 32.75 |
| BC1 cutinite | nd | 1.70 | 17.47 | 1.60 | 43.24 | 36.00 | 10.25 |

nd, Not determined.

however, may involve some error (Snape *et al.* 1989; Wind *et al.* 1993).

The coal aromaticity values, $f_a$, were calculated from the NMR spectra (Table 6). The data were plotted against the H/C atomic ratio and the light reflectance $R_o$ (Fig. 4). The former correlation, which has been commonly used, was thought by Choi *et al.* (1989), to indicate changes in the carbon aromaticity with structural properties of the organic coal components. Axelson (1987) further suggested that this dependence also expresses a decrease in the paraffinic carbon content and substitution of the aromatic ring with increasing coalification. However, no correlation between the $f_a$ values and the H/C ratio was observed by Yoshida *et al.* (1992). As light reflectance is one of the main indices of coalification, the $f_a$ and $R_o$ data can be used to infer structural differences between coal samples exhibiting nearly identical $R_o$ values (0.4). A similar correlation has been described by Kalkreuth *et al.* (1991). Our results indicate a dependence of $f_a$ on coal maceral composition; the finely dispersed liptodetrinite of samples L4 and L7 is responsible for the lower $f_a$ values and the highest aromaticity was found for the coal with a high inertinite content from the Saddle Member.

The structure of coal in the coalification series was examined by DRIFT. This technique does not require material to be mounted in a supporting matrix which can contain artifacts and enables measurement of the infra-red spectra with only minor sample preparation. The following band assignment can be made for the organic components of coal: 3500–3300 cm⁻¹, OH stretching; 3050 cm⁻¹, aromatic C—H stretching; 2920 and 2858 cm⁻¹, aliphatic C—H stretching; 1700 cm⁻¹, C=O stretching; 1600 cm⁻¹, aromatic ring vibration; 1450 cm⁻¹,

CH₂/CH₃ deformation; 1375 cm⁻¹, CH₃ deformation; 1200 cm⁻¹, C—O—C stretching; and 900–700 cm⁻¹, aromatic out of plane.

The DRIFT spectra (Fig. 5) indicate that the aromatic content increases with increasing coalification (intensity increase at 3050 cm⁻¹, 1600 cm⁻¹ and in the 700–900 cm⁻¹ range), whereas the reverse is true of the aliphatic content (3000–2800, 1450 and 1375 cm⁻¹ bands). The decrease in the intensities of bands at 3400 cm⁻¹, 1700 cm⁻¹ and about 1000 cm⁻¹ indicates a decreasing fraction of oxygen functional groups. A semi-quantitative assessment of the changes is presented in Table 6, giving the relative percentages of methyl, methylene and methine groups in the coal samples. The aromatic C=C and carbonyl contents are also included. The aromatic C=C content increases with increasing coalification, whereas the carbonyl content exhibits the opposite trend, in accordance with the decreasing oxygen content $O_d^{daf}$ (Table 2). The methylene and methyl data were used to calculate the CH₂/CH₃ ratio, allowing aliphatic chain length and branching in the organic moiety to be assessed (the ratio increases with increasing chain length and decreases with increasing branching). This parameter has been discussed by Lin & Ritz (1993) with respect to the effect of the chain length and branching on the gas/condensate proneness of coal and kerogens. Our ratios lie typically between 1.2 and 5.1 and agree with the data by Lin & Ritz for coal macerals.

The structure of the individual macerals of some coal samples was investigated by reflectance FTIR microscopy. This technique allows small sample areas (20 × 20 μm) to be analysed and gives semi-quantitative data on the chemical structure of coal macerals without the need for time-consuming isolation. The spectra of huminites in the lignites L1 and L3 and vitrinites in the bituminous coal sample BC1 and anthracite sample A are shown in Fig. 6.

The spectra for huminite/vitrinite change with increasing coalification in a manner similar to the spectra of the parent coal. Thus with increasing rank the spectra of huminite/vitrinite show an increase in the aromatic bands of C—H and C—C bonds (3050 cm⁻¹, 900–700 cm⁻¹, 1600 cm⁻¹). Furthermore, it is possible to see a decrease of carbonyl bands near 1700 cm⁻¹. The semi-quantitative structural parameters derived from the FTIR microscopy data are given in Table 6. The aromatic C=C content in the huminites/vitrinites is higher than in the initial coal, except for the lignite L3 where the two values approach each other. The oxygen

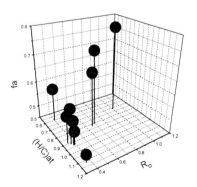

**Fig. 4.** Relationship between atomic ratio H/C, vitrinite reflectance $R_o$ and coal aromaticity $f_a$; see text for explanation.

functional group content is higher for the lignites and nearly identical with that of the parent coal for the bituminous coal and anthracite. Table 6 also shows the structural parameters of two additional macerals of the liptinite group, sporinite and cutinite, for the BC1 coal. Their spectra are shown in Fig. 7. The two macerals exhibit high aliphatic C—H contents compared with vitrinite and this is most apparent in cutinite. Sporinite displays the lowest aromatic C=C bond content of all the samples.

It is well known that substantial amounts of $CH_4$ are produced from coals of rank ranging

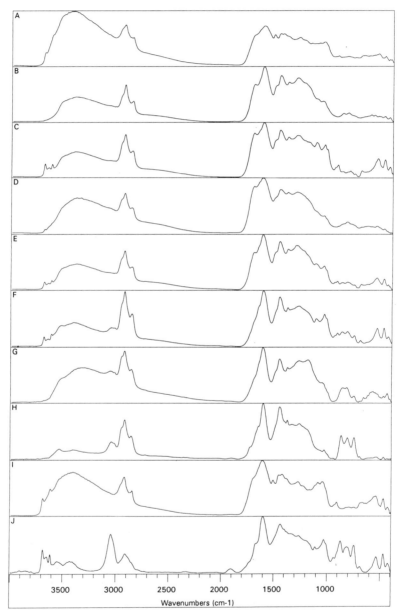

**Fig. 5.** DRIFT spectra of coals: A, L1; B, L2; C, L3; D, L4; E, L5; F, L6; G, BC1; H, BC2; I, BC3; J, BC4; and K, A; see text for explanation.

from high-volatile bituminous coals ($V_{daf} = 40\%$) to anthracite ($V_{daf} = 5\%$). Important economic deposits of gas can be found in coals with $R_o = 1-3.0\%$. In the case of the coals studied in this investigation, those with these properties are: BC3 ($R_o = 1.01\%$, $V_{daf} = 28.78\%$, H/C $= 0.71$); BC4 ($R_o = 1.24\%$, $V_{daf} = 26.12\%$, H/C $= 0.67$); and A ($R_o = 2.13\%$, $V_{daf} = 8.74\%$, H/C $= 0.47$).

These are the coals in which it has been shown that the percentages of aliphatic $-CH_3$, $-CH_2$ and $-CH$ and of aromatic C$=$C groups are highest, and in which the aromaticity value $f_a$ is also highest. Conversely, they are the coals in

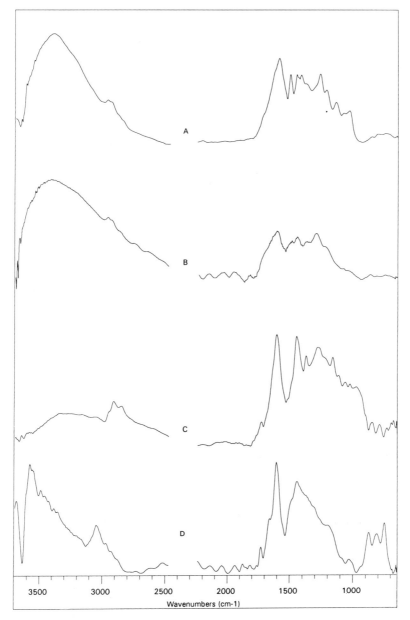

**Fig. 6.** Micro-FTIR spectra of huminites/vitrinites of coals: A, L1 huminite; B, L3 huminite; C, BC1 vitrinite; and D, A anthracite.

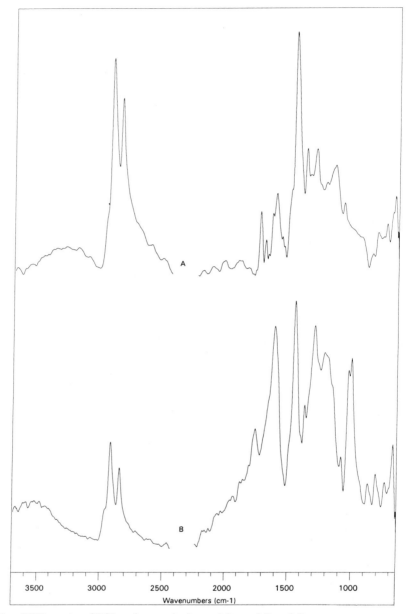

**Fig. 7.** Micro-FTIR spectra of BC1 coal macerals: A, sporinite; and B, cutinite.

which the percentage of carbonyl groups is lowest. All these factors relate to the gas proneness of the coals.

The highest incidence of methane outbursts in the Czech Republic is found in many of the seams of the Ostrava-Karviná coalfield (Dopita & Kumpera 1993). The sample BC3 comes from this area and shows the greatest proneness to $CH_4$. Further studies will consider the relationship between the structural parameters of coal from the Upper Silesian Basin and the abundance of $CH_4$.

## Conclusions

Optical and infra-red microscopy, DRIFT and CP/MAS NMR were used to determine the degree of coalification and structural characteristics of samples of lignites and bituminous coal from the Czech Republic, whose huminite/vitrinite reflectance values lie within the range of $R_o = 0.28–2.13\%$. The quantitative structural parameters obtained by FTIR microscopy, i.e. the percentages of aliphatic and aromatic structures and carbonyl contents, revealed that the huminite/vitrinite maceral group shows similar changes with increasing coalification to those of the parent coal. The structure of two macerals of the liptinite group, sporinite and cutinite, was investigated for bituminous coal and the aromatic and oxygen functional group contents were found to be higher in cutinite than in sporinite. Aromaticity $f_a$, which was determined by $^{13}$C CP/MAS NMR, was found to be a valuable parameter reflecting the degree of coalification, particularly for low and middle rank coal, and thus complements the information derived from vitrinite reflectance $R_o$ and $(H/C)_{at}$ ratio data.

The results suggest that optical microscopy and FTIR microscopy are powerful tools for the study of coal and the determination of its rank. The two techniques allow in-depth characterization of the structure of coal macerals, giving valuable information on coal geology, technology and gas proneness. Further research should be aimed at applying the methods to the study of maceral reactivity, for inertinite in particular.

# References

AXELSON, D. E. 1987. Solid state carbon-13 nuclear magnetic study of Canadian coals. *Fuel Processing Technology*, **16**, 257–278.

BÖHLMANN, W., HOFFMANN, W. D., ZAHN, A. & VOLKMANN, N. 1992. Spectroscopic and fluorescence microscopic investigation of lignite hydrogenation residues. *Fuel*, **71**, 545–552.

BOTTO, R. E., WILSON, R. & WINANS, R. E. 1987. Evaluation of the reliability of solid state $^{13}$C NMR spectroscopy for the quantitative analysis of coals: study of whole coals and maceral concentrates *Energy & Fuels*, **1**, 173–181

CHOI, C. Y., MUNTEAN, J. V., THOMPSON, A. R. & BOTTO, R. E. 1989. Characterization of coal macerals using combined chemical and NMR spectroscopic methods. *Energy & Fuels*, **3**, 528–533

DIESSEL, C. F. K. 1992. Coal facies and depositional environments *In*: GUNDERMANN, I. (ed.) *Coal-bearing Depositional Systems*. Springer, Berlin, 161–261.

DOPITA, M. & KUMPERA, O. 1993. Geology of the Ostrava-Karviná coalfield, Upper Silesian basin, Czech Republic, and its influence on mining. *International Journal of Coal Geology*, **23**, 291–321.

DYRKACZ, G. R., BLOOMQUIST, C. A. A. & SOLOMON, P. R. 1984. Fourier transform infrared study of high purity maceral types. *Fuel*, **53**, 536–542

——, ——, RUSIC, Z. & CRELLING, J. C. 1991. An investigation of the vitrinite maceral group in microlithotypes using density gradient saparation methods. *Energy & Fuels*, **5**, 155–163

GANZ, H. & KALKREUTH, W. 1990. The potential of infrared spectroscopy for the classification of kerogen, coal and bitumen. *Erdöl und Kohle*, **43**, 116–117.

HURNÍK, S. 1991. Coal (humic gel) clastic dike in the North Bohemian Basin (Miocene). *Vistník ÚÚG*, **66**, 23–30.

KALKREUTH, W., STELLER, M., WIESCHENKAMPER, I. & GANZ, S. 1991. Petrographic and chemical characterization of Canadian and German coals in relation to utilization potential 1. Petrographic and chemical characterization of feedcoals *Fuel*, **70**, 683–694

KISTER, J., GUILIANO, M., TOTINO, E. & MULLES, J. F. 1986. Analyse par spectroscopie infrarouge á transformée de Fourier (IRTF) des macéraux de charbons. *Comptes Rendu de Academie des Sciences Paris*, **302**, 527–531.

KUEHN, D. W., SNYDER, R. W., DAVIS, A. & PAWTER, P. C. 1982. Characterization of vitrinite concentrates 1. Fourier transform infrared studies. *Fuel*, **61**, 682–694.

LANDAIS, P. & ROCHDI, A. 1990. Reliability of semiquantitative data extracted from transmision microscopy–Fourier transform infrared spectra of coal. *Energy & Fuels*, **4**, 290–295

LEMOS DE SOUSA, M. J., FLORES, D., PINHEIRO, H. J. & VASCONCELDS, L. 1992. Coal classification and codification up-date on the state of the art and critical review. *Publicacoes do Museu e Laboratório Mineralógico e geológico da Faculdade de ciencias do Porto*, **2**, 1–61.

LIN, R. & RITZ, P. 1993. Studying individual macerals using i. r. microspectroscopy, and implications on oil versus gas/condensate proneness and "low-rank" generation. *Organic Geochemistry*, **20**, 695–706

MASTALERZ, M. & BUSTIN, R. M. 1993. Electron microprobe and micro-FTIR analyses applied to maceral chemistry. *International Journal of Coal Geology*, **24**, 333–345.

——WILKS, K. R. & BUSTIN, R. M. 1993. Variation in vitrinite chemistry as a function of associated liptinite content; a microprobe and FT–i.r. investigation. *Organic Geochemistry*, **20**, 555–562.

MCFARLANE, R. A., GENTZIS, T., GOODARZI, F., HANNA, J. V. & VASSALLO, A. M. 1993. Evolution of the chemical structure of Hat Creek resinite during oxidation: a combined FT–IR photoacoustic, NMR and optical study. *International Journal of Coal Geology*, **22**, 119–147

MICHAELIAN, K. H., FRIESEN, W. I. 1990. Photoacoustic FT–i.r. spectra of separated western Canadian coal macerals. Analysis of the CH stretching region by curve-fitting and deconvolution *Fuel*, **69**, 1271–1275.

PAINTER, P. C. , SNYDER, R. W., STARSINIC, M.,
COLEMAN, M. M., KUEHN, D. W. & DAVIS, A.
1981. Concerning the application of FT–IR to the
study of coal: a critical assessment of band
assignment and the application of spectral analysis
programs. *Applied Spectroscopy*, **35**, 475–485.

RETCOFSKY, H. L. & VANDERHART, D. L. 1978.
$^{13}C$–$^1H$ Cross-polarization nuclear magnetic
resonance spectra of macerals from coal. *Fuel*,
**57**, 421–423.

ROZKOŠNÝ, I., MACHOVIČ, V., PAVLÍKOVÁ, H. &
HEMELÍKOVÁ, B. 1994 Chemical structure of
migrabitumens from Silurian Crinoidea, Prague
Basin, Barrandian (Bohemia). *Organic geochemistry*, **21**, 1131–1140.

SENFTLE, J. T., KUEHN, D., DAVIS, A., BROZOSKI, B.
RHOADS, C. PAINTER, P. C. 1984. Characterization of vitrinite concentrates 3. Correlation of FT-
i.r. measurments to thermoplastic and liquefaction behaviour. *Fuel*, **63**, 245–250.

SNAPE, C. E., AXELSON, D. E., BOTTO, R. E.,
DELPUECH, J. J., TEKELY, P., GERSTEIN, B. C.,
PRUSKI, M., MACIEL, G. E. & WILSON, M. A.
1989. Quantitative reliability of aromaticity and
related measurments on coals by $^{13}C$ n.m.r. A
debate. *Fuel*, **68**, 547–560.

SNOWDON, L. R., BROOKS, P. W. & GOODARZI, F.
1986. Chemical and petrological properties of
some liptinite-rich coals from British Columbia.
*Fuel*, **65**, 459–471.

STASIUK, L. D., KYBETT, B. D. & BEND, S. L. 1993.
Reflected light microscopy and micro-FTIR of
Upper Ordovician Gloeocapsomorpha prisca
alginite in relation to paleoenvironment and
petroleum generation, Saskatchewan, Canada.
*Organic Geochemistry*, **20**, 707–719.

TEICHMÜLLER, M. & TEICHMÜLLER, R. 1982
Fundamentals of coal petrology. *In*: CHANDRA,
D., MACKOWSKY, M.-TH., STACH, T., TAYLOR,
G. H., TEICHMÜLLER, M. & TEICHMÜLLER, R.
(eds) *Stach's Textbook of Coal Petrology*, Gebrü-
der Borntraeger, Berlin, 5–82.

VASSALO, A. M., LOCKHART, N. C., HANNA,
J. V., CHAMBERLAIN, R., PAINTER, P. C. &
SOBKOWIAK, M. 1991. Infrared and nuclear
magnetic resonance spectroscopy of density
fractions from Callide coal. *Energy & Fuels*, **5**,
477–482.

WIND, R. A., MACIEL, G. E. & BOTTO, R. R. 1993.
Quantitation in $^{13}C$ NMR spectroscopy of
carbonaceous solids. *In*: BOTTO, R. E. &
SANADA, Y. (eds) *Magnetic Resonance of Carbonaceous Solids*. Advances in Chemistry Series,
**229**, 3–26.

YOSHIDA, R., YOSHIDA, T., NARITA, H. &
MAEKAWA, Y. 1992. Carbon distribution analysis
of coals by CP/MAS $^{13}C$ NMR. *Journal of Coal
Quality*, **11**, 38–43.

# Petrological coal seam accumulation model for the Zacler Formation of the Lower Silesian coal basin, southwestern Poland

GRZEGORZ J. NOWAK

*Polish Geological Institute, Lower Silesian Branch, al. Jaworowa 19, 53-122 Wroclaw, Poland*

**Abstract:** Coal-bearing strata of Westphalian A–C age occur in the Zacler Formation of the Lower Silesian coal basin of the Central Sudetes. Numerous coal seams were studied petrographically to reconstruct their depositional environments. Coal seams were sampled from several localities, predominantly from mines and boreholes in the Wałbrzych (northwestern), Nowa Ruda (eastern) and Słupiec (southeastern) regions of the Lower Silesian coal basin. The coals studied range from high- to medium-volatile bituminous coals ($R_m$ 0.72–1.35%). Petrographically, the coals from the northwestern region are characterized by low to high vitrinite contents, low to medium inertinite contents and a constant amount of liptinite. Coals in the eastern part of the basin are vitrinite dominated, with medium to high inertinite and negligible liptinite contents. All of the coals studied in the southeastern area of the basin have a moderate vitrinite content with high percentages of inertinite (semifusinite, fusinite and inertodetrinite) and with constant amounts of liptinite.

Depositional environments were determined from the petrographic maceral analysis indices. Environments in peat swamps were interpreted from triangular facies diagrams and the tissue-preservation index versus gelification index relationships. Forest moors of telmatic and limno-telmatic zones predominate in the northwestern and eastern areas. In the southeastern region coals of reed–moor facies are also present. On the basis of the maceral composition, two main petrological coal facies were identified: (1) vitrinite–fusinite and (2) densosporinite facies. The variation in spore assemblage and petrographic composition of the seams indicates that environmental changes were accompanied by changes in vegetation. The vitrinite-fusinite facies, high in vitrinite, is characteristic of the forest moor/swamp environment (the Wałbrzych and Nowa Ruda regions), where arborescent vegetation predominated. The densosporinite facies, high in inertinite and intermediate in vitrinite, is typical of reed moor/marsh environments where herbaceous plants prevail.

The main aim of this paper is to reconstruct coal facies and palaeoenvironmental conditions for the coal seams of the Zacler Formation in the Lower Silesian coal basin, southwestern Poland. The analysis is carried out on coal petrology data.

The Lower Silesian coal basin is the smallest of the three Polish bituminous coal basins. It represents an intermontane basin with a limno-fluvial character of coal deposition. It is located in the Intrasudetic Basin, which is filled with Carboniferous, Permian, Triassic and Upper Cretaceous sediments (Fig. 1). This basin is a major geological unit within the Central Sudetes, forming a peripheral zone of the Bohemian Block.

Coal-bearing strata of the Intrasudetic Basin are of Namurian and Westphalian age (Fig. 2). The lower part of the succession is the Wałbrzych Formation, up to 300 m thick, consisting of sandstones and mudstones, with rare conglomerates interlayered with coal seams. This formation is of Namurian A age and is overlain by the Biały Kamień Formation of Namurian B and C age (Górecka & Górecka-Nowak 1990), consisting

mainly of coarse conglomerates and sandstones. The youngest coal-bearing unit, the Zacler Formation, is up to 800 m thick and is composed mainly of sandstones and mudstones with numerous coal seams. This formation is of Westphalian A B and C age (Górecka-Nowak 1995). Sedimentation of phytogenic material was limited to some local troughs, such as the Wałbrzych trough, the Jugow–Nowa Ruda trough, the Słupiec trough and several other troughs in the Czech Republic (Augustyniak 1964; 1970).

## Samples and methods

Coal samples were collected from localities in the Lower Silesian coal basin, predominantly from mines and boreholes in the Wałbrzych (northwestern), Nowa Ruda (eastern) and Słupiec (southeastern) regions of the Intrasudetic Basin (Table 1; Fig. 3). Megascopic coal seam profiles have been constructed for coal seams studied underground. These include descriptions of lithotypes and the lithology of clastic partings

*From* Gayer, R. & Harris, I. (eds) 1996, *Coalbed Methane and Coal Geology*, Geological Society Special Publication No 109, pp 261–286.

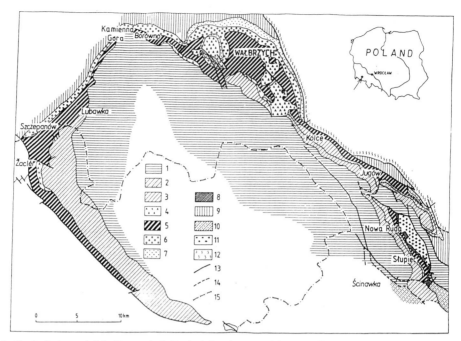

**Fig. 1.** Geological map of the Intrasudetic Basin (after Augustyniak 1970, slightly simplified). 1, Permian and younger undivided rocks; 2, sediments of the Ludikowice Formation (Stephanian); 3, deposits of the Glinik Formation (Upper Westphalian–Lowest Stephanian); 4, volcanic rocks occurring among Carboniferous strata; 5, coal-bearing strata of the Zacler Formation (Westphalian A–C); 6, conglomerates of the Biały Kamień Formation (Uppermost Namurian); 7, coal-bearing strata of the Wałbrzych Formation (Namurian A); 8, weathered rocks deposited on the gabbro–diabase massif of Nowa Ruda; 9, Lower Carboniferous deposits; 10, phyllites; 11, gabbros and diabases of the Nowa Ruda massif; 12, gneisses of the Sowie Mountains; 13, ascertained faults; 14, supposed faults; and 15, country boundaries.

| AGE | | | | LITHOSTRATIGRAPHICAL UNITS |
|---|---|---|---|---|
| CARBONIFEROUS | SILESIAN | STEPHANIAN | D | LUDWIKOWICE FORMATION |
| | | | C | |
| | | | B | |
| | | | A | |
| | | WESTPHALIAN | D | GLINIK FORMATION |
| | | | C | ZACLER FORMATION |
| | | | B | |
| | | | A | |
| | | NAMURIAN | C | BIAŁY KAMIEŃ FORMATION |
| | | | B | |
| | | | A | WAŁBRZYCH FORMATION |
| | LOWER VISEAN | | | SZCZAWNO FORMATION |

**Fig. 2.** Lithostratigraphy of Carboniferous deposits in the Intrasudetic Basin.

**Table 1.** *Analysed coal seams of the Zacler Formation*

| Location | Seam No. | No. of profiles | No. of samples |
|---|---|---|---|
| Wałbrzych region ($n = 81$) | | | |
| Victoria mine | 430 | 2 | 17 |
| GV-19 borehole | 304 | | |
| | — | — | 32 |
| | 317 | | |
| Borówno 1 borehole | 307 | | |
| | — | — | 30 |
| | 317 | | |
| Suliszów borehole | 447 | — | 2 |
| Nowa Ruda region ($n = 22$) | | | |
| Piast field of the | 405 | 1 | 7 |
| Nowa Ruda mine | — | | |
| | 415 | 4 | 15 |
| Słupiec region ($n = 25$) | | | |
| Słupiec field of the | 409 | 3 | 7 |
| Nowa Ruda mine | 410 | 3 | 6 |
| | 412/413 | 2 | 7 |
| | 414/415 | 2 | 5 |

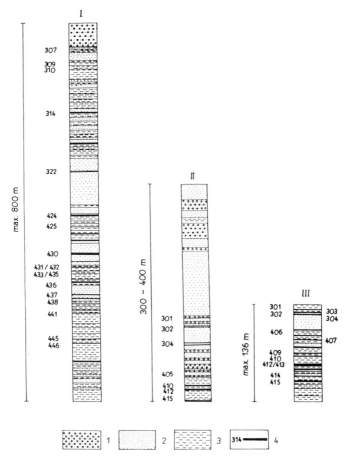

**Fig. 3.** Schematic lithological profiles of Zacler Formation of the studied areas. I, the Wałbrzych region (northwestern part of the Intrasudetic Basin); II, the Nowa Ruda region (eastern part of the Intrasudetic Basin); III, the Słupiec region (southeastern part of the Intrasudetic Basin). 1, Conglomerates; 2, sandstones; 3, claystones and siltstones; 4, coal seams (301 . . . , numbers of coal seams).

as well as of floor and roof rocks. Each coal section has been subdivided into petrographic intervals, which are often represented by a single lithotype or group of lithophytes, consisting of two or more coal types (Nowak 1993). Where it was impossible to recognize lithotypes, seams were arbitrarily divided into petrographic intervals (seams 414, 415, 412/413, 410, 409 of the Słupiec area). Channel samples were taken from each petrographic interval and coal seams from cores. Polished pellets were made from all samples according to standard ICCP preparation methods (Stach *et al.* 1982). Maceral analysis was carried out in reflected white light. The analytical procedures used in microscopic studies followed ICCP standards and the maceral terminology used was that of Stach

*et al.* (1982). Coal rank was measured as mean random telocollinite reflectance ($R_m$) at 546 nm.

## Petrological characteristics of the coal seams

### Coal rank

For the purposes of this paper, low and medium rank coal seams ($R_m = 0.72$–$1.35\%$) were selected. The highest degree of coalification is displayed by seams of the Nowa Ruda region ($R_m = 1.28$–$1.35\%$), whereas the coals of the Słupiec area show a constant $R_m$ of 0.99–1.11%. The lowest rank coals selected for this study are seams of the northwestern part of the basin

($R_m = 0.72–1.09\%$). In this latter area coals are also affected by magmatic activity, producing the highest degree of metamorphism up to meta-anthracites and natural cokes (Kwiecińska 1967, 1980).

## Lithotypes

The coal lithotype classification used is a slight modification of the Australian lithotype scheme (Diessel 1965; Marchioni 1980, Lamberson et al. 1992). The minimum lithotype thickness recorded was 5 mm.

Eight lithotypes were identified in the coal seams of the Zacler Formation, namely, bright, banded bright, semibright, semibright with fibrous coal, banded semibright, banded coal, dull and fibrous coal (Nowak 1992; 1993). In the Wałbrzych region semibright coals predominate (e.g. the 430 seam), whereas other coal types are represented sporadically. A characteristic feature of the 430 seam is a high content of clastic partings, which are intercalated with generally semibright and banded coal lithotypes. The layers of semibright coal are thick in the seam profiles studied. However, dull, fibrous coal and the other lithotypes also occur as thin bands in the profiles of the examined seam.

In the Nowa Ruda area bright and semibright types, often forming an undivided group of litho-types, prevail over the rest of lithotypes (i.e. 415 and 405 seams) (Nowak 1992; 1993). A low number of clastic partings is characteristic of these coals (Table 6). However, in the same region (seam 405) the banded semibright lithotype occurs most frequently and is interlayered with semibright and dull coal (Table 6).

In the Słupiec region the coal structure made determination of lithotypes impossible.

## Maceral and microlithotype composition

Maceral and microlithotypes analyses were conducted on all samples for which the determination of lithotypes was possible (seams occurring in mines), whereas in boreholes, where the coal structure was not observable, only maceral analyses were made. The maceral and micro-lithotypes analyses are characterized by inter-mediate to high vitrinite contents and cor-respondingly intermediate to low inertinite and constant amounts of liptinite (exinite) (Tables 2–5; Fig. 4). In seam 430 the vitrinite content is usually 47.6–71.8% (Table 2), mostly exceeding 50%, whereas in seams drilled in the boreholes GV-19, Borowno 1 and Suliszow the vitrinite

content varies from nearly 30 to 83.3% (Tables 3–5). This maceral group is represented by telocolli-nite, desmocollinite and rarely by vitrodetrinite, telinite and corpocollinite (Tables 2–5). Pseudo-vitrinite also occurs sporadically in the studied coals. Inertinite is represented mainly by semi-fusinite and fusinite, predominating over inerto-detrinite. Lesser inertinites include macrinite, sclerotinite and micrinite. However, only spor-inite makes up a significant proportion of liptinitic macerals, whereas the others are rare in the composition of coals in this area of the basin. In the coal seams studied a great variety of mineral matter is typical (Tables 2–5). Clays, carbonates and framboidal pyrite are the domi-nant inorganic constituents. These coal seams display a great variation in principal microlitho-type content (Table 2). The predominant types in seams from this area are claritic, vitrinertitic and trimaceritic coals.

Seams of the Nowa Ruda region mainly represent coals of a high vitrinite content, medium to high inertinite and negligible exinite percentages (Table 6, Fig. 4). The vitrinite content varies from 47.0 to 83.8% and the major maceral of this group is collinite, with rare vitrodetrinite. In this region inertinite does not exceed 32%. It appears as fusinite, semifusinite and inertodetrinite, with subordinate micrinite, macrinite and sclerotinite. The main maceral of the liptinite group is metasporinite. Cutinite and resinite occur sporadically. In the coal seams studied in the eastern part of the basin low amounts of inorganic components occur (Table 6). Clay minerals predominate, forming bands, lenses and dispersed mineral matter. Among the microlithotypes, vitrite and vitrinertite prevail (Table 6).

All of the coal seams studied in the Słupiec area are of intermediate vitrinite content (34.4–77.0%), high inertinite content (14.2–45.8%) and constant exinite percentages (7.6–19.4%) (Table 7, Fig. 4). Vitrinite consists mainly of desmocollinite and telocollinite, occasional vitro-detrinite and accessory telinite and corpocollinite. Semifusinite, fusinite and inertodetrinite domi-nate in the inertinite group with macrinite, sclerotinite and micrinite being of secondary import-ance. Liptinite occurs mainly as spori-nite, primarily crassisporinite and tenuisporinite. Megaspores were also observed, along with rare cutinite and resinite. Typical of coals of this area is the low content of inorganic components (Table 7), which comprise mainly clay minerals, pyrite, carbonates and quartz. In this third area trimacerite predominates over other microlitho-types (Table 7), with vitrite occurring in lesser amounts.

**Table 2.** Petrographic composition of the 430 coal seam of the Wałbrzych region

| Profile | Interval | Lithotype | Thickness (m) | Petrography (vol%) Vitrinite | | | | | Exinite | | | | Intertinite | | | | | | Mineral matter | | | | | Microlithotypes | | | | | | | |
|---|---|---|---|---|---|---|---|---|---|---|---|---|---|---|---|---|---|---|---|---|---|---|---|---|---|---|---|---|---|---|---|
| | | | | Tc | De | Co | Te | Vd | Sp | Cu | Re | Ld | Id | Mi | Fu | Sf | Sc | Ma | Il | Pi | Ga | Q | W | Vt | Lp | In | CR | Vi | Du | Tr | Ca |
| 1 | A | Semibright | 0.80 | 17.2 | 17.8 | — | 0.6 | 12.6 | 7.2 | 0.4 | 0.4 | — | 7.8 | 0.8 | 6.8 | 7.2 | 0.4 | 2.0 | 4.0 | 2.0 | — | — | 2.4 | 24.8 | — | 9.4 | 13.2 | 10.4 | 2.8 | 26.4 | 13.0 |
| | B | Semibright with fibrous coal | 0.10 | 23.4 | 16.6 | 0.4 | 0.4 | 5.4 | 9.0 | 2.0 | 0.4 | — | 12.6 | — | 11.6 | 5.4 | 0.8 | 1.6 | 5.2 | 2.0 | — | 0.4 | 0.4 | 29.0 | 0.4 | 23.0 | 15.0 | 4.8 | 2.2 | 12.6 | 13.0 |
| | C | Parting | 0.10 | — | — | — | — | — | — | — | — | — | — | — | — | — | — | — | — | — | — | — | — | — | — | — | — | — | — | — | — |
| | D | Parting | 0.10 | — | — | — | — | — | — | — | — | — | — | — | — | — | — | — | — | — | — | — | — | — | — | — | — | — | — | — | — |
| | E | Bright and semibright | 0.25 | 45.0 | 18.2 | 0.2 | — | 3.2 | 11.4 | 1.2 | 0.4 | 0.2 | 3.6 | — | 1.8 | 2.2 | 0.6 | 0.6 | 3.6 | 1.2 | — | 0.6 | 0.8 | 49.8 | 1.0 | 3.2 | 18.2 | 4.4 | 1.2 | 12.2 | 10.0 |
| | F | Parting | 0.10 | — | — | — | — | — | — | — | — | — | — | — | — | — | — | — | — | — | — | — | — | — | — | — | — | — | — | — | — |
| | G | Semibright | 0.60 | 35.8 | 29.2 | 0.6 | — | 1.6 | 10.6 | 0.6 | 0.2 | 0.2 | 5.0 | 1.4 | 2.8 | 7.0 | 0.4 | 1.4 | 0.4 | 1.2 | 0.8 | — | 0.4 | 40.8 | 0.2 | 6.6 | 13.4 | 7.2 | 2.4 | 19.6 | 9.8 |
| 2 | A | Semibright | 0.30 | 37.4 | 24.8 | 0.6 | — | 0.8 | 13.4 | 0.8 | — | — | 3.8 | 3.6 | 3.6 | 3.0 | 1.0 | 1.4 | 2.8 | 0.2 | 0.4 | — | 0.4 | 44.4 | 2.0 | 4.2 | 14.6 | 4.8 | 0.8 | 21.2 | 8.0 |
| | B | Banded semibright | 0.10 | 32.2 | 34.0 | — | — | — | 10.0 | 1.2 | 0.2 | — | 6.4 | 0.6 | 3.2 | 3.6 | 0.8 | 1.0 | 1.6 | 0.2 | 0.2 | 0.2 | 0.2 | 39.2 | — | 5.0 | 25.8 | 6.6 | — | 15.6 | 7.8 |
| | C | Semibright | 0.30 | 37.0 | 20.0 | 0.6 | — | 1.6 | 10.8 | 0.6 | — | 0.6 | 4.8 | 1.0 | 3.0 | 5.2 | 0.2 | 1.0 | 5.6 | 3.4 | 1.4 | — | 2.0 | 41.4 | 0.2 | 5.0 | 18.6 | 4.0 | 3.0 | 14.8 | 13.0 |
| | D | Parting | 0.10 | — | — | — | — | — | — | — | — | — | — | — | — | — | — | — | — | — | — | — | — | — | — | — | — | — | — | — | — |
| | E | Banded | 0.10 | 26.0 | 19.6 | — | — | 2.2 | 8.2 | 1.6 | — | 2.6 | 6.6 | — | 7.0 | 9.4 | 0.6 | 0.8 | 6.6 | 2.4 | 2.4 | — | 1.0 | 27.8 | 0.8 | 18.2 | 21.8 | 2.0 | 5.8 | 6.6 | 17.0 |
| | F | Parting | 0.20 | — | — | — | — | — | — | — | — | — | — | — | — | — | — | — | — | — | — | — | — | — | — | — | — | — | — | — | — |
| | G | Banded | 0.10 | 34.8 | 19.8 | — | — | 3.0 | 13.4 | 0.3 | 0.3 | — | 5.5 | 0.3 | 5.2 | 1.2 | 1.2 | 0.6 | 9.1 | 2.1 | — | 1.5 | 1.7 | 36.3 | 0.3 | 7.6 | 22.8 | 4.2 | — | 8.2 | 20.6 |
| | H | Parting | 0.40 | — | — | — | — | — | — | — | — | — | — | — | — | — | — | — | — | — | — | — | — | — | — | — | — | — | — | — | — |
| | O | Group of lithotypes | 0.25 | 21.4 | 18.8 | 0.4 | — | 8.6 | 11.0 | 0.2 | 0.4 | 0.6 | 5.0 | — | 6.4 | 5.8 | 1.0 | 1.4 | 10.8 | 2.2 | 0.2 | — | — | 31.2 | 0.2 | 11.6 | 13.8 | 5.8 | 5.2 | 16.2 | 16.0 |

Vitrinite: Tc, telocollinite; De, desmocollinite; Co, corpocollinite; Te, telinite; and Vd, vitrodetrinite.
Exinite: Sp, sporinite; Cu, cutinite; Re, resinite; and Ld, liptodetrinite.
Inertinite: Id, inertodetrinite; Mi, micrinite; Fu, fusinite; Sf, semifusinite; Sc, sclerotinite; and Ma, macrinite.
Mineral matter: Il, clay minerals; Pi, pyrite; Ga, galena; Q, quartz; and W, carbonates.
Microlithotypes: Vt, vitrite; Lp, liptite; In, inertite; CR, clarite; Vi, vitrinertite; Du, durite; Tr, trimacerite; and Ca, carbominerite.

**Table 3.** *Maceral composition (vol%) of coals in borehole GV-19*

| Depth (m) | Vitrinite (vol%) | | | | | | Exinite (vol%) | | | | Inertinite (vol%) | | | | | | Mineral matter (vol%) | | | | |
|---|---|---|---|---|---|---|---|---|---|---|---|---|---|---|---|---|---|---|---|---|---|
| | Tc | De | Co | Te | Vd | Ps | Sp | Cu | Re | Ld | Id | Mi | Fu | Sf | Sc | Ma | Il | Pi | Ga | Q | W |
| 17.25 | 30.4 | 21.0 | — | 1.6 | 0.2 | 0.8 | 8.8 | — | — | 0.8 | 4.2 | 0.4 | 18.8 | 6.6 | 1.2 | 1.8 | 2.2 | 1.8 | — | 0.2 | — |
| 30.00 | 11.8 | 34.0 | 0.4 | — | — | — | 8.2 | — | 0.6 | 0.8 | 4.4 | — | 16.6 | 15.6 | 1.0 | 1.6 | 2.6 | 2.0 | — | 0.2 | 0.2 |
| 31.00 | 53.0 | 22.0 | 0.4 | 0.4 | 0.4 | — | 7.2 | 0.2 | 0.2 | — | 2.8 | 0.2 | 5.6 | 3.2 | 0.4 | 1.4 | 0.6 | 1.8 | — | 0.2 | — |
| 38.70 | 48.0 | 22.4 | 0.4 | 0.4 | 2.8 | — | 5.4 | — | 0.2 | 0.4 | 2.2 | 0.2 | 4.6 | 2.6 | 0.6 | 0.8 | 5.4 | 2.0 | — | 0.8 | 0.6 |
| 52.45 | 26.6 | 35.6 | — | — | 1.2 | — | 10.4 | 0.4 | 0.6 | 1.6 | 3.6 | 0.4 | 2.6 | 6.6 | 1.0 | 1.6 | 3.0 | 4.0 | 0.2 | 0.6 | — |
| 74.30 | 30.4 | 16.0 | 1.0 | 0.6 | 1.2 | — | 13.6 | 0.6 | 0.4 | — | 3.0 | 0.6 | 14.4 | 11.6 | 1.2 | 2.6 | 1.8 | 0.4 | 0.2 | — | 0.4 |
| 85.15 | 46.2 | 20.6 | 0.2 | — | 2.6 | — | 5.2 | 0.2 | — | — | 0.2 | — | 5.6 | 4.8 | 0.4 | 1.0 | 9.8 | 1.6 | — | 0.8 | 0.8 |
| 88.30 | 49.4 | 19.0 | 0.8 | 1.2 | 1.6 | 0.8 | 4.0 | — | 0.4 | — | 1.2 | 0.6 | 5.6 | 5.6 | 0.4 | 1.2 | 5.4 | 1.8 | 0.2 | 0.8 | 0.8 |
| 91.20 | 44.4 | 20.6 | 0.4 | 1.4 | 2.8 | 0.2 | 5.4 | 0.2 | 0.6 | — | 2.8 | — | 4.2 | 3.6 | 0.8 | 1.2 | 6.4 | 1.4 | — | 0.4 | 2.6 |
| 103.60 | 59.2 | 22.8 | 0.2 | 1.2 | 0.2 | — | 4.0 | 0.2 | 0.2 | — | 0.6 | 1.0 | 2.0 | 0.4 | 0.2 | 1.0 | 2.2 | 3.4 | 0.4 | — | 0.6 |
| 117.50 | 21.8 | 25.0 | — | — | 1.8 | — | 19.0 | 0.4 | 0.4 | 1.2 | 3.0 | 0.4 | 10.6 | 7.0 | 0.8 | 1.2 | 6.2 | 0.6 | 0.2 | 0.4 | — |
| 142.70 | 50.6 | 16.6 | — | — | 0.8 | 0.2 | 6.4 | 0.2 | 0.2 | 0.2 | 2.4 | 0.6 | 8.2 | 10.0 | 0.6 | 0.8 | 1.0 | 0.6 | — | 0.6 | — |
| 155.65 | 19.8 | 26.6 | 0.8 | 0.8 | 2.8 | 0.2 | 8.8 | 0.4 | 0.6 | 0.4 | 2.6 | 0.2 | 11.6 | 12.2 | 0.6 | 1.2 | 7.0 | 2.6 | 0.2 | 0.6 | — |
| 156.95 | 34.8 | 27.5 | 0.2 | — | 0.8 | — | 7.4 | 0.1 | 0.2 | 0.2 | 2.3 | 0.6 | 3.3 | 7.5 | 0.3 | 0.5 | 11.1 | 2.2* | 0.2 | 1.1 | 0.1 |
| 170.03 | 17.6 | 13.8 | — | — | 1.6 | — | 20.2 | 0.2 | — | 2.4 | 4.4 | — | 13.8 | 17.8 | 0.8 | 0.6 | 5.2 | 0.6 | — | 0.8 | 0.2 |
| 172.40 | 35.2 | 17.6 | — | 0.4 | — | — | 5.6 | 0.2 | 0.2 | 1.4 | 3.0 | — | 7.2 | 5.8 | 0.4 | 0.4 | 14.4 | 4.0 | — | 1.8 | — |
| 180.45 | 17.0 | 24.4 | — | 0.4 | 2.4 | — | 9.4 | 0.6 | 0.4 | — | 4.2 | 1.0 | 18.2 | 18.4 | 1.0 | 1.6 | 1.0 | 1.2 | — | 0.4 | 0.2 |
| 194.10 | 15.0 | 33.6 | — | 0.6 | 0.6 | 0.6 | 13.0 | 0.2 | 0.4 | — | 2.6 | 0.6 | 16.0 | 13.8 | 1.4 | 0.8 | — | 1.2 | — | 0.2 | — |
| 205.90 | 26.0 | 24.0 | — | — | — | — | 15.4 | 0.4 | 0.2 | 0.2 | 6.4 | — | 13.0 | 11.4 | 0.6 | 0.4 | 0.2 | 1.6 | — | — | — |
| 217.10 | 39.4 | 25.8 | 0.6 | 0.2 | 0.2 | — | 8.6 | 0.2 | 0.2 | — | 2.0 | 0.2 | 6.6 | 8.0 | 0.6 | 1.2 | 3.2 | 1.0 | — | 0.6 | — |
| 226.90 | 35.0 | 18.6 | — | 1.0 | 1.6 | — | 14.0 | — | — | 2.0 | 4.4 | — | 6.8 | 7.4 | 0.6 | 1.2 | 4.0 | 0.4 | — | 2.0 | — |
| 229.60 | 6.4 | 31.6 | — | — | 2.6 | — | 18.0 | 0.6 | 0.2 | 0.2 | 7.4 | 0.4 | 10.6 | 18.6 | 1.8 | 1.6 | 0.4 | 1.6 | 0.6 | — | — |
| 231.30 | 11.0 | 22.6 | 0.2 | 0.2 | 2.6 | — | 11.6 | 0.2 | 0.2 | 5.8 | 8.4 | — | 11.4 | 15.6 | 1.0 | 0.8 | 3.2 | 4.8 | 0.2 | 0.2 | — |
| 235.00 | 37.0 | 25.6 | 0.8 | 0.4 | 2.4 | — | 10.6 | — | — | 1.6 | 3.0 | 0.6 | 3.8 | 3.4 | 0.2 | 0.6 | 7.4 | 2.2 | 0.4 | — | — |
| 236.40 | 15.4 | 20.0 | — | 0.2 | 1.6 | 0.4 | 18.4 | 0.2 | 0.2 | 3.2 | 6.4 | 0.2 | 11.6 | 8.2 | 1.6 | 2.2 | 8.0 | 1.6 | — | — | — |
| 254.00 | 45.4 | 18.8 | 0.2 | 1.2 | 0.8 | 0.4 | 9.4 | — | 0.2 | — | 5.2 | 0.8 | 8.2 | 5.8 | 0.6 | 1.0 | — | 0.8 | 0.8‡ | 0.8 | — |
| 262.35 | 39.4 | 27.0 | — | — | 2.4 | 0.4 | 3.8 | — | — | 1.6 | 3.6 | — | 7.0 | 1.6 | 0.2 | 0.8 | 11.0 | 1.2 | — | 0.2 | — |
| 266.50 | 25.6 | 10.0 | — | 0.2 | — | 0.4 | 15.8 | — | — | — | 6.0 | — | 19.2 | 20.2 | 1.0 | 1.4 | — | 0.2 | — | — | — |

Tc, Telocollinite; De, desmocollinite; Co, corpocollinite; Te, telinite; Vd, vitrodetrinite; Ps, pseudovitrinite; Sp, sporinite; Cu, cutinite; Re, resinite; Ld, liptodetrinite; Id, inertodetrinite; Mi, micrinite; Fu, fusinite; Sf, semifusinite; Sc, sclerotinite; Ma, macrinite; Il, clay minerals; Pi, pyrite; Ga, galena; Q, quartz; and W, carbonates.
* Pyrite + chalcopyrite (0.2%).
† Galena + hematite (0.6%).

**Table 4.** *Maceral composition (vol%) of coals in borehole Borówno 1*

| Depth (m) | Vitrinite (vol%) | | | | | | Exinite (vol%) | | | | Inertinite (vol%) | | | | | | Mineral matter (vol%) | | | | |
|---|---|---|---|---|---|---|---|---|---|---|---|---|---|---|---|---|---|---|---|---|---|
| | Tc | De | Co | Te | Vd | Ps | Sp | Cu | Re | Ld | Id | Mi | Fu | Sf | Sc | Ma | Il | Pi | Ga | Q | W |
| 50.50 | 23.6 | 35.4 | 0.4 | 0.4 | 0.4 | — | 10.4 | 0.2 | — | 0.4 | 4.2 | 0.6 | 3.2 | 11.6 | 0.8 | 1.8 | 4.2 | 1.6 | — | 0.8 | — |
| 57.30 | 17.0 | 40.4 | 0.6 | 0.2 | 1.0 | — | 8.4 | 0.2 | 0.2 | 1.2 | 2.0 | 0.6 | 2.8 | 13.8 | 0.6 | 1.6 | 5.4 | 3.2 | — | 0.8 | — |
| 59.55 | 58.8 | 19.2 | 0.1 | 1.6 | 0.8 | 0.4 | 8.0 | — | — | 0.8 | 0.4 | 0.4 | 1.6 | 3.2 | 0.4 | 0.6 | 1.4 | 1.8 | — | 0.8 | — |
| 76.90 | 23.7 | 33.7 | — | 0.7 | 0.7 | — | 14.8 | — | 0.2 | 0.7 | 5.8 | 0.8 | 5.0 | 5.0 | 1.4 | 1.3 | 3.7 | 1.9 | 0.1 | 0.4 | — |
| 82.60 | 32.4 | 22.2 | 0.8 | — | 6.6 | — | 6.6 | — | — | 1.2 | 1.6 | — | 1.4 | 4.0 | 0.2 | 1.0 | 20.6 | 0.8 | — | 0.6 | — |
| 87.50 | 45.0 | 23.0 | 0.2 | 0.2 | 0.4 | 0.2 | 8.8 | — | 0.2 | 0.2 | 0.6 | 0.6 | 7.6 | 6.4 | 0.2 | 0.6 | 3.8 | 0.2 | — | 0.8 | — |
| 94.85 | 29.2 | 26.0 | — | 0.4 | 0.2 | — | 10.4 | 0.2 | 0.2 | — | 7.4 | 0.2 | 11.8 | 15.4 | 0.8 | 1.6 | 1.8 | 2.8 | 0.2 | 0.6 | — |
| 95.00 | 22.6 | 46.4 | 0.2 | 0.2 | 0.4 | — | 11.6 | — | 0.2 | — | 2.2 | 1.0 | 3.0 | 1.8 | 0.2 | 0.8 | 8.2 | 0.8 | — | — | 0.2 |
| 98.60 | 61.2 | 15.0 | — | 0.8 | — | 1.4 | 6.2 | — | 0.2 | 1.6 | 4.8 | 0.2 | 3.0 | 3.6 | 0.4 | 1.4 | 0.6 | 1.2 | — | — | — |
| 104.85 | 19.0 | 12.4 | — | — | 0.8 | 0.4 | 10.8 | — | — | 1.0 | 2.2 | — | 17.6 | 23.0 | 0.4 | 0.2 | 9.6 | 1.4 | — | — | — |
| 110.40 | 24.0 | 17.6 | 0.4 | 0.4 | 0.2 | — | 22.8 | 0.2 | 0.2 | — | 2.4 | — | 18.4 | 7.0 | 1.4 | 1.2 | 1.6 | 0.4 | — | 1.2 | — |
| 133.50 | 4.6 | 20.8 | 0.2 | 1.2 | — | 0.4 | 16.2 | 0.2 | 0.2 | 0.2 | 1.4 | — | 29.2 | 20.8 | 0.8 | 1.6 | 1.4 | 0.8 | 0.2 | 0.6 | 5.6 |
| 153.90 | 26.2 | 20.0 | 0.2 | 0.4 | 2.4 | — | 12.4 | — | — | 0.4 | 1.0 | 0.6 | 7.0 | 7.0 | 0.6 | 1.2 | 11.0 | 3.4 | — | 0.6 | — |
| 169.80 | 22.4 | 21.0 | — | — | 0.4 | — | 7.8 | — | — | — | 5.6 | 0.2 | 17.6 | 10.6 | 0.6 | 1.0 | 12.2 | 0.2 | — | — | 1.0 |
| 174.80 | 14.2 | 31.4 | 0.4 | 0.6 | 0.8 | — | 21.0 | 0.6 | — | — | 2.8 | 0.8 | 6.2 | 13.8 | 1.2 | 1.2 | 5.0 | 0.6 | — | — | — |
| 183.60 | 29.0 | 19.6 | — | — | 0.4 | 1.4 | 7.2 | 0.2 | 0.4 | — | 1.6 | 0.8 | 19.4 | 10.6 | 0.4 | 0.4 | 8.2 | 0.8 | — | — | — |
| 198.70 | 41.8 | 28.4 | 0.4 | — | 0.6 | — | 5.6 | 0.4 | 0.2 | — | 0.6 | 0.2 | 3.8 | 7.0 | — | 0.8 | 5.4 | 3.4 | — | 0.2 | 1.0 |
| 199.20 | 40.8 | 29.2 | 0.2 | — | 0.6 | — | 14.8 | 0.2 | 0.2 | 1.4 | 0.8 | 1.0 | 1.2 | 4.4 | 0.6 | 1.0 | 2.4 | 1.0 | — | — | — |
| 223.30 | 25.8 | 37.8 | — | 2.6 | 1.0 | 0.6 | 9.8 | — | — | — | 4.8 | 0.6 | 1.2 | 6.6 | 1.2 | 1.0 | 8.6 | 1.2 | 0.2 | 0.2 | — |
| 226.60 | 18.4 | 16.2 | 0.2 | — | 0.2 | — | 18.6 | — | 0.2 | 0.2 | 1.6 | — | 7.8 | 26.6 | 1.0 | 1.2 | 1.6 | 2.4 | — | — | 1.0 |
| 230.60 | 23.8 | 29.7 | — | 1.0 | 1.3 | — | 9.2 | — | 0.6 | — | 1.0 | 0.4 | 5.6 | 15.0 | 0.6 | 1.2 | 1.2 | 0.8 | — | — | — |
| 236.80 | 23.0 | 28.6 | — | 0.8 | 0.2 | — | 15.0 | — | — | — | 3.4 | 0.8 | 7.4 | 12.8 | 0.8 | 2.4 | 1.8 | 2.0 | — | 0.2 | — |
| 237.60 | 41.0 | 28.4 | — | 0.4 | — | — | 11.6 | — | 0.8 | — | 1.0 | — | 11.2 | 2.6 | 1.4 | 1.0 | 0.2 | 0.2 | — | — | — |
| 238.40 | 40.8 | 30.4 | 0.4 | 0.4 | — | — | 6.0 | — | 0.2 | 0.8 | 3.8 | 0.6 | 6.2 | 7.0 | 0.4 | 1.2 | 0.4 | 1.6 | — | — | — |
| 238.80 | 26.2 | 35.2 | 0.4 | — | 0.6 | — | 12.8 | — | — | 3.8 | 1.4 | — | 8.2 | 5.2 | 0.6 | 1.2 | 3.6 | 2.4 | — | — | 0.8 |
| 261.50 | 18.8 | 26.2 | — | — | 0.4 | — | 17.0 | — | — | — | 7.8 | 0.4 | 14.0 | 8.8 | 0.4 | 1.0 | 0.8 | 0.6 | — | — | 0.8 |
| 264.95 | 43.6 | 31.2 | 0.2 | — | — | — | 4.6 | 0.2 | 0.2 | — | 0.8 | 0.2 | 5.4 | 7.8 | 0.4 | 0.4 | 1.2 | 2.2 | — | 0.6 | 1.0 |

Key as for Table 3.

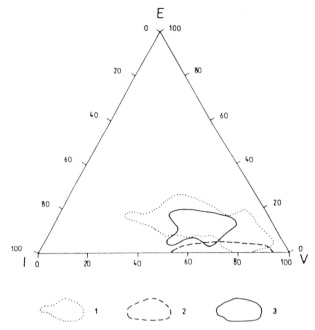

**Fig. 4.** Maceral composition of the seams studied, (mineral matter free). V, Vitrinite; E, exinite (liptinite); and I, inertinite. Coal seams studied of: 1, the Wałbrzych region; 2, the Nowa Ruda region; and 3, the Słupiec region.

**Table 5.** *Maceral composition (vol%) of coals in Suliszów borehole*

| Depth (m) | Vitrinite (vol%) | | | | | | Exinite (vol%) | | | |
|---|---|---|---|---|---|---|---|---|---|---|
| | Tc | De | Co | Te | Vd | Ps | Sp | Cu | Re | Ld |
| 394.00 | 39.1 | 31.9 | — | 0.8 | 1.1 | — | 3.9 | 0.1 | 0.3 | 0.1 |

| Depth (m) | Inertinite (vol%) | | | | | | Mineral matter (vol%) | | | | |
|---|---|---|---|---|---|---|---|---|---|---|---|
| | Id | Mi | Fu | Sf | Sc | Ma | Il | Pi | Ga | Q | W |
| 394.00 | 4.1 | 0.3 | 9.2 | 5.0 | 0.3 | 1.0 | 1.3 | 1.1 | — | — | 0.4 |

Key as for Table 3.

## Coal facies analysis

### Principles

Coal petrographical and palynological investigations can be related to palaeo-conditions in the original mires (Hacquebard & Donaldson 1969; Diessel 1982; 1986; 1992; Marchioni 1980; Marchioni *et al.* 1993; Calder *et al.* 1991; Eble & Grady 1993; Eble *et al.* 1994*a*; 1994*b*; Hower *et al.* 1994). Teichmüller (1962) and Hacquebard & Donaldson (1969) considered that different environments of coal accumulation are related to water depth in the peat swamp, affecting both the type of vegetation and the mode of preservation of petrographic entities. They applied Osvald's (1937) terminology of zones characterizing swamps. Hacquebard & Donaldson (1969) constructed a facies diagram based on microlithotype composition, in which seams are assigned to an origin in one of the four moor environments (Teichmüller 1950). A slightly modified version of this diagram is used for interpretation of the depositional environments of the coal seams of the Zacler Formation from the Lower Silesian coal basin (Fig. 5).

The palaeoenvironmental conditions of swamps are determined from petrographic facies indices calculated from maceral analyses. The ratios used for the facies analysis of coal

**Table 6.** *Petrographic composition of coal seams 405 and 415 in the Nowa Ruda region*

| Seam | Profile | Interval | Lithotype | Thickness (m) | Vitrinite | | | Exinite | | | Intertinite | | | | | | Mineral matter | | | | | Microlithotypes | | | | | | | |
|---|---|---|---|---|---|---|---|---|---|---|---|---|---|---|---|---|---|---|---|---|---|---|---|---|---|---|---|---|---|
| | | | | | Cl | Te | Vd | Sp | Cu | Re | Id | Mi | Fu | Sf | Sc | Ma | Il | Pi | Ga | Q | W | Vt | Lp | In | CR | Vi | Du | Tr | Ca |
| 415 | 1 | A | Bright | 0.40 | 79.6 | — | — | 0.6 | — | — | 3.8 | 8.2 | 2.0 | 1.6 | 0.6 | 2.0 | 1.2 | — | 0.2 | — | 0.2 | 67.4 | — | 4.0 | 0.6 | 26.4 | — | — | 1.6 |
| | | C1 | Bright + semibright | 0.40 | 67.4 | — | 1.0 | 3.0 | — | — | 7.2 | 3.8 | 3.2 | 10.2 | 0.8 | 3.0 | 0.2 | 0.2 | — | — | — | 52.8 | — | 12.6 | 0.6 | 31.4 | 0.8 | 1.4 | 0.4 |
| | | C2 | Bright + semibright | 0.40 | 68.4 | — | 3.0 | 1.8 | 0.2 | 0.4 | 8.6 | 2.4 | 3.8 | 9.4 | 0.8 | 2.2 | 1.2 | — | — | — | 0.8 | 48.6 | 0.2 | 12.4 | 0.8 | 32.8 | 1.2 | 2.2 | 1.8 |
| | 2 | A1 | Bright | 0.60 | 83.8 | — | — | 0.6 | — | — | 2.4 | 7.6 | 1.6 | 2.8 | — | 0.6 | 0.4 | — | — | 0.2 | — | 64.8 | — | 4.8 | 0.6 | 28.2 | — | 0.8 | 0.8 |
| | | A2 | Bright + semibright | 0.60 | 77.6 | — | 0.4 | 1.6 | — | 0.4 | 4.4 | 3.6 | 1.4 | 7.6 | 0.8 | 1.2 | 1.0 | — | — | — | — | 60.6 | — | 10.0 | 0.8 | 25.0 | — | 1.0 | 2.6 |
| | 3 | B | Semibright | 0.25 | 61.0 | — | 8.8 | 2.4 | — | 0.4 | 2.2 | 2.6 | 5.2 | 2.8 | 0.6 | 0.4 | 10.8 | 2.4 | — | — | 0.4 | 61.8 | — | 8.4 | 3.6 | 9.8 | — | — | 16.4 |
| | | C | Bright | 0.20 | 69.0 | 0.2 | 3.4 | 5.2* | 0.4 | — | 7.0 | 3.4 | 2.8 | 5.6 | 0.6 | 2.2 | 0.4 | — | — | — | — | 51.8 | — | 7.2 | 0.8 | 26.4 | 6.2 | 7.2 | 2.4 |
| | | D | Semibright | 0.05 | 63.0 | — | 0.2 | 7.2 | 0.2 | — | 6.6 | 5.0 | 5.2 | 7.8 | 0.2 | 1.6 | 1.4 | 0.6 | 0.2 | — | — | 47.6 | — | 12.6 | 1.2 | 33.2 | — | 3.0 | 2.4 |
| | | E + F | Semibright + fibrous | 0.60 | 62.4 | — | 1.8 | 2.4 | 0.2 | — | 8.0 | 3.0 | 4.0 | 14.6 | 0.6 | 1.4 | 10 | — | — | 0.2 | 0.4 | 44.4 | — | 15.2 | 0.8 | 33.4 | 1.2 | 3.2 | 1.8 |
| | | G | Parting | 0.05 | — | | | | | | | | | | | | | | | | | | | | | | | | |
| | | H | Undivided sequence of dull and bright | 0.20 | 65.4 | — | 7.2 | 3.4 | — | 0.2 | 3.2 | 2.4 | 4.4 | 2.8 | 0.2 | 0.2 | 10.0 | 0.6 | — | 0.2 | — | 55.0 | — | 6.4 | 1.2 | 16.8 | 0.4 | 4.2 | 46.0 |
| | 4 | A1 A2 | Bright + semibright | 1.60 | 61.8 | — | 4.2 | 1.8 | 0.2 | 0.2 | 10.0 | 2.4 | 5.4 | 7.2 | 1.2 | 1.2 | 1.4 | 0.2 | — | 0.4 | 2.4 | 44.0 | — | 16.0 | 1.0 | 28.8 | 0.8 | 0.6 | 8.8 |
| 405 | 5 | C | Semibright | 0.40 | 75.8 | — | 2.2 | 3.0 | — | — | 4.8 | 3.2 | 2.2 | 5.8 | 0.8 | 1.4 | 0.4 | 0.4 | — | — | — | 64.4 | — | 7.0 | 2.4 | 18.0 | 1.0 | 5.4 | 1.8 |
| | | D | Parting | 0.10 | — | | | | | | | | | | | | | | | | | | | | | | | | |
| | | E + F | Banded, semibright, dull | 0.30 0.10 | 62.8 | — | — | 5.0 | — | 0.2 | 8.0 | 2.8 | 3.4 | 9.6 | 0.4 | 2.0 | 0.4 | — | — | — | — | 53.0 | 0.2 | 19.2 | 2.8 | 21.4 | 2.6 | 9.2 | 0.6 |
| | | G | Banded semibright | 0.30 | 74.4 | 0.2 | — | 5.8 | — | 0.8 | 2.4 | 3.4 | 4.6 | 6.6 | 0.2 | 0.8 | 0.4 | — | — | — | 0.8 | 65.4 | — | 7.0 | 4.4 | 17.4 | 1.6 | 3.0 | 1.2 |
| | | H | Parting | 0.05 | — | | | | | | | | | | | | | | | | | | | | | | | | |
| | | I | Semibright | 0.20 | 67.2 | — | 3.6 | 2.4 | — | — | 9.0 | 2.2 | 5.2 | 5.6 | 1.4 | 1.2 | 1.6 | 0.2 | — | — | 0.4 | 51.2 | — | 9.6 | 1.6 | 30.2 | 1.0 | 3.8 | 2.6 |
| | | J | Dull | 0.07 | 24.2 | — | 22.8 | 5.8 | — | — | 7.8 | 1.0 | 2.4 | 14.2 | 1.2 | 1.0 | 18.2 | — | — | 1.2 | 0.2 | 21.2 | 0.4 | 15.2 | 3.0 | 28.6 | 2.8 | 4.4 | 24.0 |
| | | K | Banded semibright | 0.60 | 39.6 | 0.2 | 21.4 | 1.6 | — | — | 7.4 | 5.6 | 5.6 | 12.4 | 0.6 | 0.8 | 9.4 | 0.2 | 0.2 | — | 0.4 | 36.4 | — | 17.2 | 0.6 | 29.0 | 1.0 | 1.6 | 14.2 |

Vitrinite: Cl, collinite; Te, telinite; and Vd, vitrodetrinite.
Exinite: Sp, sporinite; Cu, cutinite; Re, resinite; Sp*, sporinite + Lipodetrinite (1.2%).
Inertinite: Id, inertodetrinite; Mi, micrinite; Fu, fusinite; Sf, semifusinite; Sc, sclerotinite; and Ma, macrinite.
Mineral matter: Il, clay minerals; Pi, pyrite; Ga, galena; Q, quartz; and W, carbonates.
Microlithotypes: Vt, vitrite; Lp, liptite; In, inertite; CR, clarite; Vi, vitrinertite; Du, durite; Tr, trimacerite; and Ca, cabominerite.

**Table 7.** *Petrographic composition of coal seams in the Słupiec area*

| Seam | Profile | Interval | Thickness (m) | Vitrinite | | | | | Exinite | | | Intertinite | | | | | | Mineral matter | | | | | Microlithotypes | | | | | | | |
|---|---|---|---|---|---|---|---|---|---|---|---|---|---|---|---|---|---|---|---|---|---|---|---|---|---|---|---|---|---|---|
| | | | | Cl | Co | Te | Vd | Sp | Cu | Re | Ld | Id | Mi | Fu | Sf | Sc | Ma | Il | Pi | Ga | Q | W | Vt | Lp | In | CR | Vi | Du | Tr | Ca |
| 414/415 | 1 | A | 0.38 | 46.2 | — | 0.4 | 7.0 | 5.8 | — | 0.4 | 2.8 | 9.0 | 1.4 | 9.4 | 13.4 | 0.6 | 1.2 | 0.8 | 0.6 | — | 0.4 | 0.6 | 23.0 | 0.4 | 19.2 | 2.4 | 14.4 | 1.6 | 35.8 | 3.2 |
| | | B* | 0.05 | — | — | — | — | — | — | — | — | — | — | — | — | — | — | — | — | — | — | — | — | — | — | — | — | — | — | — |
| | | C | 0.38 | 44.2 | — | 0.4 | 4.4 | 8.6 | — | 0.4 | 1.6 | 8.4 | 0.8 | 10.0 | 17.2 | 0.8 | 1.2 | 1.0 | 0.4 | — | 0.2 | 0.4 | 22.4 | — | 21.8 | 1.2 | 17.2 | 1.8 | 33.0 | 2.6 |
| | 2 | A | 1.40 | 56.2 | — | 1.0 | 2.0 | 7.0 | — | 0.4 | 0.2 | 7.8 | 1.0 | 7.8 | 11.2 | 0.8 | 0.8 | 2.2 | 1.0 | — | 0.2 | 0.4 | 26.6 | 0.2 | 13.4 | 1.6 | 26.0 | 3.0 | 26.0 | 3.0 |
| | | B* | 0.06 | — | — | — | — | — | — | — | — | — | — | — | — | — | — | — | — | — | — | — | — | — | — | — | — | — | — | — |
| 412/413 | 3 | A | 0.80 | 49.0 | 0.8 | 0.2 | 3.0 | 10.8 | 0.2 | 0.6 | 2.4 | 6.2 | 1.2 | 7.2 | 8.2 | 0.8 | 1.4 | 5.8 | 1.0 | — | 0.2 | 1.0 | 21.0 | 0.8 | 13.0 | 8.6 | 8.4 | 3.6 | 29.4 | 15.2 |
| | | B* | 0.10 | — | — | — | — | — | — | — | — | — | — | — | — | — | — | — | — | — | — | — | — | — | — | — | — | — | — | — |
| | | C | 0.80 | 29.4 | 0.2 | — | 4.8 | 12.6 | 0.4 | 0.2 | 2.6 | 8.0 | 1.6 | 11.6 | 21.2 | 1.0 | 2.4 | 1.0 | 0.8 | 1.0 | 0.2 | 1.0 | 11.6 | 1.2 | 17.2 | 1.8 | 5.4 | 11.8 | 47.2 | 3.8 |
| 4 | 4 | C (banded semibright) | 0.58 | 50.8 | — | 0.2 | 4.6 | 11.2 | 0.6 | 0.6 | 0.8 | 4.8 | 2.6 | 7.4 | 9.2 | 1.0 | 2.0 | 2.6 | — | 0.2 | 0.6 | 0.8 | 21.2 | 0.4 | 15.0 | 13.2 | 6.6 | 2.0 | 36.8 | 4.8 |
| | | D* | 0.09 | — | — | — | — | — | — | — | — | — | — | — | — | — | — | — | — | — | — | — | — | — | — | — | — | — | — | — |
| | | E (semibright with fibrous) | 0.30 | 42.2 | 0.2 | 0.4 | 3.8 | 11.2 | 0.2 | 1.0 | 2.4 | 7.8 | 1.0 | 9.4 | 10.4 | 0.6 | 1.2 | 5.0 | 1.0 | — | 0.2 | 1.4 | 19.2 | 1.2 | 14.4 | 5.6 | 3.0 | 1.0 | 40.4 | 15.2 |
| | | Eb (semibright) | 0.27 | 58.8 | 0.4 | 0.2 | 2.4 | 11.0 | 0.4 | 0.8 | 1.4 | 4.0 | 1.0 | 7.6 | 7.0 | 0.6 | 1.0 | 2.2 | 0.6 | — | 0.2 | 0.4 | 26.6 | — | 9.4 | 13.4 | 6.4 | 6.0 | 33.4 | 4.8 |
| | | F* | 0.20 | — | — | — | — | — | — | — | — | — | — | — | — | — | — | — | — | — | — | — | — | — | — | — | — | — | — | — |
| | | G (semibright + bright) | 0.58 | 68.0 | — | — | 9.0 | 3.8 | 0.2 | 0.2 | 1.6 | 7.8 | 0.6 | 3.8 | 1.0 | 0.6 | 0.4 | 0.6 | 1.0 | — | 0.4 | 1.8 | 47.0 | — | 3.8 | 6.4 | 16.8 | 1.6 | 19.2 | 5.0 |
| 410 | 5 | A | 0.90 | 43.0 | 0.2 | — | 3.4 | 10.4 | — | 0.4 | 0.6 | 8.4 | 1.8 | 5.8 | 16.6 | 1.0 | 1.8 | 5.0 | 0.6 | 0.6 | — | 0.6 | 19.2 | 0.6 | 14.6 | 4.4 | 8.0 | 5.8 | 38.8 | 8.6 |
| | 6 | A | 1.00 | 61.6 | 0.4 | 0.2 | 3.0 | 10.2 | — | 0.6 | 1.0 | 3.4 | 2.0 | 4.8 | 3.6 | 0.8 | 1.4 | 6.0 | 0.6 | 0.2 | — | 0.2 | 43.0 | — | 7.8 | 12.4 | 5.0 | 0.6 | 22.8 | 8.4 |
| | 7 | A | 0.60 | 39.0 | 0.4 | 0.6 | 6.2 | 7.2 | 0.6 | 0.2 | 1.8 | 6.8 | 1.0 | 10.0 | 19.2 | 1.4 | 1.2 | 3.6 | 0.4 | — | 0.2 | 0.2 | 21.4 | — | 25.4 | 4.8 | 7.8 | 11.0 | 22.6 | 7.2 |
| | | B* | 0.10 | — | — | — | — | — | — | — | — | — | — | — | — | — | — | — | — | — | — | — | — | — | — | — | — | — | — | — |
| | | C1 | 0.38 | 41.2 | 0.6 | 0.6 | 11.0 | 13.4 | 0.2 | — | 5.0 | 9.4 | 0.8 | 3.4 | 9.8 | 0.8 | 2.2 | 0.8 | 0.8 | — | — | — | 17.6 | — | 8.6 | 2.0 | 6.8 | 2.8 | 59.2 | 3.2 |
| | | C2 | 0.36 | 48.8 | — | 0.2 | 3.4 | 9.0 | — | 1.6 | 1.0 | 8.4 | 3.0 | 5.4 | 13.2 | 1.4 | 1.4 | 2.6 | 0.2 | — | — | — | — | — | — | — | — | — | — | — |
| 409 | 8 | A | 1.80 | 55.0 | 0.2 | — | 3.2 | 14.8 | 0.6 | — | — | 6.0 | 1.2 | 4.6 | 7.4 | 0.6 | 2.8 | 2.4 | 0.4 | — | — | — | 30.4 | 1.0 | 6.4 | 7.4 | 10.8 | 8.0 | 32.4 | 3.6 |
| | 9 | A | 0.50 | 62.2 | 0.6 | 0.2 | 1.0 | 8.6 | — | 0.2 | 0.6 | 3.0 | 1.0 | 5.6 | 9.6 | 0.6 | 1.0 | 4.0 | 1.2 | — | — | — | 38.2 | 2.4 | 11.4 | 3.8 | 9.2 | 6.4 | 17.6 | 11.0 |
| | | B* | 0.10 | — | — | — | — | — | — | — | — | — | — | — | — | — | — | — | — | — | — | — | — | — | — | — | — | — | — | — |
| | | C | 0.40 | 54.4 | 0.6 | 0.4 | 3.0 | 9.0 | 0.8 | 0.8 | 1.4 | 6.2 | 1.8 | 7.2 | 9.6 | 1.2 | 1.4 | 1.8 | — | — | 0.2 | 0.2 | 39.2 | 0.6 | 15.6 | 6.4 | 9.4 | 4.0 | 22.0 | 2.8 |
| | | D* | 0.05 | — | — | — | — | — | — | — | — | — | — | — | — | — | — | — | — | — | — | — | — | — | — | — | — | — | — | — |
| | | E | 0.38 | 52.0 | 0.2 | — | 5.0 | 9.6 | 0.6 | 0.4 | 1.2 | 8.4 | 1.6 | 4.0 | 12.0 | 0.2 | 1.4 | 2.6 | 0.8 | — | — | — | 30.0 | 0.4 | 13.4 | 6.4 | 14.8 | 2.2 | 27.4 | 5.4 |
| 10 | 10 | A | 0.80 | 45.2 | — | 0.2 | 3.0 | 19.0 | 0.2 | — | 0.2 | 9.4 | 1.8 | 3.8 | 12.4 | 0.4 | 1.4 | 2.6 | 0.2 | — | — | — | 14.6 | 4.0 | 7.6 | 5.4 | 4.4 | 12.6 | 48.2 | 3.2 |
| | | A1* | 0.08 | — | — | — | — | — | — | — | — | — | — | — | — | — | — | — | — | — | — | — | — | — | — | — | — | — | — | — |
| | | B | 0.66 | 60.6 | — | 0.4 | 2.8 | 10.8 | 0.4 | 1.2 | 0.6 | 5.0 | 1.2 | 7.0 | 6.0 | 0.6 | 2.0 | 0.8 | 0.4 | — | 0.4 | 0.2 | 41.8 | 1.0 | 6.6 | 12.2 | 7.0 | 2.4 | 27.2 | 1.8 |

\* Intervals of clastic partings.

Cl, colinite; Co, corpocollinite; Te, telinite; Vd, vitrodetrinite; Sp, sporinite; Cu, cutinite; Re, resinite; Ld, liptodetrinite; Id, inertodetrinite; Mi, micrinite; Fu, fusinite; Sf, semifusinite; Sc, sclerotinite; Ma, macrinite; Il, clay minerals; Pi, pyrite; Ga, galena; Q, quartz; W, carbonates; Vt, vitrite; Lp, liptite; In, inertite; CR, clarite; Vi, vitrinertite; Du, durite; Tr, trimacerite; and Ca, carbominerite.

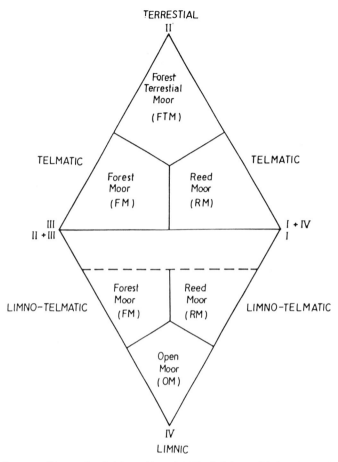

**Fig. 5.** Facies diagram of Hacquebard & Donaldson (1969), slightly modified. I, Sporoclarite + duroclarite + vitrinertoliptite; II, Vitrinertite II + inertite; III, Clarite V + Vitrite + Vitrinertite V, Cuticuloclarite; and IV, Clarodurite + durite + carbominerite.

seams in Zacler Formation include: the SF/F ratio = semifusinite/fusinite (Kalkreuth & Leckie 1989); the VA/VB ratio = vitrinite A/vitrinite B (Kalkreuth & Leckie 1989); the VI/I ratio = total vitrinite/total inertinite (Harvey & Dillon 1985); the T/F ratio = total vitrinite/(fusinite + semifusinite) (Diessel 1982); the W/D ratio = (vitrinite A + fusinite + semifusinite)/(alginite + sporinite + inertodetrinite) (Diessel 1982); the IR ratio = (semifusinite + fusinite)/(inertodetrinite + macrinite + micrinite) (Kalkreuth & Leckie 1989); the S/D ratio = (vitrinite A + fusinite + semifusinite)/(alginite + sporinite + inertodetrinite + vitrinite B + vitrodetrinite) (Kalkreuth & Leckie 1989); the TPI ratio = (vitrinite A + fusinite + semifusinite)/(vitrinite B + macrinite + inertodetrinite) (Diessel 1982); and GI ratio = (total vitrinite + macrinite)/(semifusinite + fusinite + inertodetrinite) (Diessel 1982).

The SF/F ratio can be considered as a relative measure of the degree of oxidation which has taken place in the palaeo-swamp (Kalkreuth & Leckie 1989). The VA/VB ratio indicates the original contribution of preserved woody tissues (VA) versus detrital woody matter (VB) (Kalkreuth & Leckie 1989). High percentages of vitrinite B and/or vitrodetrinite suggest some transportation of organic material before final accumulation. The VI/I ratio increases in coals formed in the vicinity of channels and decreases away from the channels, and is related to a lowering of water level. The T/F ratio characterizes the relative amounts of gelified vitrinite over fusinized and partly fusinized organic material. If the T/F ratios are high (>4), they are considered to indicate wet forest–moor conditions, whereas low values (<0.25) are typical of dry forest–moor

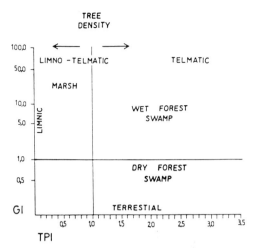

**Fig. 6.** Facies diagram for seams: tissue preservation index (TPI) versus gelification index (GI), according to Diessel (1986).

environments (Diessel 1982). According to Diessel (1982) the W/D ratio indicates the humidity of a peat bog, whereas the IR ratio relates to the degree of degradation that took place in the swamp area before burial of the phytogenic material (Kalkreuth & Lechie 1989).

The interrelationship between the tissue preservation index (TPI) and gelification index (GI) is determined by the depositional environment of the coal (Fig. 6) (Diessel 1982; 1983; 1986).

Using the results of the maceral analyses, it has been possible to establish two petrological coal facies and eight subfacies according to a slight modification of the Strehlau (1990) nomenclature. Table 8 summarizes the characterization of this scheme. The boundaries between the particular subfacies or main facies are often transitional.

Palynological studies (Górecka-Nowak unpublished data) have complemented this facies analysis of the coals of the Zacler Formation by determining the parent plants of the coal and thus allowing a more precise interpretation of the environments of coal deposition.

Facies interpretations of individual seams of the Zacler Formation in the Intrasudetic Basin are based on both maceral analyses and microlithotype data for coals from the mines, but only on maceral analyses for the boreholes.

## Results

### The Walbrzych region

The 430 seam has a variable composition, but with a predominance of vitrite and clarite as microlithotypes (Table 2). This results in coals of the 430 seam plotting in varying positions on the facies diagram, representing forest moors of both the telmatic and limno-telmatic zones (Fig. 7). The sedimentation of phytogenic material under

**Table 8.** *Summary of characteristics of main petrological facies and subfacies distinguished in the studied seams of the Zacler Formation [nomenclature after Strehlau (1990), slightly modified]*

| Facies | Subfacies | Description |
|---|---|---|
| 1. Vitrinite–fusinite | 1a. Telocollinitic | High in vitrinite; more than 50% of vitrinite is telocollinite |
| | 1b. Desmocollinitic | High in vitrinite; more than 50% of vitrinite is desmocollinite |
| | 1c. Teleocollinitic–desmocollinitic | High in vitrinite; content of telocollinite and desmocollinite more than 50%; sporinite, cutinite and corpocollinite may be the common macerals of this subfacies |
| | 1d. Trimaceritic | Medium in vitrinite; enrichment of inertinite and exinite; fusinite, semifusinite, intertodetrinite, tenuisporinite may be the common macerals of this subfacies |
| | 1e. Fusinitic–semifusinitic | Higher in fusinite and semifusinite; other inertinite macerals may be also enriched |
| | 1f. Duritic | Relatively low in vitrinite; sporinite (tenuisporinite), megasporinite and the rest of exinite macerals are typical; semifusinite and fusinite are the dominating interinities; clay minerals may occur |
| 2. Densosporinite | 2a. Trimaceritic (vitrinite content 35–50%) | Crassisporinite and oval-shaped densosporinite are typical; megasporinite and other exinites may be also enriched; intertodetrinite and semifusinite are dominating intertinites, but fusinite is also common in this facies; sometimes vitrinite is slightly enriched. |
| | 2b. Duritic (low in vitrinite, high in intertinite and eximite) | |

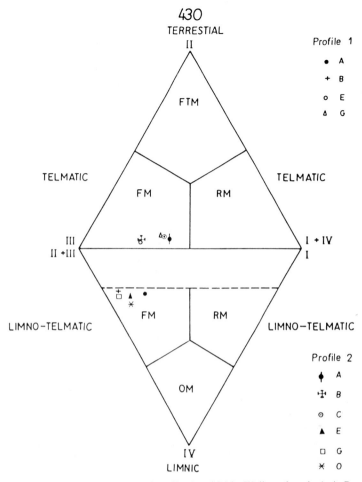

**Fig. 7.** Facies diagram for the 430 coal seam of profiles 1 and 2 (the Wałbrzych region). A, B..., Coal lithotypes.

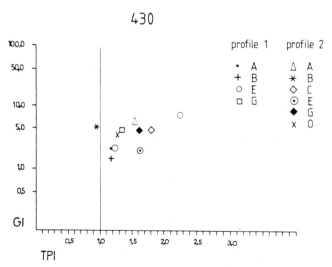

**Fig. 8.** Relationship between TPI and GI indices for the 430 coal seam of profiles 1 and 2. A, B,..., Coal lithotypes.

these conditions was interrupted from time to time by short periods of flooding, recorded by partings of clastic rocks in the seam. Petrographic facies indices such as W/D (1.87–3.27) and TPI suggest that the structural components of the coal are well preserved (Nowak 1992). The VA/VB ratio (0.59–2.10) displays a dominance of 'structural' vitrinite over its unstructured forms. The VI/I ratio for profile 1 of seam 430 gives values (1.44–7.57) typical of wet conditions during the deposition of phytogenic material; it may indicate that the coal was formed under conditions of high water level, adjacent to channels (flood plains). Very similar or higher values of this indicator are observed in profile 2 (1.9–4.24). The dominance and relatively high content of vitrinite also indicated a pH typical of

an acid environment (Cecil *et al.* 1981). This is confirmed by a T/F ratio of 2.72–16.65. The source of inertinite may be periodic desiccation of the swamp as well as the influence of well oxygenated water. The S/D ratio (0.69–1.35) displays variable values in the two profiles, which suggests that conditions favoured tree development. Relationships between TPI and GI (Fig. 8) put the coal of seam 430 in the forest swamp of the telmatic zone. Medium TPI and the high GI values suggest wet conditions and arboreal vegetation in the palaeo-swamp.

In the profiles of seam 430 one petrological facies is distinguished: vitrinite–fusinite, which is represented by four subfacies (telocollinitic, telocollinitic–desmocollinitic, fusinitic–semifusinitic and trimaceritic). In profile 1 (Fig. 9) the

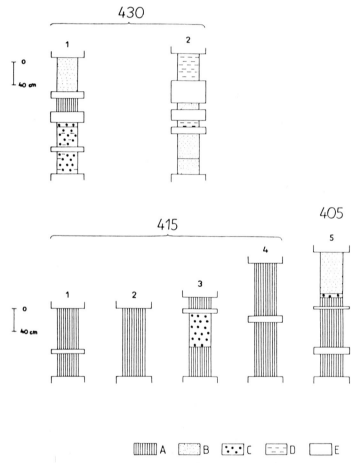

**Fig. 9.** Petrological coal facies (vitrinite–fusinite) distinguished for the 430 seam in profiles 1 and 2 (the Wałbrzych region) and for the 415 and 405 seams of profiles 1–5 from the Nowa Ruda area. A, tellocollinitic subfacies; B, telocollinitic–desmocollinitic subfacies; C, fusinitic–semifusinitic subfacies; D, trimaceritic subfacies; and E, partings.

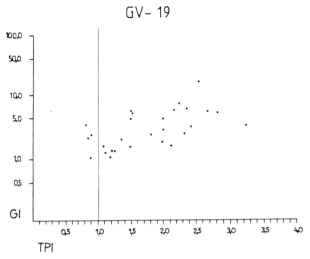

**Fig. 10.** Relationship beetwen TPI and GI indices for seams of the GV-19 borehole (the Wałbrzych region).

following sequence (from the floor to the roof) is observed: fusinitic–semifusinitic; telocollinitic; and telocollinitic–desmocollinitic.

This sequence displays an increase in vitrinite content towards the roof of the seam and a simultaneous decrease in inertinite macerals. This reflects increasingly moist conditions during the deposition of the seam—from relatively dry (fusinitic–semifusinitic) through moist (telocollinitic) to wet (telocollinitic–desmocollinitic).

In profile 2 of seam 430, the following sequence of subfacies of the vitrinite–fusinite facies is observed (from the floor to the roof; Fig. 9): telocollinitic–desmocollinitic; trima-ceritic; telocollinitic–desmocollinitic; and trima-ceritic. This succession shows variable, but moist, conditions of coal deposition.

Palynological studies of seam 430 show a predominance of *Lycospora* (Górecka-Nowak unpublished data), which is a typical spore of forest swamps (Hacquebard & Donaldson 1969; Teichmüller 1989) of Westphalian age (Philips & DiMichele 1981; 1985). Thus the petrological and palynological analyses of seam 430 indicate a forest swamp character of coal depositional environment.

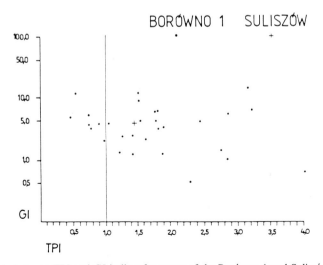

**Fig. 11.** Relationship between TPI and GI indices for seams of the Borówno 1 and Suliszów wells.

**Table 9.** *Petrological coal facies for seams of the GV-19 borehole*

| Depth (m) | Vitrinite–fusinite | | | | |
|---|---|---|---|---|---|
| | Tc | Tc–De | Fu–Sf | Tr | Du |
| 17.25 | — | — | + | — | — |
| 30.00 | — | — | + | — | — |
| 31.00 | + | — | — | — | — |
| 38.70 | + | — | — | — | — |
| 52.45 | — | + | — | — | — |
| 74.30 | — | — | + | — | — |
| 85.15 | + | — | — | — | — |
| 88.30 | + | — | — | — | — |
| 91.20 | + | — | — | — | — |
| 103.60 | + | — | — | — | — |
| 117.50 | — | — | — | + | — |
| 142.70 | + | — | — | — | — |
| 155.65 | — | — | + | — | — |
| 156.95 | — | + | — | — | — |
| 170.30 | — | — | — | — | + |
| 172.40 | — | + | — | — | — |
| 180.45 | — | — | + | — | — |
| 194.10 | — | — | + | — | — |
| 205.90 | — | — | + | — | — |
| 217.10 | — | + | — | — | — |
| 226.90 | — | — | — | + | — |
| 229.60 | — | — | — | — | + |
| 231.30 | — | — | — | — | — |
| 235.00 | — | + | — | — | — |
| 236.40 | — | — | — | — | + |
| 254.00 | — | + | — | — | — |
| 262.35 | — | + | — | — | — |
| 266.50 | — | — | + | — | — |

Subfacies: Tc, telocollinitic; Tc–De, telocollinitic–desmocollinitic; Fu–Sf, fusinitic–semifusinitic; Tr, trimaceritic; and Du, duritic.

**Table 10.** *Petrological coal facies for seams of the Borówno 1 borehole*

| Depth (m) | Vitrinite–fusinite | | | | |
|---|---|---|---|---|---|
| | Tc | Tc–De | Fu–Sf | Tr | Du |
| 50.50 | — | — | — | + | — |
| 57.30 | — | — | — | + | — |
| 59.55 | + | — | — | — | — |
| 76.90 | — | — | — | + | — |
| 82.60 | — | + | — | — | — |
| 87.50 | — | + | — | — | — |
| 94.85 | — | — | + | — | — |
| 95.00 | + | — | — | — | — |
| 98.60 | + | — | — | — | — |
| 104.85 | — | — | + | — | — |
| 110.40 | — | — | — | — | + |
| 133.50 | — | — | + | — | — |
| 153.90 | — | — | — | + | — |
| 169.80 | — | — | + | — | — |
| 174.80 | — | — | — | — | + |
| 183.60 | — | — | + | — | — |
| 198.70 | — | + | — | — | — |
| 199.20 | — | + | — | — | — |
| 223.30 | — | + | — | — | — |
| 226.60 | — | — | — | — | + |
| 230.60 | — | — | — | + | — |
| 236.80 | — | — | — | + | — |
| 237.60 | — | + | — | — | — |
| 238.40 | — | + | — | — | — |
| 238.80 | — | + | — | — | — |
| 261.50 | — | — | — | — | + |
| 264.95 | — | + | — | — | — |

Key as for Table 9.

**Table 11.** *Petrological coal facies for seam 447 of the Suliszów borehole*

| Depth (m) | Vitrinite–fusinite | | | | |
|---|---|---|---|---|---|
| | Tc | Tc–De | Fu–Sf | Tr | Du |
| 394.00 | — | + | — | — | — |

Key as for Table 9.

The coals in boreholes GV-19 and Borówno 1 show variable values of the petrographic indices (Nowak 1992), suggesting changing conditions of coal accumulation. The TPI–GI relationships show that these coals represent both telmatic forest swamps and limno-telmatic marsh (Figs 10 and 11). However, seam 447 of the Suliszów well was formed in a wet forest swamp area (Fig. 11).

Only the vitrinite–fusinite facies is present in these coals. The seams in the GV-19 well represent five subfacies, with the fusinitic-semifusinitic subfacies, occurring most frequently, followed by the telocollinitic, telocollinitic–desmocollinitic, duritic and trimaceritic subfacies (Table 9). In contrast, seams of the Borówno 1 well (Table 10) show a predominance of telocollinitic–desmocollinitic subfacies, and in diminishing order trimaceritic and fusinitic–semifusinitic subfacies are represented by seam 447 in the Suliszów borehole (Table 11).

## Nowa Ruda region

The triangular coal facies diagram (Fig. 12) shows the microlithotype compositions for each lithotype sample in the seam profiles for coals 415 and 405. The bright and semibright coals, comprising predominantly vitrite with smaller contents of vitrinertite (Table 6), plot in the upper triangle at or near the component III apex or adjacent to the axes of components II + III. This represents the forest–moor environment in the telmatic zone. The exception is the one dull

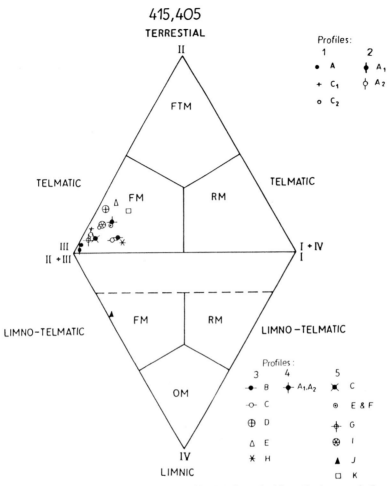

**Fig. 12.** Facies diagram of seams 415 and 405 in profiles 1–5 from the Nowa Ruda area. A, B,...,
Coal lithotypes.

band in seam 405, which consists mainly of vitrinertite and carbominerite (Table 6), and plots in the lower triangle near the axes of components II + III representing the forest–moor facies in the limno-telmatic zone (Fig. 12). High values of the petrographic indices (VA/VB, 1.03–67.4; S/D, 1.1–31.5; and TPI, 4.55–29.4) reflect the presence of structural components. The SF/F ratio (0.54–5.92) indicates that semi-fusinite dominates over fusinite, whereas the IR ratios (0.26–2.2) display the changing proportions of the structural and non-structural components of inertinite. The wet conditions of the swamp and its periodic drying out are seen in the magnitudes of the VI/I ratios (1.70–5.59). The values of the T/F ratio (2.83–22.11) reflect

the VI/I ratios and indicate wet conditions during peat deposition. These data are confirmed by the TPI–GI diagram (Fig. 13), which shows that the coals of seams 415 and 405 have formed in areas of wet forest swamp with a high tree density.

The abundance of vitrinite, often combined with medium to high inerinite contents (Table 6), defines the vitrinite–fusinite facies as the one main petrological coal facies in these seams. As previously indicated, these seams represent in general forest palaeoenvironments. Within this facies it is possible to determine the three sub-facies: telocollinitic and occasionally fusinitic–semifusinitic or telocollinitic–desmocollinitic subfacies (Fig. 9).

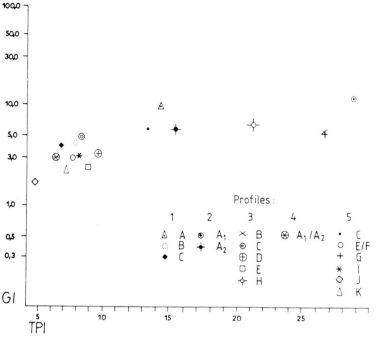

**Fig. 13.** Relationship between TPI and GI indices for seams 415 and 405. A, B,..., Coal lithotypes.

*Slupiec region*

In contrast with the facies deduced for the coal beds in the two regions just described, the coal seams 414/415, 412/413, 410 and 409 of the southeastern part of the basin represent not only forest moors but also reed moors of telmatic and limno-telmatic zones (Figs 14, 16, 18 and 20).

Seam 415/414 represents mires of the telmatic zone. Profile 1 of this seam plots on the facies diagram (Fig. 14) in the area of the reed moor and along the boundary of reed and terrestial moors. This corresponds to the abundance of inertinite (Table 7). The coal in profile 2 is located in the forest moor area of Fig. 14. The petrographic indices such VA/VB (0.42–2.55), W/D (1.03–4.17) and S/D (0.37–2.06) display moderate values (Nowak 1992). The S/F ratio shows that semifusinite prevails over fusinite. The values of the VI (1.28–2.01) and T/F (1.80–3.12) ratios suggest that sedimentation of phytogenic material took place distally from the channels in medium wet conditions. Because of the enrichment of inertinite, the points representing seam 414/415 are located near the boundary between dry and wet forest swamp on the TPI-GI diagram (Fig. 15) and they represent the second type of peat-forming facies listed here.

Seam 412/413 represents a wide range of coal depositional environments from telmatic forest moor to reed moor (Fig. 16). The microlithotype composition of the coal with the highest vitrite content (Table 7) represents forest moor in the telmatic zone. However, the coals with lower vitrinite contents were also formed in forest moor environments, but in the limno-telmatic zone, and some plot near the boundary with the reed moor field on this same diagram. The banded semibright coal of profile 4 represents the reed moor environment of the telmatic zone (Fig. 16). The deductions from this triangular facies diagram are clearly reflected in the TPI–GI diagram (Fig. 17), where the points representing particular petrographic intervals (lithotypes in profile 4) plot in both the forest swamp and marsh fields.

In the coals of seam 410, a convergence between the data plotted on the triangular facies diagram (Fig. 18) and the TPI–GI diagram (Fig. 19) is observed. These coals, representing equally high contents of vitrite and trimacerite, are typical of forest moors as well as the telmatic and limno-telmatic zones (Fig. 18). The interval ($C^1$) of profile 7 is characterized by the highest amount of trimacerite (Table 7), representing telmatic reed moor. A similar

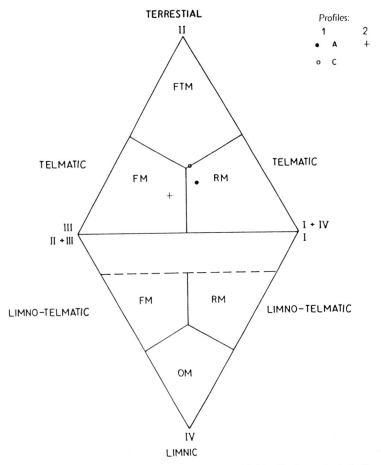

**Fig. 14.** Facies diagram of the 414/415 coal seam of profiles 1 and 2 (the Słupiec region). A, B, ...,
Petrographic intervals.

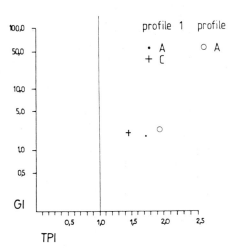

**Fig. 15.** Relationship between TPI and GI indices for
the 414/415 seam. A, B, ..., petrographic intervals.

picture of the coal facies can be interpreted from
the TPI–GI relationships, where the moderate
values of tissue preservation and gelification
indices suggest forest swamp and marsh. $C^1$ is
situated in the marsh area and suggests wetter
palaeoenvironmental conditions.

Seam 409, as with seams 414/415, 412/413 and
410, represents a wide range of peat-forming
facies. The microlithotypes of this seam are
dominated by vitrite and trimacerite (Table 7).
They represent palaeo-mires of the limno-telmatic reed moor type, the transitional between
telmatic forest and reed moors, and forest moors
of the both telmatic and limno-telmatic zones
(Fig. 20). The GI (2.24–3.27) and TPI (0.71–2.91)
(Nowak 1992) indicate changeability of both
wetness and forest material input during the
deposition of this seam. On the TPI–GI diagram
the coals of seam 409 represent both marsh and
forest swamp palaeoenvironments (Fig. 21).

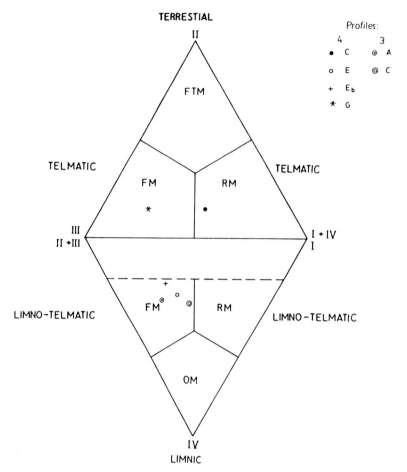

**Fig. 16.** Facies diagram of the 412/413 coal seam of profiles 3 and 4 (the Słupiec region). A, B, ...,
Petrographic intervals.

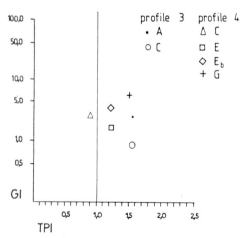

**Fig. 17.** Relationship between TPI and GI indices for
the 412/413 seam. A, B, ..., Petrographic intervals.

The petrographic indices of seams 414/415,
412/413, 410 and 409 display typical values for
these palaeoenvironments (Nowak 1992 in press).
The VA/VB (0.42–2.55), W/D (1.03–4.17) and
S/D (0.37–2.06) ratios are characterized by low
values.

The petrographic data for the seams of the
Słupiec region (Table 7) show medium values of
vitrinite, with the enrichment of inertinite and
constant liptinite percentages. The characteristic
and diagnostic feature of these coals is the
occurrence of the thick-walled, oval-shaped
crassisporinites of the densosporinite group.
Taking into account the proportions between
macerals and the presence of the thick-wall
sporinite, this gives a basis for the determination
of petrological coal facies. The macerals of these
coal seams indicate the presence of both
vitrinite–fusinite and densosporinite main facies.

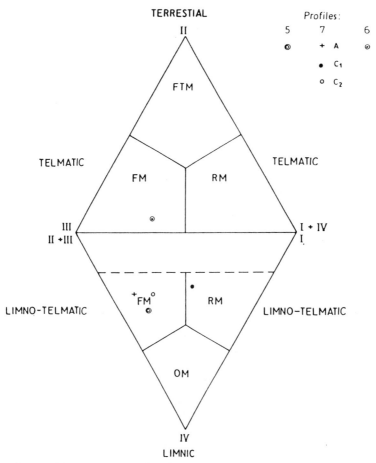

**Fig. 18.** Facies diagram of the 410 coal seam of profiles 5–7 (the Słupiec region). A, B, . . ., Petrographic intervals.

In the first of these the telocollinitic–desmo-collinitc, fusinitic–semifusinitic and trimaceritic subfacies occur. However, the densosporinite facies is represented by the trimaceritic and duritic subfacies. In the seam profiles a changing succession of the coal facies can be observed (Fig. 22). Seam 414/415 represents only the fusinitic–semifusinitic subfacies, indicating relatively dry conditions during peat deposition. However, variable conditions of coal accumulation are visible in the rest of the profiles studied. The lower part of seam 412/413 in profile 3 represents the trimaceritic subfacies of the vitrinite–fusinite facies. Above the clastic parting in the same profile, the coals of the trimaceritic subfacies of the densosporinite facies occur, and may suggest a transitional change in coal facies, whereas in profile 4 coals of the vitrinite–fusinite facies are observed as interlayered trimaceritic,

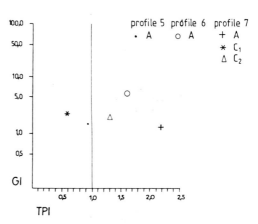

**Fig. 19.** Relationship between TPI and GI indices for the 410 seam. A, B, . . ., Petrographic intervals.

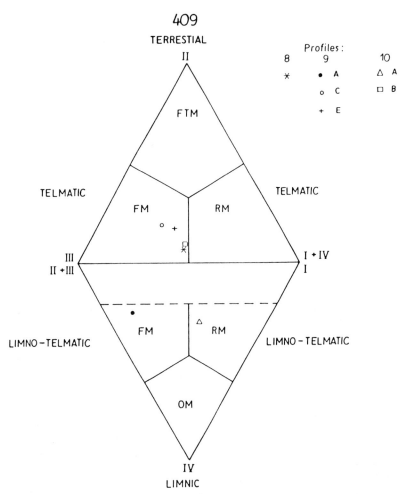

**Fig. 20.** Facies diagram of the 409 coal seam of profiles 8–10 (the Słupiec region). A, B, . . . , Petrographic intervals.

fusinitic–semifusinitic and telocollinitic–desmo-collinitic subfacies (Fig. 22). Seam 410 also shows two main coal facies. In profile 7 of this seam the following upwards succession has been recorded: trimaceritic subfacies of the densosporinite facies; clastic parting; and trimaceritic and fusinitic–semifusinitic subfacies of vitrinite–fusinite facies. This succession indicates a change in the peat depositional conditions from wet, through drier to relatively dry. In seam 409 the coals represent both vitrinite–fusinite and desosporinite facies. The former is represented by the telocollinitic–desmocollinitic and trimaceritic subfacies, whereas in profile 10 of this seam the lower interval displays typical features of the duritic subfacies of the densosporinite facies (the occurrence of densosprinite and a high content of inertinite and liptinite with moderate percentages

of vitrinite). The successions of coal petrological facies occurring in these profiles reflect relatively wet conditions of peat-forming palaeoenvironments.

These facies interpretations can be related to palynological data (Górecka-Nowak unpublished data), which indicate that *Densosporites*, a miospospore with thick walls representing herbaceous plants, is the main palynological taxon in the coals of the southeastern area of the Intrasudetic Basin.

## Discussion

In this study special attention has been paid to the problem of how a specific coal petrographic composition is related to palaeoenvironmental

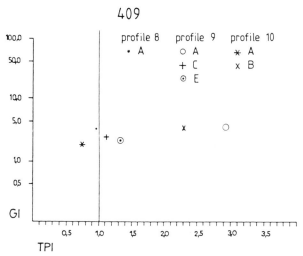

**Fig. 21.** Relationship between TPI and GI indices for the 409 seam. A, B, ..., Petrographic intervals.

conditions. Several types of facies analysis have been applied to support the interpretations (triangular facies diagram, TPI–GI relationships, petrographic indices and petrological coal facies). The methods used are based on different criteria. One aim of this study was to test the different methods by comparing the results. The coal facies analysis presented here has allowed the following relationships to be established.

The coal seams of the Zacler Formation in the Intrasudetic Basin accumulated in palaeo-mires of the telmatic and limno-telmatic zones. These facies are thought to represent peats that were deposited under mainly anoxic conditions, which favoured humification of organic material. A high amount of vitrinite indicates wet conditions and may be related to the geochemical situation in more oxygenated (relatively high pH) water with abundant nutrients. However, the high proportion of inertinite (especially fusinite and semifusinite) within the coals may have been caused by frequent fires (formation of charcoal) or other oxidation processes. A high content of inertodetrinite may be related to the fusinization process of herbaceous vegetation and/or may have been derived from larger plants (as a result of fragmentation of fusinite and semifusinite). The situation favourable for inertinite formation is related to a temporary lowering of the groundwater table. In some instances enrichment in inertinite has been connected to the occurrence of pseudovitrinite related to periodic desiccation of swamps (Hagemann & Wolf 1989; Wolf & Wolff-Fischer 1984).

The application of triangular facies diagrams and TPI–GI relationships to the coals of the Zacler Formation suggests that two general types of coal depositional environment were present: forest moor/swamp and reed moor/marsh. The two types can thus be regarded as mire ecosystems.

Forest palaeo-swamps are represented by coals of medium to high vitrinite contents, where vitrite, poor sporinite clarite, vitrinertite and trimacerite are the main microlithotypes. As and when the groundwater sank below the peat surface, oxidation took place and led to the formation of durites rich in inertinite, (Teichmüller 1989). This situation is observed in seam 405 in the Nowa Ruda region. The Carboniferous forest swamps discussed here occurred in the limno-telmatic zone, where trees were standing in deep water, and also in the drier telmatic zone. The typical spore of the Carboniferous forest swamps is *Lycospora*. This conclusion is supported by palynological data (Górecka-Nowak unpublished data).

The coals of the reed moor/marsh are characterized by a moderate content of vitrinite and a high amount of inertinite. The palaeo-peats of these facies were formed by herbaceous vegetation, which is confirmed by palynological studies (Górecka-Nowak unpublished data), *Densosporites* being a characeristic spore of this facies.

The petrographic results indicate that two main petrological coal facies are present [after the Strehlau (1990) nomenclature]: vitrinite–fusinite and densosporinite facies. The facies analyses presented here suggest that the vitrinite–fusinite

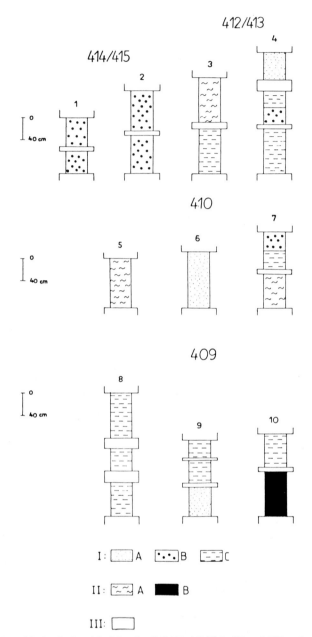

**Fig. 22.** Petrological coal facies distinguished for the 414/415, 412/413, 410 and 409 coal seams from the Słupiec area. I, Vitrinite–fusinite facies: A, tellocollinitic–desmocollinitic subfacies; B, fusinitic–semifusinitic subfacies; and C, trimaceritic subfacies. II, Densosporinite facies: A, trimaceritic subfacies; and B, duritic subfacies. III, Partings.

facies corresponds with forest peat environments and the densosporinite facies is related to reed moor/marsh environments.

There may be problems in reconstructing the facies relationships for Carboniferous bituminous coals and thus in determining a depositional model for the seams of the Zacler Formation in the Intrasudetic Basin. For example, Crosdale (1993), using maceral ratios to compare Tertiary brown coals and sub-bituminous coals with modern peat bogs, cast doubt on the methods used. However, these actualistic methods should be applied carefully for palaeoenvironmental considerations because of the drastic changes in plant associations since the Carboniferous (Scott 1979). The validity of using maceral ratios, maceral and microlithotype compositions for the interpretation of the mire type has been established by petrographic studies and supported by palynological data.

## Conclusions

The facies analyses of coal seams of the Zacler Formation in the Lower Silesian coal basin described in this paper have been based on petrological studies. These methods have shown that the forest moors of the telmatic and limno-telmatic zones, determined from the facies diagram of Hacquebard & Donaldson (1969) are equivalent to the forest swamps of the telmatitic zone indicated by the TPI–GI diagram, and both represent the vitrinite–fusinite petrological coal facies. These environments of coal sedimentation are characterized by a predominance of arborescent vegetation, for which *Lycospora* is the main palynological taxon. Reed moors of the telmatic and limno-telmatic zones (the facies diagram of Hacquebard & Donaldson 1969) find their equivalent on the TPI–GI diagram as marshes of the limno-telmatic zone. For these environments the densosporinite facies is typical, which is characterized by the prevalence of herbaceous plants, of which *Densosporites* is the main miospore.

The results indicate some general conclusions about the environments of coal accumulation. The coal seams of the Zacler Formation originated from autochthonous plant material deposited in palaeo-swamps. Frenzel (1983) and Boron *et al.* (1987) suggested that two major processes lead to the origin of swamps: (1) terrestialization by infilling of shallow lakes; and (2) paludification, swamping of badly drained areas, e.g. flood plains of rivers.

Sedimentological studies of the rocks of the Zacler Formation from the Wałbrzych trough demonstrate that the areas on which swamps developed were flood plains (Mastalerz 1990). Thus it is plausible that, at least for this part of the basin, paludification led to the formation of palaeo-swamps, which were connected with fluvial environments. However, the zonal distribution of coal facies (Teichmuller 1962; 1989; Hacquebard & Donaldson 1969; Strehlau 1990) suggests that coals originating in reed moors (marshes) are the result of mires formed by terrestrialization. Fluvial influences are marked as clastic partings in seams, especially among

coals of the vitrinite–fusinite facies representing forest swamps.

In summary, the vitrinite–fusinite facies, high in vitrinite, is characteristic of forest swamps (northwestern and eastern areas of the Lower Silesian coal basin), where arborescent vegetation predominated. The densosporinite facies, high in inertinite and intermediate in vitrinite, is typical of reed moor (marsh) environments, where herbaceous plants prevailed.

## References

AUGUSTYNIAK, K. 1964. Uwagi na temat sedymentacji westfalu w niecce srodsudeckiej. *Przeglad Geologiceny*, **7/8**, 339–342.
——1970. *Geological Atlas of the Lower Silesian Coal Basin Part II.* Instytut Geologiczny, Warszawa.
BORON, D. J., EVANS, E. W. & PETERSON, J. M. 1987. An overview of peat research, utilization and environmental considerations. *International Journal of Coal Geology*, **8**, 1–31.
CALDER, J. H., GIBLING, M. R. & MUKHOPADHYYAY, P. K. 1991. Peat formation in a Westphalian B piedmont setting, Cumberland basin, Nova Scotia: implications for the maceral-based interpretation of rheotrophic and raised paleomires. *Bulletin de la Societé Geologique de France*, **162**, 283–298.
CECIL, C. B., RANDAL III, A. H., STANTON, R. W. DULONG, F. T. & RUPPERT, L. 1981. A geochemical model for origin of low-ash and low-sulfur coal. *In: Geological Society of America, Cincinnatti, Field Trip No. 4*, 175–177.
CROSDALE, P. J. 1993. Coal maceral ratios indicators of environment of deposition: do they work for ombrogenous mires? An example from the Miocene of New Zealand. *Organic Geochemistry*, **20**, 797–809.
DIESSEL, C. F. K. 1965. Correlation of macro- and micropetrography of some New South Wales coals. *In: Proceedings General, 6 8th Commonwealth Mineralogy and Metallurgy Congress, Melbourne*, 669–677.
——1982. An appraisal of coal facies based on maceral characteristics. *Australian Coal Geology*, **4**, 474–483.
——1983. Macerals as coal facies indicators. *In:* 10 *Congr. Int. Stratigr. Geol. Carbonifer., Madrit, 1983. Comptes Rendues*, **3**, 367–373.
——1986. On the correlation between coal facies and deopsitional environments. *In: Proceedings 20th Symposium, Depositional Geology, University of Newcastle, NSW*, 19–22.
——1992. *Coal-bearing Depositional Systems.* Springer, Berlin.
EBLE, C. F. & GRADY, W. C. 1993. *Palynologic and Petrographic Characteristics of Two Middle Pennsylvanian Coal Beds and a Probable Modern Analogue.* Geological Society of America, Special Paper, **286**, 119–138.

——, GASPALDO, R. A., DEMKO, T. M. & LIU, Y. 1994a. Coal compositional changes along a mire interior to mire margin transection in the Mary Lee coalbed, Warrior Basin, Alabama, USA. *International Journal of Coal Geology*, **26**, 43–62.

——, HOWER, J. C. & ANDREWS, JR, W. M. 1994b. Palaeoecology of the Fire Clay coal bed in portion of the Eastern Kentucky Coal Field. *Palaeogeography, Palaeoclimatology, Palaeoecology*, **106**, 287–305.

DiMICHELE, W. A. & PHILLIPS, T. 1985. Arborescent Lycopod reproduction and paleoecology in coal-swamp environment of late Middle Pennsylvanian age (Herrin Coal, Illinois, USA). *Reviews in Palaebotany and Palynology*, **44**, 1–26.

FRENZEL, B. 1983. Mires—repositorries of climatic information on self-perpetuating ecosystems. *In*: GORE, A. J. P. (ed.) *Mires: Swamp, Bog and Moor, General Studies*. Elsevier, Amsterdam, 63–85.

GÓRECKA, T. & GÓRECKA-NOWAK, A. 1989. Palynostratigraphic studies of Uper Carboniferous deposits from the Intra-Sudetic Basin, southwestern Poland. *Reviews in Palaeobotany and Palynology*, **65**, 287–292.

GÓRECKA-NOWAK, A. 1995. Palynostratigaphy of Westphalian Rocks in Northwestern Part of the Intrasudetic Basin. *Prace Mineralogiczno-Geologiczne*, **40** [in Polish, with English translation].

HACQUEBARD, P. A. & DONALDSON, J. R. 1969. Carboniferous coal associated with flood-plain and limnic environments in Nova Scotia. *In*: HOPKINS, M. E. (ed.) *Environment of Caol Deposition*. Geological Society of America, Special Paper, **114**, 143–191.

HAGEMANN, H. W. & WOLF, M. 1989. Palaeoenvironments of lacustrine coals—the occurrence of algae in humic coals. *International Journal of Coal Geology*, **12**, 511–522.

HARVEY, R. D. & DILLON, J. W. 1985. Maceral distributions in Illinois coals and their paleoenvironmental implications. *International Journal of Coal Geology*, **5**, 141–165.

HOWER, J. C., EBLE, C. F. & RATHBONE, R. F. 1994. Petrology and palynology of the No. 5 Block caol bed, northeastern Kentucky. *International Journal of Coal Geology*, **25**, 171–193.

KALKREUTZ, W. & LECKIE, D. A. 1989. Sedimentaological and petrographical characteristics of Cretaceous strandplain coals: model for coal accummulation from the North American Western Interior Seaway. *International Journal of Coal Geology*, **12**, 381–424.

KWIECIŃSKA, B. 1967. Węgle skoksowane z Zagłębia Wałbrzyskiego. *Prace Mineralogiczne PAN*, **9**.

——1980. Mineralogy of natural graphites. *Prace Mineralogiczne*, **67**, 87.

LAMBERSON, M. N., BUSTLIN, R. M. & KALKAREUTH, W. 1992. Lithotype (maceral) composition and variation as correlated with paleo-wetlandenvironments, Gates Formation, northeastern British Columbia, Canada. *International Journal of Coal Geology*, **18**, 87–124.

MARCHIONI, D. L. 1980. Petroigraphy and deposition environment of the Liddell seam, Upper Hunter Valley, New South Wales. *International Journal of Coal Geology*, **1**, 35–61.

——, KALKREUTH, W., UTTING, J. & FOWLER, M. 1994. Petrographical, palynological and geochemical analyses of the Hub and Harbour seams, Sydney Coalfield, Nova Scotia, Canada—implications for facies development. *Palaeogeography, Palaeoclimatology, Palaeoecology*, **106**, 241–270.

MASTALERZ, K. 1990. *Sedymentacja warstw żaclerskich (dolny westfal) w niece wałbrzyskiej*. PhD Thesis, University of Wrocław.

NOWAK, G. J. 1992. Petrologia pokładów węgla warstw żaclerskich w depresji śródsudeckiej (Donnoslaskie Załębie Węglowe). PhD Thesis, Polish Geological Institute.

——, 1993. Lithotype variation and petrography of coal semas from Zacler Formation (Westphalian) in the Intrsudetic Basin, southwestern Poland. *Organic Geochemistry*, **20**, 295–313.

——, Petrology of Coal Seams of the Zacler Formation in the Intrasudetic Basin. *Prace Minerologiczno–Geologiczne*. [in Polish, with English summary]. In press.

OSVALD, H. 1937. *Myar och Myrodling* [Peats and their cultivation]. Kooperativa Förbundels Bokoforlag, Stockholm.

PHILLIPS, T. L. & DiMICHELE, W. A. 1981. Paleoecology of Middle Pennsylvanian age coal swamps in southern Illinois, Herrin coal member at Sahara Mine No. 6. *In*: NICLAR, K. J. (ed.) *Paleobotany, Paleoecology and Evolution*. Praeger, New York, Vol. 1, 231–284.

SCOTT, A. C. 1979. The ecology of Coal Measure floras from northern Britain. *Proceedings of the Geologists' Association*, **90**, 97–116.

STACH, E., MACKOWSKI, M.-Th., TEICHMÜLLER, M., TAYLOR, G. H., CHANDRA, D. & TEICHMÜLLER, R. 1982. *Coal Petrology*, 3rd edn. Gerbijder Borntraeger, Berlin.

STREHLAU, K. 1990. Facies of Carboniferous coal seams of Northwest Germany. *International Journal of Coal Geology*, **15**, 245–292.

TEICHMÜLLER, M. 1950. Zum petrographischen Aufbau und Wedegang der Weichbraunmkohle (mit Berücksichtigung genetische Fragen der Steikohlenpetrographie). *Geologisches Jahrbuch*, **64**, 429–488.

——1962. Die Genese der Kohle. *In*: *Comptes Rendus 4 ieme Cngr. Int. Stratigr. Geol. Carbonifere, Heerlen, 1958*, **3**, 607–722.

——1989. The genesis of coal from viewpoint of coal petrology. *International Journal of Coal Geology*, **12**, 1–87.

WOLF, M. & WOLFF-FISCHER, E. 1984. Alginit in Humuskohlen karbonischen Alters und sein Einfluss auf die optischen Eigen schaften des beleitenden vitrinits. *Glückouf-Forschungshefte*, **45**, 243–246.

# Minerals and major elements in density-separated coal fractions from Point of Ayr coal, Wales, UK

J. BARRAZA, A. GILFILLAN, M. CLOKE & D. CLIFT

*Coal Technology Research Group, Chemical Engineering Department,*
*University of Nottingham, Nottingham NG7 2RD, UK*

**Abstract:** Density separated coal fractions from 1.30 floats to 1.70 sinks of Point of Ayr coal for the particle sizes $-3350 + 2800$, $-1000 + 850$, $-300 + 212$ and $-38 + 20\,\mu m$, have been studied for major element and mineral content. Low temperature ash residues were analysed using Fourier transform infra-red and X-ray diffraction spectrometry. Major elements (Al, Ca, Fe, K Mg, Mn, Na and Si) were determined in high-temperature ash residues using atomic absorption spectrometry and atomic emission spectrometry. The minerals identified in the samples were kaolinite, illite, montmorillonite, quartz, dolomite, anhydrite, calcite, gypsum, mica, bassanite and interlayer smectite. The study suggests the origin of the minerals, on the basis of their association with the organic or mineral matter fractions, and indicates that there is a correlation between some major elements and mineral distribution in the coal fractions. It is shown that both the organic affinity of the elements and the abundance of specific minerals are related to the particle size.

Minerals and major elements have been evaluated in coal and coal fractions separated by different methods (Gluskoter *et al.* 1977; Palmer 1983; Valkovic 1983; Conzemius *et al.* 1988; Norton & Markuszewski 1989; Martinez-Tarazona *et al.* 1992*a*). It is well known that minerals and major elements present a number of problems in coal utilization processes. For example, in the hydrocracking of coal liquids there is a deactivation of the catalyst due to the deposition of certain major elements (Coleman *et al.* 1977; Hellgeth *et al.* 1984; Cloke *et al.* 1987; Cloke & Aquino 1991). In combustion, minerals produce abrasion in the coal handling circuits and there are problems with slagging and fouling (Garcia & Martinez-Tarazona 1993). In Coalbed methane (CBM) extraction, reduction in the production of gas due to the presence of minerals filling the cleat/fracture systems in the coal has been reported (Gamson *et al.*, this volume). Environmental problems may also arise from sulphur dioxide emissions generated from pyrite combustion (Clarke & Sloss 1992).

Coal beneficiation provides a means of separating the mineral matter from the more reactive macerals. However, the removal of minerals in the coal beneficiation process depends on their origin, as well as on their particle size. It has been reported (Murchison & Westoll 1968) that the syngenetic minerals (detrital and authigenic) are more difficult to remove than epigenetic minerals because they are intimately associated with the coal. On the other hand, the particle size plays an important part in the coal beneficiation process, having an influence on the economy of the coal

crushing process, the liberation and separation of mineral phases and the velocity of sedimentation of the density-separated coal fractions.

In this work, density-separated coal fractions of Point of Ayr coal were characterized by elemental and mineralogical analysis to identify the types of minerals, to determine their distribution, to suggest their origin on the basis of their organic or mineral association and to evaluate the influence of the particle size on both the liberation of the mineral phases and the organic affinity of the elements.

## Experimental

### Methods

The coal used was a bituminous coal from Point of Ayr (North Wales) supplied by British Coal.

Coal fraction samples were obtained by the float (F)/sink (S) process using specific gravities of <1.30 (1.30F), 1.30–1.35 (1.35F), 1.35–1.40 (1.40 F), 1.45–1.55 (1.55F), 1.55–1.70 (1.70 F) and >1.70 (1.70S). For good separations using the float–sink techniques, reasonably tight size fractions are required. To cover a range of coal sizes, without generating a large number of samples, separations were carried out using ground coal of the following sizes (in $\mu m$): $-3350 + 2800$ (coarse), $-1000 + 850$ (medium), $-300 + 212$ (small) and $-38 + 20$ (fines). Solutions of sodium polytungstate (SPT), of various specific gravities, were used as the dense liquid medium. The medium was adjusted to the required specific gravity by concentration or dilution. A surfactant, Brij 35 (supplied by ICI

*From* Gayer, R. & Harris, I. (eds) 1996, *Coalbed Methane and Coal Geology*,
Geological Society Special Publication No 109, pp 287–299.

Chemicals Ltd) was added to a final concentration of 8 g/l to prevent agglomeration of the fine coal particles. Separations were carried out under gravity in 1 l separating cylinders. The major element concentrations were determined by atomic absorption spectrometry (AAS) and atomic emission spectrometry (AES). The procedure used to dissolve the coal fractions has been reported by Hamilton (1986). Standards and controls used to determine the major element concentrations were made from BDH Spectrosol solutions.

Mineral matter forms in the coal fractions were determined by X-ray diffraction (XRD) and Fourier transform infra-red (FTIR) spectrometry of the appropriate low-temperature ash (LTA). Low temperature ash residues were prepared in a Nanotech Plasmaprep 100 plasma furnace. A radiofrequency forward power of 30 W was used for 48–72 h. X-Ray diffractograms were obtained on a Phillips PW/1130/00 spectrometer. Samples were mixed with acetone, spread on a glass slide, dried and scanned from $6°$ $2\theta$ to $65°$ $2\theta$. FTIR spectra were recorded on a

**Table 1.** *Analysis of coal fractions of particle size $-3350 + 2800\,\mu m$*

| Coal fraction | Analysis (wt%) | | Petrographic analysis (%v, mmf) | | |
|---|---|---|---|---|---|
| | Moisture | Ash (dB) | Vitrinite | Liptinite | Inertinite |
| Feed  | 2.2 | 9.4  | 70.4 | 19.6 | 9.9  |
| 1.30F | 2.3 | 1.9  | 87.3 | 7.0  | 5.6  |
| 1.35F | 1.9 | 6.6  | 79.3 | 8.0  | 12.7 |
| 1.40F | 1.9 | 13.0 | 66.1 | 10.3 | 23.1 |
| 1.45F | 1.8 | 16.2 | 68.7 | 11.3 | 20.0 |
| 1.55F | 1.8 | 22.4 | 70.8 | 9.7  | 19.5 |
| 1.70F | 1.7 | 34.7 | 71.0 | 7.7  | 21.2 |
| 1.70S | 1.2 | 73.0 | 74.7 | 3.3  | 22.0 |

**Table 2.** *Analysis of coal fractions of particle size $-1000 + 850\,\mu m$*

| Coal fraction | Analysis (wt%) | | Petrographic analysis (%v, mmf) | | |
|---|---|---|---|---|---|
| | Moisture | Ash (dB) | Vitrinite | Liptinite | Inertinite |
| Feed  | 1.8 | 10.8 | 77.5 | 8.3  | 14.2 |
| 1.30F | 2.9 | 1.5  | 80.2 | 10.8 | 9.0  |
| 1.35F | 2.5 | 4.5  | 67.0 | 15.4 | 17.6 |
| 1.40F | 2.1 | 9.0  | 66.0 | 11.4 | 22.6 |
| 1.45F | 1.8 | 13.5 | 68.7 | 10.7 | 20.6 |
| 1.55F | 1.8 | 18.3 | 67.4 | 11.1 | 24.7 |
| 1.70F | 1.9 | 30.3 | 63.7 | 14.1 | 22.2 |
| 1.70S | 0.9 | 74.0 |      |      |      |

**Table 3.** *Analysis of coal fractions of particle size $-300 + 212\,\mu m$*

| Coal fraction | Analysis (wt%) | | Petrographic analysis (%v, mmf) | | |
|---|---|---|---|---|---|
| | Moisture | Ash (dB) | Vitrinite | Liptinite | Inertinite |
| Feed  | 2.4 | 12.4 | 81.3 | 9.7  | 9.0  |
| 1.30F | 3.1 | 1.0  | 84.3 | 8.9  | 6.8  |
| 1.35F | 2.2 | 3.0  | 68.5 | 8.6  | 22.9 |
| 1.40F | 2.4 | 7.2  | 64.6 | 10.6 | 24.8 |
| 1.45F | 2.3 | 19.9 | 61.8 | 8.2  | 30.0 |
| 1.55F | 2.1 | 18.3 | 62.5 | 9.7  | 27.7 |
| 1.70F | 2.7 | 29.3 | 64.6 | 9.5  | 25.9 |
| 1.70S | 1.5 | 74.2 |      |      |      |

Perkin-Elmer Model 1710 spectrometer by co-adding 32 scans at a resolution of $4\,cm^{-1}$. Samples were introduced as 1% w/w dispersions in KBr discs. Spectra were obtained for the light (1.30F), medium (1.55F) and heavy (1.70S) coal fractions of coarse and fine particle sizes.

## Results and discussion

Proximate and petrographic analyses of the feed and coal fractions at different particle size are reported in Tables 1–4. In general, the light coal fractions (1.30F, 1.35F) show high vitrinite concentration and low ash content for all the particle sizes. The coal fractions 1.30F and 1.35F of fine particle size show the highest vitrinite content. In contrast, the heavy coal fractions (1.70F, 1.70S) show a low vitrinite content and a high concentration of ash, whereas the coal fractions of medium specific gravities (1.40F, 1.45F, 1.55F) exhibit medium levels of vitrinite and ash. These differences suggest that some minerals were liberated and separated effectively from the organic matter

**Table 4.** *Analysis of coal fractions of particle size* $-38 + 20\,\mu m$

| Coal fraction | Analysis (wt%) | | Petrographic analysis (%v, mmf) | | |
|---|---|---|---|---|---|
| | Moisture | Ash (dB) | Vitrinite | Liptinite | Inertinite |
| Feed | 1.7 | 6.5 | 81.3 | 4.0 | 14.7 |
| 1.30F | 0.8 | 1.0 | 91.4 | 4.4 | 4.2 |
| 1.35F | 0.9 | 1.7 | 90.9 | 2.2 | 6.9 |
| 1.40F | 2.3 | 4.4 | 70.9 | 3.6 | 25.5 |
| 1.45F | 2.2 | 8.5 | 50.2 | 2.2 | 47.6 |
| 1.55F | 2.1 | 13.2 | 34.8 | 3.0 | 62.2 |
| 1.70F | 2.5 | 18.1 | 27.1 | 1.6 | 71.3 |
| 1.70S | 3.4 | 69.1 | 47.4 | 5.3 | 47.4 |

**Table 5.** *Weight distribution in coal fractions*

| Samples | Particle size ($\mu$m) | | | | | | | |
|---|---|---|---|---|---|---|---|---|
| | $-3350 + 2800$ | | $-1000 + 850$ | | $-300 + 212$ | | $-38 + 20$ | |
| | Wt% | Cum. wt% | Wt% | Cum. wt% | Wt% | Cum. wt% | Wt% | Cum. wt% |
| 1.30F | 76.67 | 76.67 | 68.40 | 68.40 | 69.10 | 69.10 | 73.75 | 73.75 |
| 1.35F | 7.85 | 84.52 | 11.90 | 80.30 | 8.10 | 77.20 | 11.18 | 84.93 |
| 1.40F | 3.80 | 88.32 | 4.40 | 84.70 | 5.50 | 82.70 | 3.40 | 88.33 |
| 1.45F | 1.52 | 89.84 | 2.20 | 86.90 | 1.90 | 84.60 | 1.70 | 90.03 |
| 1.55F | 1.72 | 91.56 | 2.10 | 89.00 | 2.20 | 86.80 | 2.08 | 92.11 |
| 1.70F | 1.21 | 92.77 | 1.30 | 90.30 | 1.20 | 88.00 | 1.72 | 93.83 |
| 1.70S | 7.22 | 99.99 | 9.80 | 100.10 | 12.10 | 100.10 | 6.17 | 100.00 |

**Table 6.** *Ash distribution in coal fractions, dry basis*

| Samples | Particle size ($\mu$m) | | | | | | | |
|---|---|---|---|---|---|---|---|---|
| | $-3350 + 2800$ | | $-1000 + 850$ | | $-300 + 212$ | | $-38 + 20$ | |
| | Ash (%) | Cum. ash (%) | Ash (%) | Cum. ash (%) | Ash (%) | Cum. ash (%) | Ash (%) | Cum. ash (%) |
| 1.30F | 1.90 | 1.90 | 1.49 | 1.49 | 1.00 | 1.00 | 1.03 | 1.03 |
| 1.35F | 6.57 | 2.34 | 4.54 | 1.95 | 3.00 | 1.21 | 1.68 | 1.12 |
| 1.40F | 13.02 | 2.80 | 9.00 | 2.31 | 7.20 | 1.61 | 4.37 | 1.24 |
| 1.45F | 16.17 | 3.02 | 13.47 | 2.59 | 19.90 | 2.02 | 8.52 | 1.38 |
| 1.55F | 22.40 | 3.39 | 18.27 | 2.96 | 18.30 | 2.43 | 13.15 | 1.65 |
| 1.70F | 34.73 | 3.80 | 30.35 | 3.36 | 29.30 | 2.80 | 18.13 | 1.95 |
| 1.70S | 73.05 | 8.12 | 74.04 | 9.96 | 74.20 | 11.44 | 69.07 | 6.08 |

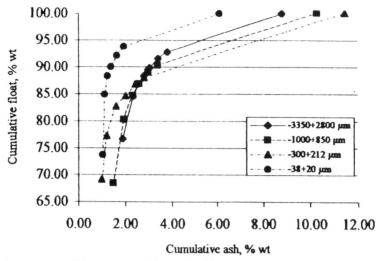

Fig. 1. Cumulative float (wt%) versus cumulative ash (wt%).

during the grinding and the float/sink processes and they may be a basis to suggest their possible origin.

The mass yield, cumulative mass, ash and cumulative ash of the coal as a function of the particle size are given in Tables 5 and 6. Some changes appear in the cumulative mass and cumulative ash contents of the coal fractions at different particle size. To examine these changes the results are plotted as washability curves. Figures 1 and 2 show the washability curves of the cumulative mass of float (wt%) versus cumulative ash (wt%), and the cumulative mass of float (wt%) versus specific gravity of float. In general, the curves exhibit similar behaviour; however, the main point to note is that high mass yield is produced in light coal fractions with low ash content and fine particle size. Technically, these washability curves have a great importance in coal preparation to separate coal fractions at the required density and ash content.

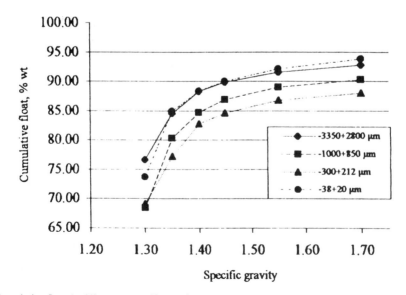

Fig. 2. Cumulative float (wt%) versus specific gravity.

## Elemental concentration in high-temperature ash of coal fractions

The concentration of the major elements in the high-temperature ash of the density-separated coal fractions as a function of the particle size are shown in Tables 7–10. In general, silicon and iron show the highest concentrations for all the density-separated coal fractions and particle sizes, followed by (in decreasing order) Al, Ca, Mg, K Ti, Na and Mn. The high concentrations of silicon and aluminium suggest an abundance of clay minerals, whereas the levels of iron present indicate an abundance of pyrite.

The distribution of the major elements in high-temperature ash of the coal fractions show several trends that may be identified from the data presented in Tables 7–10. Calcium and magnesium show similar patterns of distribution and in general these elements show enrichment in both the lightest and heaviest fractions. Martinez-Tarazona et al. (1992a) reported that preferential association of carbonates with the organic material in coal has been found in previous studies of high-volatile bituminous coals, which could explain the higher concentrations of both calcium and magnesium in the lightest fraction. However, analysis of the mineral forms present

**Table 7.** Elemental concentration in ash of coal fractions for particle size $-3350 + 2800\,\mu m$

| Coal fraction | Element concentration in ash (wt%) | | | | | | | | |
|---|---|---|---|---|---|---|---|---|---|
| | Al | Ca | Fe | Mg | Mn | K | Si | Na | Ti |
| Feed | 9.70 | 5.85 | 10.42 | 2.65 | 0.13 | 1.17 | 22.90 | 0.48 | 0.65 |
| 1.30F | 8.02 | 9.35 | 18.45 | 1.73 | 0.04 | 0.11 | 18.02 | 0.96 | 0.88 |
| 1.35F | 10.19 | 2.88 | 24.64 | 0.89 | 0.05 | 0.44 | 14.33 | 0.29 | 0.60 |
| 1.40F | 10.85 | 2.02 | 20.15 | 0.69 | 0.03 | 0.72 | 20.29 | 0.37 | 0.68 |
| 1.45F | 12.15 | 1.27 | 17.75 | 0.84 | 0.04 | 0.99 | 21.09 | 0.47 | 0.58 |
| 1.55F | 12.40 | 2.02 | 13.09 | 1.13 | 0.06 | 1.23 | 24.82 | 0.38 | 0.59 |
| 1.70F | 15.70 | 3.99 | 12.46 | 2.58 | 0.08 | 1.79 | 26.28 | 0.37 | 0.68 |
| 1.70S | 10.00 | 6.56 | 6.30 | 3.66 | 0.20 | 1.55 | 27.31 | 0.43 | 0.66 |

**Table 8.** Elemental concentration in ash of coal fractions for particle size $-1000 + 850\,\mu m$

| Coal fraction | Element concentration in ash (wt%) | | | | | | | | |
|---|---|---|---|---|---|---|---|---|---|
| | Al | Ca | Fe | Mg | Mn | K | Si | Na | Ti |
| Feed | 9.07 | 5.92 | 10.47 | 3.00 | 0.16 | 1.18 | 21.70 | 0.43 | 0.56 |
| 1.30F | 6.60 | 10.00 | 17.00 | 2.08 | 0.03 | 0.14 | 18.20 | 0.76 | 0.62 |
| 1.35F | 10.10 | 2.74 | 24.14 | 0.65 | 0.02 | 0.45 | 17.17 | 0.53 | 0.55 |
| 1.40F | 10.48 | 2.51 | 25.78 | 0.66 | 0.02 | 0.40 | 27.31 | 0.37 | 0.57 |
| 1.45F | 11.12 | 0.96 | 19.61 | 0.32 | 0.02 | 0.77 | 20.10 | 0.56 | 0.68 |
| 1.55F | 12.23 | 0.54 | 15.12 | 0.32 | 0.01 | 1.02 | 24.10 | 0.50 | 0.60 |
| 1.70F | 13.69 | 0.50 | 10.80 | 0.54 | 0.02 | 1.39 | 24.10 | 0.44 | 0.59 |
| 1.70S | 9.16 | 6.78 | 7.53 | 3.86 | 0.22 | 1.51 | 23.20 | 0.39 | 0.57 |

**Table 9.** Elemental concentration in ash of coal fractions for particle size $-300 + 212\,\mu m$

| Coal fraction | Element concentration in ash (wt%) | | | | | | | | |
|---|---|---|---|---|---|---|---|---|---|
| | Al | Ca | Fe | Mg | Mn | K | Si | Na | Ti |
| Feed | 9.70 | 3.65 | 9.00 | 2.30 | 0.10 | 1.70 | 26.70 | 0.26 | 0.50 |
| 1.30F | 6.22 | 17.64 | 12.17 | 3.47 | 0.02 | 0.38 | 11.82 | 0.65 | 0.56 |
| 1.35F | 8.80 | 3.70 | 18.57 | 0.81 | 0.01 | 0.50 | 21.51 | 0.41 | 0.70 |
| 1.40F | 10.40 | 1.18 | 18.32 | 0.40 | 0.02 | 0.72 | 21.51 | 0.29 | 0.68 |
| 1.45F | 10.40 | 1.98 | 11.02 | 1.23 | 0.05 | 1.27 | 24.82 | 0.25 | 0.67 |
| 1.55F | 12.65 | 0.29 | 13.38 | 0.45 | 0.02 | 1.24 | 27.57 | 0.35 | 0.58 |
| 1.70F | 13.14 | 0.36 | 10.28 | 0.63 | 0.02 | 1.39 | 27.31 | 0.22 | 0.49 |
| 1.70S | 10.49 | 3.30 | 7.89 | 2.27 | 0.12 | 1.90 | 30.81 | 0.25 | 0.47 |

**Table 10.** *Elemental concentration in ash of coal fractions for particle size* $-38 + 20\,\mu m$

| Coal fraction | Element concentration in ash (wt%) | | | | | | | | |
|---|---|---|---|---|---|---|---|---|---|
| | Al | Ca | Fe | Mg | Mn | K | Si | Na | Ti |
| Feed | 6.50 | 8.45 | 16.30 | 3.56 | 0.16 | 0.68 | 16.40 | 0.27 | 0.54 |
| 1.30F | 5.90 | 11.14 | 5.13 | 2.49 | 0.02 | 0.68 | 11.60 | 0.63 | 0.75 |
| 1.35F | 9.00 | 7.15 | 10.13 | 1.64 | 0.01 | 0.90 | 18.20 | 0.48 | 1.06 |
| 1.40F | 11.86 | 1.48 | 13.40 | 0.52 | 0.01 | 1.08 | 21.09 | 0.29 | 0.59 |
| 1.45F | 10.88 | 2.09 | 19.12 | 1.20 | 0.05 | 0.93 | 21.09 | 0.21 | 0.79 |
| 1.55F | 9.90 | 1.82 | 15.10 | 0.95 | 0.05 | 0.90 | 17.84 | 0.28 | 0.46 |
| 1.70F | 10.10 | 0.91 | 14.77 | 0.70 | 0.03 | 1.01 | 21.72 | 0.30 | 0.59 |
| 1.70S | 6.00 | 9.73 | 21.65 | 4.29 | 0.23 | 0.67 | 17.84 | 0.22 | 0.48 |

indicates high carbonate concentrations (as dolomite) in the 1.70S fractions, together with enrichment of the minerals montmorillonite and anhydrite in the 1.30F fractions. It therefore appears that high concentrations of calcium at 1.30F may be due to the presence of both montmorillonite and anhydrite, with magnesium being accounted for in montmorillonite.

Iron concentrations show maxima between 1.35F and 1.45F for the coarse, medium and small particle sizes, with the lowest concentrations being found in the fraction at 1.70S. By contrast, the fine particle size exhibits the highest concentration at 1.70S, closely followed by a similar level at 1.45F. Iron is associated with the mineral pyrite, which, because of its high density, is expected to show the highest concentration in the heaviest fraction of each size. It appears that particle size thus plays an important part in the liberation of pyrite. However, a high concentration of iron remains in the lighter fractions (1.45F) of fine particle size. Pyrite has been shown to occur as very fine particles (Martinez-Tarazona *et al.* 1992*b*), which would account for this high concentration found in the fraction 1.45F of fine particle size.

Aluminium shows the same distribution pattern for the coarse, medium and small particle sizes. The lowest concentration is found in the 1.30F fraction, gradually increasing to a maximum at 1.70F. The coal fractions of fine particle size show a maximum at 1.40F, with the lowest concentration occurring at 1.70S. Aluminium occurs in clay minerals such as illite and kaolinite and as these minerals contain silicon, we might expect that the distribution of silicon would vary in sympathy with aluminium. However, the results of this study indicate a more irregular distribution of silicon than aluminium. This will be due to the presence of silicon associated with quartz and analysis of the minerals indicates the presence of quartz in all the coal fractions.

The alkali metals sodium and potassium show disimilar distribution patterns, despite their similar chemistries. Sodium concentrations remain more or less constant for all the particle sizes and fractions, showing a maximum at 1.30F. The higher levels of sodium at 1.30F can be correlated with the greater abundance of montmorillonite found in this fraction. Potassium shows the highest concentrations in the coal fractions of 1.70F and 1.70S for coarse, medium and small particle sizes, with corresponding minima in the 1.30F fraction, whereas concentrations remain fairly constant for all the coal fractions of fine particle size. The lower potassium content in the lighter fractions may be attributed to a decrease in the abundance of the mineral illite, such as is shown in the analysis of minerals (Fig. 4).

*Element washability curves*

Element washability curves (cumulative element concentration, wt% versus cumulative ash, wt%) are important in coal preparation in determining the optimum conditions for the elimination of the major elements and for the study of the organic affinity of the elements (Martinez-Tarazona *et al.* 1992*a*). The cumulative element concentrations of the coal fractions at different particle size are reported in Tables 11–14. As illustrated, Fig. 3 shows typical washability curves for aluminium, calcium and titanium. Some differences were found in the elements studied when the cumulative element concentration is analysed in the cumulative ash range 0–4 wt%. The particle size does not appear to affect the cumulative concentration of aluminium, magnesium, manganese and silicon. As can be seen in the aluminium curve, most points lie on the same line independent of the particle size. However, in general, calcium, iron and sodium show an increase in their

**Table 11.** *Cumulative element concentration in coal fractions for particle size −3350 + 2800 μm*

| Coal fraction | Cumulative element concentration (wt%) | | | | | | | | |
|---|---|---|---|---|---|---|---|---|---|
| | Al | Ca | Fe | Mg | Mn | K | Si | Na | Ti |
| Feed | 0.91 | 0.55 | 0.98 | 0.25 | 0.01 | 0.11 | 2.15 | 0.05 | 0.06 |
| 1.30F | 0.15 | 0.18 | 0.35 | 0.03 | 0.00 | 0.00 | 0.34 | 0.02 | 0.02 |
| 1.35F | 0.20 | 0.18 | 0.47 | 0.04 | 0.00 | 0.00 | 0.40 | 0.02 | 0.02 |
| 1.40F | 0.25 | 0.18 | 0.56 | 0.04 | 0.00 | 0.01 | 0.49 | 0.02 | 0.02 |
| 1.45F | 0.28 | 0.18 | 0.60 | 0.04 | 0.00 | 0.01 | 0.54 | 0.02 | 0.02 |
| 1.55F | 0.33 | 0.19 | 0.64 | 0.04 | 0.00 | 0.02 | 0.64 | 0.02 | 0.03 |
| 1.70F | 0.40 | 0.20 | 0.69 | 0.05 | 0.00 | 0.02 | 0.75 | 0.02 | 0.03 |
| 1.70S | 0.89 | 0.53 | 0.97 | 0.24 | 0.01 | 0.10 | 2.14 | 0.04 | 0.06 |

**Table 12.** *Cumulative element concentration in coal fractions for particle size −1000 + 850 μm*

| Coal fraction | Cumulative element concentration (wt%) | | | | | | | | |
|---|---|---|---|---|---|---|---|---|---|
| | Al | Ca | Fe | Mg | Mn | K | Si | Na | Ti |
| Feed | 0.98 | 0.64 | 1.13 | 0.32 | 0.02 | 0.13 | 2.34 | 0.05 | 0.06 |
| 1.30F | 0.10 | 0.15 | 0.25 | 0.03 | 0.00 | 0.00 | 0.27 | 0.01 | 0.01 |
| 1.35F | 0.15 | 0.15 | 0.38 | 0.03 | 0.00 | 0.00 | 0.35 | 0.01 | 0.01 |
| 1.40F | 0.19 | 0.15 | 0.48 | 0.03 | 0.00 | 0.01 | 0.46 | 0.01 | 0.01 |
| 1.45F | 0.23 | 0.15 | 0.53 | 0.03 | 0.00 | 0.01 | 0.51 | 0.02 | 0.02 |
| 1.55F | 0.27 | 0.15 | 0.59 | 0.03 | 0.00 | 0.01 | 0.60 | 0.02 | 0.02 |
| 1.70F | 0.33 | 0.15 | 0.63 | 0.04 | 0.00 | 0.02 | 0.70 | 0.02 | 0.02 |
| 1.70S | 0.96 | 0.62 | 1.11 | 0.31 | 0.02 | 0.13 | 2.32 | 0.05 | 0.06 |

**Table 13.** *Cumulative element concentration in coal fractions for particle size −300 + 212 μm*

| Coal fraction | Cumulative element concentration (wt%) | | | | | | | | |
|---|---|---|---|---|---|---|---|---|---|
| | Al | Ca | Fe | Mg | Mn | K | Si | Na | Ti |
| Feed | 1.20 | 0.45 | 1.33 | 0.29 | 0.02 | 0.21 | 3.31 | 0.03 | 0.06 |
| 1.30F | 0.06 | 0.18 | 0.12 | 0.03 | 0.00 | 0.00 | 0.12 | 0.01 | 0.01 |
| 1.35F | 0.08 | 0.17 | 0.17 | 0.03 | 0.00 | 0.00 | 0.17 | 0.01 | 0.01 |
| 1.40F | 0.13 | 0.16 | 0.24 | 0.03 | 0.00 | 0.01 | 0.26 | 0.01 | 0.01 |
| 1.45F | 0.17 | 0.17 | 0.29 | 0.04 | 0.00 | 0.01 | 0.37 | 0.01 | 0.01 |
| 1.55F | 0.23 | 0.17 | 0.34 | 0.04 | 0.00 | 0.02 | 0.49 | 0.01 | 0.02 |
| 1.70F | 0.27 | 0.17 | 0.38 | 0.04 | 0.00 | 0.02 | 0.59 | 0.01 | 0.02 |
| 1.70S | 1.18 | 0.44 | 1.04 | 0.24 | 0.01 | 0.19 | 3.28 | 0.03 | 0.06 |

**Table 14.** *Cumulative element concentration in coal fractions for particle size −38 + 20 μm*

| Coal fraction | Cumulative element concentration (wt%) | | | | | | | | |
|---|---|---|---|---|---|---|---|---|---|
| | Al | Ca | Fe | Mg | Mn | K | Si | Na | Ti |
| Feed | 0.42 | 0.55 | 0.98 | 0.23 | 0.01 | 0.04 | 1.07 | 0.02 | 0.04 |
| 1.30F | 0.06 | 0.11 | 0.05 | 0.03 | 0.00 | 0.01 | 0.12 | 0.01 | 0.01 |
| 1.35F | 0.07 | 0.12 | 0.07 | 0.03 | 0.00 | 0.01 | 0.14 | 0.01 | 0.01 |
| 1.40F | 0.09 | 0.11 | 0.09 | 0.03 | 0.00 | 0.01 | 0.17 | 0.01 | 0.01 |
| 1.45F | 0.11 | 0.11 | 0.12 | 0.03 | 0.00 | 0.01 | 0.20 | 0.01 | 0.01 |
| 1.55F | 0.13 | 0.12 | 0.16 | 0.03 | 0.00 | 0.01 | 0.25 | 0.01 | 0.01 |
| 1.70F | 0.16 | 0.12 | 0.21 | 0.03 | 0.00 | 0.02 | 0.32 | 0.01 | 0.01 |
| 1.70S | 0.41 | 0.53 | 1.12 | 0.21 | 0.01 | 0.04 | 1.06 | 0.02 | 0.03 |

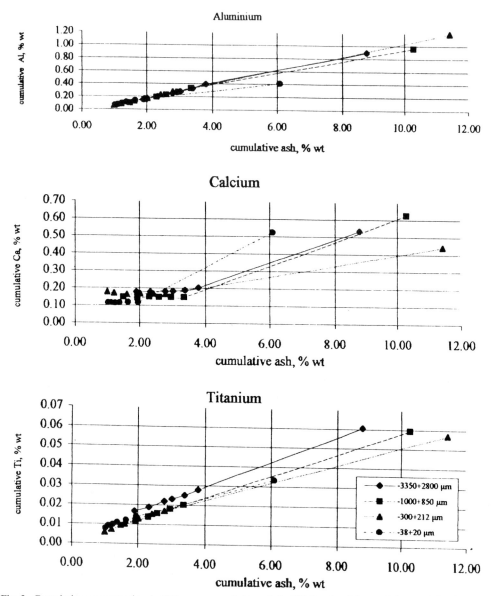

**Fig. 3.** Cumulative concentration (wt%) versus cumulative ash (wt%) for aluminium, calcium and titanium as a function of the particle size.

cumulative concentration with an increase in particle size. In contrast, potassium shows an increase in its cumulative concentration from large to small particle sizes. Titanium shows different behaviour again, with a high cumulative concentration in the coarse particle size, low in the medium particle size and medium in the fine particle size. These results suggest that the association of elements in the coal may depend on particle size. Martinez-Tarazona *et al.* (1992*a*) studied the association of elements in coal with respect to the organic or mineral matter, according to the value of slope of the element washability curves. In this study, it has been found that for all the particle sizes most of elements (Al, Fe, Mn, Si, Na, Ti and K) show an association with the mineral matter (positives slopes in all the ash cumulative range). However,

Particle size -3350+2800 um

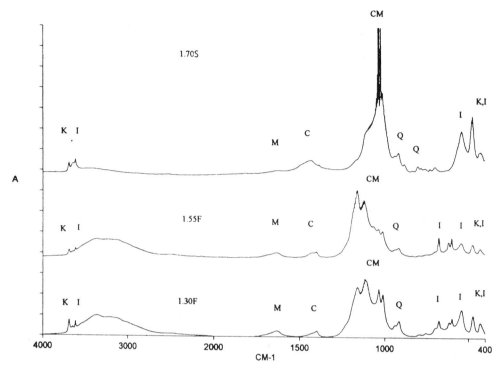

**Fig. 4.** FTIR of LTA residues of density separated fractions, particle size $-3350 + 2800\,\mu$m.

Ca and Mg show negative slope in the cumulative range 0–4 wt%, at medium and small particle sizes, whereas they have a positive slope at coarse and fine particle sizes. According to these results it may be considered that Ca and Mg show an affinity towards organic matter at medium and small particle sizes, but not at coarse and fine sizes. These findings verify that the organic affinity of some elements depends on both the specific gravity and the particle size, which is also confirmed from the mineralogical analysis conducted for this study.

### Minerals present in low-temperature ash of the coal fractions

To ascertain the association of the major elements with mineral matter present in the coal fractions, a mineralogical analysis was performed on the low-temperature ash of each coal fraction using FTIR spectrometry and XRD. Figures 4 and 5 show the FTIR spectra

of low-temperature ash for the density-separated coal fractions 1.30F, 1.55F and 1.70S, for coarse and fine particle sizes. Bands were assigned to minerals according to the wavenumbers reported by Gadsden (1975). The spectra are dominated by absorptions due to the clay minerals illite and kaolinite. Kaolinite shows characteristic absorptions in the wavenumber ranges 475–468, 542–535, 1035–1030 and 3630–3624 cm$^{-1}$, whereas absorptions due to illite are found at 470–465, 537–525, 1025–1010, 1150 and 3630 cm$^{-1}$. The variation in peak intensities is proportional to the change in concentrations of the minerals present. The intensity of these absorptions varies according to particle size and coal fraction, the most intense peaks occurring in the 1.70S fraction for the coarse particle size and in the 1.30F and 1.55F fractions for fine particle sizes. These findings show that the clay minerals were not liberated from the organic matter (high concentration in the coal fraction 1.30F of small particle size) by the grinding process, which suggests that the clay minerals

Particle size -38+20 μm

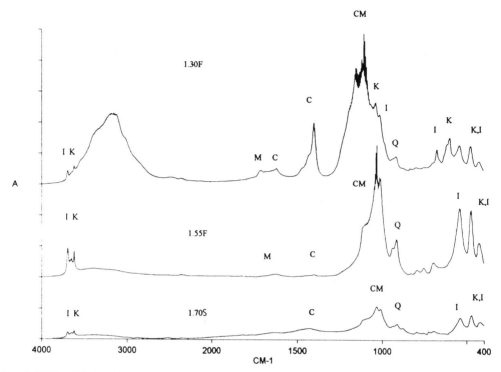

**Fig. 5.** FTIR of LTA residues of density separated fractions, particle size −38 + 20 μm.

had a syngenetic–detrital origin. Similar findings have been reported in the work of Martinez-Tarazona *et al.* (1992*b*) using Spanish coals. However, some clay minerals, mainly illite, may also have an epigenetic origin, as is shown by their high concentration in the coal fraction 1.70S of coarse particle size. For the clay minerals illite and kaolinite, a good correlation is obtained between the variation of peak intensities and the distribution of elemental aluminium. Other minerals that may be identified from the FTIR spectra include quartz and montmorillonite. Quartz shows absorptions at 915 and 805–796 cm$^{-1}$, whereas montmorillonite absorbs around 1640–1635 cm$^{-1}$.

To verify the data obtained from the FTIR spectra and to identify the other minerals present, the same LTA residues were analysed by XRD. The scans obtained of the LTA residues for the two particle sizes are shown in Figs 6 and 7. The minerals identified by XRD in the samples were clay minerals (kaolinite, illite), carbonates (dolomite, calcite), sulphide (pyrite), sulphates

(gypsum, anhydrite, bassanite), oxides (quartz, anatase), mica and interlayer smectite.

Quartz and dolomite show enrichment in the 1.70S fraction for both particle sizes. These findings show that quartz and dolomite are easily liberated, suggesting a predominantly epigenetic origin. In agreement with our results, both minerals have been reported (Murchison & Westoll 1968) as being of epigenetic origin and may be formed by deposition in peat bogs, or by solutions penetrating cracks and fissures of the coalbed. The presence of dolomite in the 1.70S fractions explains the higher concentrations of calcium and magnesium found in these fractions. The high elemental concentrations of calcium found in the 1.30F fractions may be related to the presence of calcium-rich minerals such as anhydrite, gypsum and calcite. Anhydrite was identified in the 1.30F and 1.55F fractions for the coarse particle size and at 1.30F for the fine particle size. Gypsum was found in the 1.30F fraction of fine particle size and calcite in the same fraction for the coarse particle size.

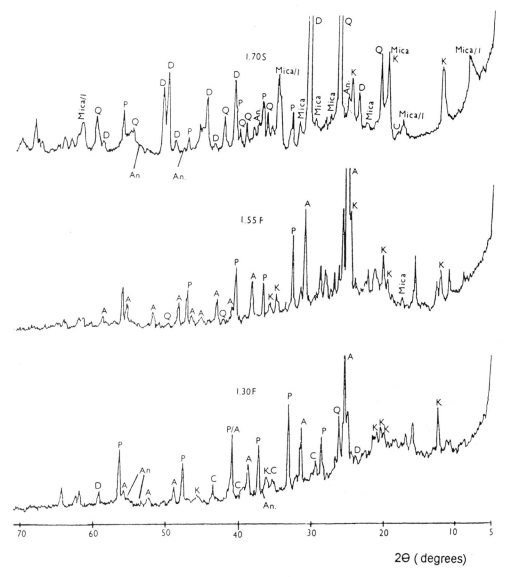

**Fig. 6.** X-ray diffractograms of LTA residues of density separated coal fractions, particle size $-3350 + 2800\,\mu m$.

It should be noted that both gypsum and anhydrite may be generated during the low-temperature ashing procedure from calcite, so that the abundance of calcite in the actual coal fractions may be higher than is indicated from the XRD spectra of the low-temperature ashes.

The distribution of kaolinite in the XRD spectra is in agreement with that of the FTIR spectra, and once again there is a correlation with the elemental aluminium distribution. Pyrite exhibits relatively low concentrations in the 1.30F fraction of fine particle size and in the 1.70S fraction of coarse particle size, whereas high concentrations are found in the fractions 1.70S and 1.30F for fine and coarse particle sizes, respectively. These results suggest an epigenetic origin for the pyrite, as it was liberated from the organic matter when the coal was ground to a fine particle size. The distribution of pyrite correlates well with that of elemental iron and confirms that the smaller particle size favours pyrite liberation.

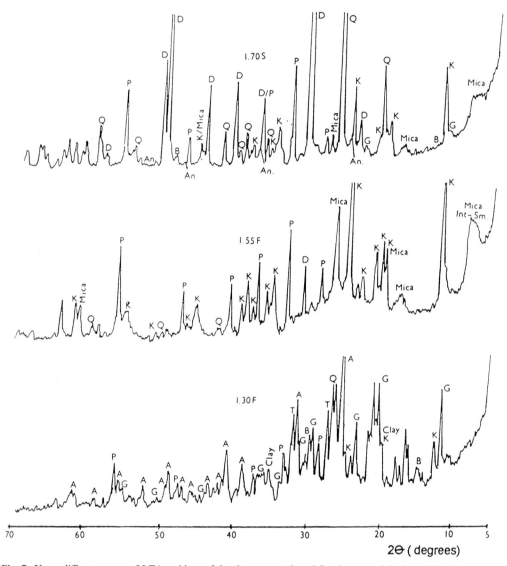

**Fig. 7.** X-ray diffractograms of LTA residues of density separated coal fractions, particle size $-38 + 20\,\mu$m.

## Conclusions

Data on the distribution of major elements and minerals in density-separated coal fractions as a function of the particle size in Point of Ayr coal may be used to reveal the coal fraction most convenient for processing. The analytical techniques allowed the evaluation of the distribution of major elements; these data were then used to identify and evaluate semi-quantitatively the minerals present in the density-separated coal fractions and to suggest their possible origin on the basis of their organic or mineral association.

The distribution of minerals in the density-separated coal fractions shows that dolomite and quartz are concentrated in the heavier fractions of the particle sizes. Pyrite is enriched in the lighter fractions for all but the fine particle size, in which high concentrations are found in the heaviest (1.70S) fraction. The smaller particle size favours mineral liberation.

The authors thank colleagues in the Department of Mineral Resources, University of Nottingham, for the supply of the density-separated coal fraction samples. The financial support provided by the Colombian

Institute of Science and Technology (COLCIEN-CIAS), the University of Valle (Colombia) and the ECSC is gratefully acknowledged.

# References

CLARKE, L. B. & SLOSS, L. L. 1992. *Trace Element Emissions from Coal Combustion and Gasification.* IEA Coal Research, London.

CLOKE, M. & AQUINO, Z. N. 1991. Production of filtered coal extract solution with very low levels of mineral matter. *In: Coal Science II.* American Chemical Society, Symposium Series, **461**, 250–259.

——, HAMILTON, S. & WRIGHT, P. 1987. Changes in trace element concentrations and catalyst activity during the hydrocracking of a coal extract solution. *Fuel,* **66**, 678–682.

COLEMAN, W. M., SZOOBE P., WOOTON, D. L. & DORN, H. C. 1977. Minor and trace metal analysis of a solvent-refined coal by flameless atomic absorption. *Fuel,* **56**, 195–198.

CONZEMIUS, R., CHRISWELL, C. D. & JUNK, G. 1988. The partitioning of elements during physical cleaning of coals. *Fuel Processing Technology,* **19**, 95–106

GADSDEN, J. A. 1975. *Infrared Spectra of Minerals and Related Inorganic Compounds.* Butterworth, London.

GAMSON, P., BEAMISH, B. & JOHNSON, D. 1996. Coal microstructure and secondary mineralization: their effect on methane recovery. *This volume.*

GARCIA, A. B. & MARTINEZ-TARAZONA, R. 1993. Trace element removal from spanish coals by flotation. *Fuel,* **72**, 15–16.

GLUSKOTER, H. J., RUCH, R. R., MILLER, W. E. & CAHILL, R. A. 1977. *Minor Elements in American Coals.* Illinois State Geological Survey, Circular, **399**.

HAMILTON, S. 1986. *Mineral forms in coal and coal derived liquids.* M. Phil. Thesis, University of Nottingham.

HELLGETH, J. W., BROWN, R. S. & TAYLOR, L. T. 1984. Effect of coal liquefaction conditions on the trace element content of the non-volatile product. *Fuel,* **63**, 453–462.

MARTINEZ-TARAZONA, M. R., SPEARS, D. A. & TASCON, J. M. 1992a. Organic affinity of trace elements in Asturian bituminous coals. *Fuel,* **71**, 909–917.

——, ——, PALACIOS, J. M., MARTINEZ-ALONSO, A. & TASCON, J. M. D. 1992b. Mineral matter in coals of different rank from the Asturian Central basin. *Fuel,* **71**, 367–372.

MURCHISON D. & WESTOLL, T. S. 1968. *Coal and Coal Bearing Strata.* Oliver and Boyd, Glasgow.

NORTON, G. A. & MARKUSZEWSKI, R. 1989. Trace element removal during physical and chemical coal cleaning. *Coal Preparation,* **7**, 55–68.

PALMER, C. A. 1983. Determination of mode of occurrence of trace elements in the upper freeport coal bed using size and density separation procedures. *In: Proceedings, International Conference on Coal Science, Pittsburgh, PA,* 365–369.

VALKOVIC, V. 1983. *Trace Elements in Coal,* Vols 1 and 2. CRC Press, Boca Raton.

# Mineralogy, geochemistry and pyrite content of Bulgarian subbituminous coals, Pernik Basin

IRENA KOSTOVA[1], OGNYAN PETROV[1] & JORDAN KORTENSKI[2]

[1] *Institute of Applied Mineralogy, Bulgarian Academy of Sciences,*
*92 Rakovski Street, 1000 Sofia, Bulgaria*
[2] *University of Mining and Geology, 1156 Sofia, Bulgaria*

**Abstract:** The mineralogy and geochemistry of Pernik subbituminous coals (coal bed A) and some genetic peculiarities related to the mineral formation were studied. The mineral matter of the coal consists chiefly of pyrite, kaolinite, siderite, quartz and calcite. Other minerals (dolomite, ankerite, plagioclase and some sulphates) are present in minor amounts, some occurring as accessory single crystals. Pyrite is the main mineral in these coals and exhibits a large array of textures and morphology. Isolated and clustered euhedral, bacterial and inorganic framboidal, cluster-like, homogeneous and microconcretional massive, infilling and replacing anhedral, and cleat-filling and fracture-filling infiltrational pyrite types were observed. Four stages of mineralization were distinguished: pyrite–kaolinite, pyrite, pyrite–siderite and sulphate stages. The amount of pyrite present in two sections of coal bed A was determined by quantitative powder X-ray diffraction analysis. The concentrations of 37 trace elements were determined. As, Cu, Co, Ni, Zn, Pb, V Ti, Mo, Rb, Cr and Mn are typomorphic for this coal. On the basis of their relation to organic or inorganic matter, four groups of trace elements were subdivided; and on the basis of cluster analysis four associations were differentiated.

The Pernik coal basin is situated in the western part of Bulgaria, about 30 km from Sofia (Fig. 1). The coal-bearing sediments have been determined as Upper Oligocene and Lower Miocene (Chernyavska 1970) on the basis of characteristic spore pollen spectra. From a lithostratigraphic viewpoint the Pernik Coal Basin consists of a sequence of coarse-grained, clayey–sandy and clayey–marly sediments subdivided into five formations (from the base upwards): conglomerate–sandy, clay–bituminous, mollasic, coal-bearing and clay–marly. The total thickness of the coal-bearing formation exceeds 1000–2000 m. The coal beds belong to the fourth of the five formations (Kamenov 1964). Five coal beds have been named (from bottom to top) as:

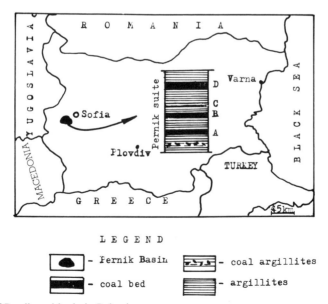

**Fig. 1.** Location of Pernik coal basin in Bulgaria.

*From* Gayer, R. & Harris, I. (eds) 1996, *Coalbed Methane and Coal Geology*,
Geological Society Special Publication No 109, pp 301–314.

shaley, A B, C and D. Only A B and D (Pernik suite, Fig. 1) are of industrial interest. The coal beds thicknesses are: A, 0.5–3 m; B, 2.5–3 m; and D, 1.5–6 m. Coal bed A has a high ash content and the fourth, coal bed C is thin. Tectonically, the Pernik Basin is a syncline with an east–west orientation. The coals range in rank from subbituminous to bituminous (Siskov *et al.* 1982). The petrological composition of the Pernik coals has been de-scribed by Konstantinova (1964) and Valcheva & Milchev (1979; 1981). These workers described the macerals and their distribution in detail, as well as the properties of the coals, but mineralogically they identified only pyrite, illite, kaolinite and siderite (in beds A and B) and detailed descriptions have not been given. Kortenski (1990) investigated the coals in bed D and determined the principal forms of the minerals present and distinguished four stages of mineralization.

The purpose of this paper is to characterize the mineralogy and geochemistry of Pernik brown coals, especially coal bed A, and to describe unusual features related to the mineral formation. Pyrite, the major sulphur-bearing mineral, was also determined quantitatively by X-ray diffraction (XRD).

## Materials and methods

Coal bed A varies in thickness from 0.5 to 3 m (1.5 m average). To determine the mineralogical and pyrite content of coal A, 17 samples at 10 cm vertical intervals were collected from two sections within this bed. Section I is representative of the thin parts of the bed and section II is typical of the average thickness of the bed.

Optical microscopy was performed under reflected light with dry and oil immersion objectives on an NU-2 Universal microscope. The morphological and compositional studies were carried out with a scanning electron microscope, SEM 515 Philips, equipped with an energy dispersive X-ray spectrometer, EDAX 9100/72, and the minor mineral phase determinations using a transmission electron microscope, TEM 420 Philips, operated at 120 keV and equipped with an EDAX X-ray microanalyser with a Si(Li) detector. The samples were prepared from alcohol suspension and were deposited on a copper grid covered with a carbon film. The mineralogical composition of the coal samples was determined by qualitative powder XRD analysis. A powder XRD approach was also applied for quantitative

determination of pyrite in the samples from which pyritic sulphur contents were calculated. The qualitative and quantitative XRD analyses were performed on a DRON 3M powder diffractometer with a Ni-filtered CuK$\alpha$ radiation and a scintilation detector. Diffraction patterns used for the phase mineralogical determinations were registered in the $2\theta$ interval 9–50° . For determination of the trace elements, a total of 44 samples, including 34 coal samples and 10 coal shales, interbeded with the coal layers, were ashed at 800°C. All samples were analysed using instrumental neutron activation analysis (INAA) and inductively coupled plasma atomic emission spectrometry (ICP-AES).

## Mineralogy

Kortenski (1990) described three stages of mineralization in coal D of the Pernik subbituminous coal basin, namely: 'pyrite–kaolinite–illite', 'pyrite' and 'pyrite–siderite' stages. Our observations on coal A from the same basin showed a similar sequence of mineral formation in the two coal beds. In coal A the same three stages of mineral deposition were observed, but in first mineral formation stage kaolinite is the only clay mineral that has formed and no illite has been detected. Our mineralogical studies showed that in coal A, an additional fourth 'sulphate' stage can be recognized.

The first 'pyrite–kaolinite' stage of mineralization occurred during peat genesis. The mineralization is controlled by conditions in the peat environment (pH 3–6.5). Framboidal, euhedral and massive pyrite, and kaolinite, were precipitated during this stage and calcite and clastic quartz were mechanically introduced by incoming waters.

The second stage, the 'pyrite' stage, is related to early diagenesis and is marked by the formation of anhedral pyrite. This type of pyrite is formed either by pore filling or by replacement processes. The first and second stages are closely related because the pyrite in both instances is formed by feeding solutions during the peat genesis.

The third stage, 'pyrite–siderite' stage, resulted in the precipitation of mineral matter during late diagenesis or katagenesis. Concordant and discordant fractures were formed in the coal beds as a result of fault movements in the Pernik graben. The pyrite and siderite were precipitated from two types of infiltrating solutions within the two distinctive fracture systems (Kortenski 1990).

Gypsum, bassanite, anhydrate, jarosite, melanterite, copiapite, szomolnoquite, aluminocopiapite, ammoniojarosite, Mg-copiapite and halotrichite have also been detected. They mark the fourth 'sulphate' stage of mineralization. Most of these minerals are of secondary origin related to the weathering of minerals, especially pyrite. Wiese *et al.* (1987) and Wiese & Fyfe (1986) described the mechanism of spontaneous oxidation of pyrite to form various hydrated iron sulphates. These workers have observed on the polished and rough-cut surfaces of Ohio and Utah coal samples stored under normal atmospheric conditions growths melanterite, rozenite, szomolnoquite and halotrichite on iron sulphides. They suggested that melanterite and rozenite were earlier formed but that szomolnoquite and halotrichite, which are more stable sulphides, formed later. Porous or spongy-textured, pyrite, high humidity and the

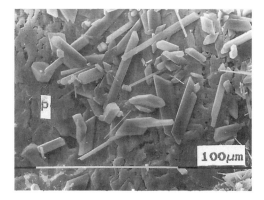

**Fig. 4.** Sulphate minerals grown on porous-textured pyrite (p) (scanning electron microscopy).

**Fig. 2.** Well-shaped crystals of pseudocubic jarosite (j) and gypsum (g) (scanning electron microscopy).

**Fig. 3.** Aggregates of secondary sulphates developed on porously textured pyrite (p) (scanning electron microscopy).

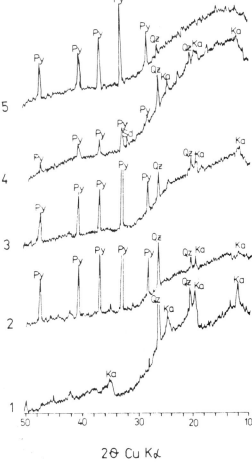

**Fig. 5.** X-Ray diffraction patterns of coal samples from section I of coal A. Py, pyrite; Qz, quartz; Ka, kaolinite; and Sd, siderite.

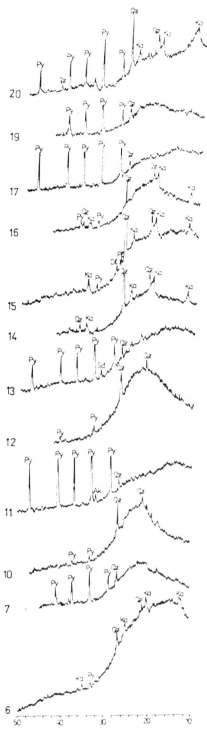

**Fig. 6.** X-Ray diffraction patterns of coal samples from section II of coal A. Py, pyrite; Qz, quartz; Ka, kaolinite; Sd, siderite; Dl, dolomite; Pl, plagioclase; and Ak, ankerite.

presence of clays associated with pyrite are all factors which influence the process of autoxidation of the sulphides and stimulate the formation of secondary sulphates. Our results confirm these suggestions. In the samples from the Pernik coals we have also observed a close association between kaolinite and pyrite and the development of sulphate minerals on pyrite with a porous texture. Figure 2 shows well-shaped crystals of pseudo-cubic jarosite (j) and gypsum (g). Aggregates and separate crystals of secondary sulphates that have developed on porous textured pyrite are shown in Figs 3 and 4.

The mineral matter of the Pernik subbituminous coal (coal A) consists chiefly of pyrite, kaolinite, siderite and quartz (Figs 5 and 6). The very low energy depositional environment, the restricted income of terrigeneous material and the greater acidity of the environment are reasons for the formation of kaolinite during peat genesis. Kaolinite forms separate fine interbeds and often associates with other minerals, mainly pyrite. It also fills cell openings in the structures of some macerals.

The formation of siderite is connected with the final stage of diagenesis and the beginning of katagenesis, when the strongly fractured top of the coal bed has allowed mineral solutions to circulate. The result is the deposition or infiltration of siderite and pyrite. Marcasite has not been detected. Quartz is predominantly terrigeneous and detrital. Single grains rarely exceeding 100–200 $\mu$m are observed.

The acidity of the environment was not unfavourable for the formation of carbonate minerals. In the less acidic levels of the peat bed, under more favourable conditions, cryptocrystalline calcite in association with kaolinite has crystallized.

Other minerals such as calcite, dolomite, ankerite, plagioclase and gypsum are present in minor amounts. The other sulphates (bassanite, anhydrate, jarosite, melanterite, copiapite, szomolnoquite, aluminocopiapite, ammoniojarosite, Mg-copiapite and halotrichite) are present in very small amounts.

Because of the significant content of pyrite, the variety of its forms and its significance with respect to the ecology and coal quality were studied in detail. A classification scheme (Table 1) describes the forms of pyrite in the Pernik subbituminous coals.

*Framboidal pyrite*

Framboids represent spherical aggregates of euhedral pyrite crystals. The globules resulting from massive mineralization have been considered as pyritized bacteria. Framboids showing

**Table 1.** *Classification of forms of pyrite in coals*

| Place of formation | Time of formation | Morphology | Pyrite forms |
|---|---|---|---|
| Peat bog | Peat genesis | Euhedral | Isolated |
| | | | Clustered |
| | | Fromboidal | Bacterial |
| | | | Inorganic |
| | | Massive | Cluster-like |
| | | | Homogeneous |
| | | | Microconcretional |
| | Early diagenesis | Anhedral | Infilling |
| | | | Replacing |
| Coal bed | After peat genesis | Infiltration | Infilling |
| | | | Cleat-filling |
| | | | Fracture-filling |
| | | | Massive |
| | | | Euhedral |
| | | | Replacing |

After Kortenski (1991) and improved by Kortenski (in press) and Kortenski & Kostova (in press).

well-shaped crystals are likely to be a result of crystallization from mineral solutions in the organic substance. Bacterial framboidal pyrite is found in all the samples under study. Bacterial framboids are characterized by the usually non-uniform mineralization of globules (Figs 7 and 10). Some of the globules are compact-pyritized (Fig. 8), whereas others exhibit only partial pyritization (Fig. 10). Bacterial framboidal pyrite preserves the independence of the separate globules even when they form aggregates (Figs 8 and 10). Some of the aggregates are pyritized bacterial colonies (Figs 9 and 11). This type of pyrite occasionally occurs as single bodies (Fig. 12), but more often can be found as aggregates of different sized globules, irregular or lenticular (Figs 8 and 10) or, less frequently, beaded in shape (Fig. 7). Bacterial framboidal pyrite is not associated with euhedral pyrite, but is often associated with clay minerals, especially kaolinite (Fig. 11).

Inorganic framboidal pyrite represents a relatively uniform mineralization of the globules (Figs 13 and 19). Only in a few instances do the framboids contain solitary crystals (Fig. 14) and the binding substance (organic matter and kaolinite) dominates (Fig. 15). This form of pyrite exhibits a symmetrical, mostly concentric ordering of the crystals in the globules (Fig. 13). In many of the inorganic framboidal pyrite globules the crystals are densely intergrown and overgrow the earlier formed crystals in the centre of the globules (Fig. 16). In contrast with the bacterial pyrite, inorganic framboidal pyrite is sometimes characterized by the coalescence of globules (Figs 14 and 17). The pyrite crystals in individual framboidal bodies are

usually equal in size (Figs 14 and 18). The crystals are octahedral, pentagonal dodecahedral and cubic (Figs 14 and 16). In almost all

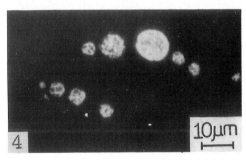

**Fig. 7.** Beaded aggregate of bacterial framboidal pyrite (oil immersion).

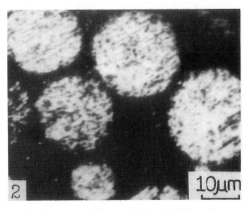

**Fig. 8.** Aggregate of bacterial framboidal pyrite (oil immersion).

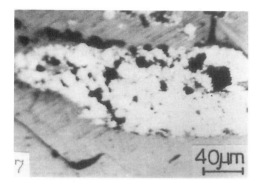

**Fig. 9.** Lenticular aggregate of bacterial framboidal pyrite (pyritized bacterial colony).

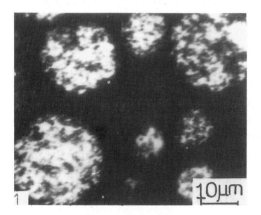

**Fig. 10.** Irregularly mineralized bacterial framboidal pyrite (oil immersion).

**Fig. 11.** Aggregate of bacterial framboidal pyrite (pyritized bacterial colony) associated with kaolinite (scanning electron microscopy). Scale bar, 0.1 mm.

instances inorganic framboidal pyrite is associated with isolated crystals or aggregates of euhedral pyrite (Figs 18 and 19) and kaolinite. This suggests similar mechanisms of formation for the two pyrite types.

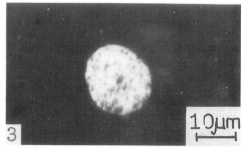

**Fig. 12.** Bacterial framboidal pyrite with oval shape (oil immersion).

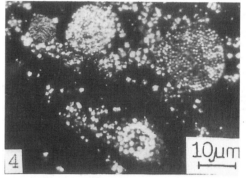

**Fig. 13.** Inorganic framboidal pyrite in association with chaotic euhedral pyrite (oil immersion).

**Fig. 14.** Aggregate of inorganic framboidal pyrite with pentagonal dodecahedral and octahedral crystals (scanning electron microscopy).

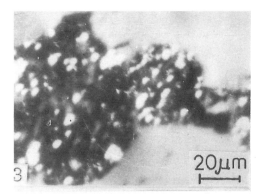

**Fig. 15.** Inorganic framboidal pyrite with solitary pyrite crystals.

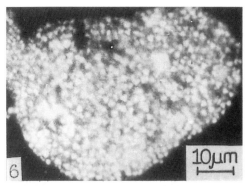

**Fig. 17.** Coalescence of two globules of inorganic framboidal pyrite (oil immersion).

**Fig. 16.** Aggregate of inorganic framboidal pyrite (scanning electron microscopy).

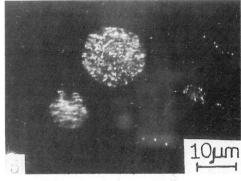

**Fig. 18.** Inorganic framboidal pyrite in association with chaotic and clustered euhedral pyrite (oil immersion).

## Euhedral pyrite

Euhedral pyrite is represented by well-shaped pyrite (Figs 20 and 21) with crystals showing pentagonal dodecahedral, octahedral or cubic crystal forms. Isolated crystals and clustered euhedral pyrite were often observed in all samples investigated. They were generally associated with kaolinite or inorganic framboidal pyrite (Figs 13 and 19). The structure of the aggregates of euhedral pyrite is reminiscent of inorganic framboidal pyrite. They can be distinguished by the absence of the typical spherical shape. In some aggregates the crystals are close to one another and with a relatively uniform distribution (Fig. 21). Euhedral pyrite was often associated with inorganic framboidal pyrite. The separate crystals were dispersed chaotically around the globules or were in the form of aggregates (Fig. 13). These aggregates surrounded spherulites of inorganic framboidal pyrite and almost coalesced with them. In the Pernik subbituminous coals, crystals of euhedral pyrite are small, ranging from submicrometre to 1–2 $\mu$m.

## Massive pyrite

The massive pyrite is represented by pyrite particles displaying a wide variety of shapes and sizes. This pyrite form is found in most of the investigated samples as irregular grains with sizes ranging from 4–5 to 10–20 $\mu$m. In the Pernik subbituminous coals homogeneous and cluster-like massive pyrite are present. The homogeneous massive pyrite is predominantly porous (Fig. 22) due to the inclusion of relict organic matter and clay minerals during the crystallization processes. Homogeneous massive pyrite is lenticular or irregular (Fig. 22). The cluster-like massive pyrite is represented by

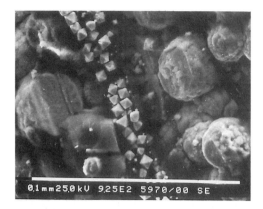

**Fig. 19.** Framboidal and euhedral pyrite (scanning electron microscopy). Scale bar, 0.1 mm.

**Fig. 21.** Clustered euhedral pyrite associated with kaolinite (scanning electron microscopy).

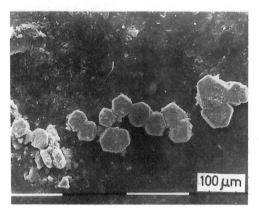

**Fig. 20.** Clusterred euhedral pyrite (scanning electron microscopy).

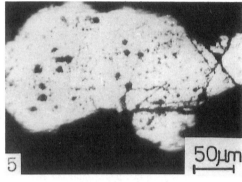

**Fig. 22.** Homogeneous massive pyrite with irregular shape.

aggregates of fine pyrite grains united by kaolinite or organic substance (Fig. 23).

### Anhedral pyrite

The shape of the anhedral pyrite depends on the shape of the plant debris associated with its deposition. Two varieties of anhedral pyrite were observed in Pernik subbituminous coals. Replacement anhedral pyrite results from either the mineralization of cell walls (Wiese & Fyfe 1986) or massive pyritic replacement of organic matter (Querol *et al.* 1989). The other variety is infilling anhedral pyrite, which fills cell lumens and pores in the plant debris. It is deposited mainly in the lumens of fusinite (Fig. 24) and textinite, taking their shape. Anhedral pyrite is less common than replacement anhedral pyrite

in the coal samples studied. In some instances the replacing anhedral pyrite, associates with anhedral pyrite which crystallizes in the internal part of the lumens (Fig. 24).

### Infiltrational pyrite

Infiltrational pyrite is deposited from the penetrating solutions along fractures and cleats in the coal beds. This type of pyrite is often seen in the Pernik subbituminous coals. Two varieties can be differentiated: infilling and replacing infiltrational pyrite. The infilling infiltrational pyrite has two varieties: fracture-filling and cleat-filling. The fracture-filling infiltrational pyrite is observed in endogeneous and exogenic fractures, filling them completely or partially (Fig. 25). The infilling infiltrational pyrite sometimes causes corrosion of earlier deposited minerals, especially siderite (Fig. 21) in fractures and cleats. Fracture-filling and cleat-filling infiltrational pyrite is usually homogeneous and massive, but in some instances it includes organic matter or other mineral grains. Pyrite which densely fills the fractures is termed massive fracture-filling pyrite. Well-shaped pyrite crystals are observed in some fractures (Fig. 15). This variety can be classified as euhedral fracture-filling pyrite. Replacing infiltrational pyrite can encrust and even fully pyritize tissues, taking on their form.

## Determination of pyritic sulphur

Querol *et al.* (1993) developed a powder XRD method for the indirect determination of the pyritic sulphur content of coals. However, although considerably less time consuming than classical chemical methods (ASTM Standard D2491-84) used for this purpose, the XRD method used here is slightly complicated by the presence of both pyrite and marcasite.

In our samples no marcasite was detected in the qualitative XRD analyses. This allowed the direct determination of pyrite by quantitative XRD methods, from which pyritic sulphur can be determined by stoichiometric calculations.

Averaged intensity values (peak heights) of the three strongest lines of pyrite (hkl 200, 210 and 211) triply registered in the $2\theta$ diffraction range, $32-42°$, with a detector speed of $1°(2\theta)/$min were used in the determinations.

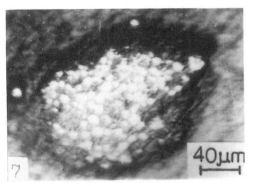

**Fig. 23.** Cluster-like massive pyrite.

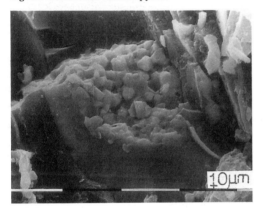

**Fig. 24.** Infilling anhedral pyrite ($Py^{an}$) and euhedral pyrite ($Py^e$) in the lumens of fusinite (scanning electron microscopy).

**Fig. 25.** Fracture-filling infiltrational pyrite and siderite.

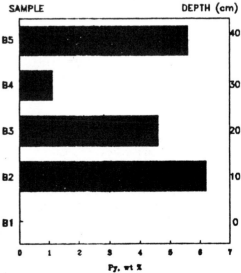

**Fig. 26.** Pyrite content in coal samples from section I determined by powder XRD.

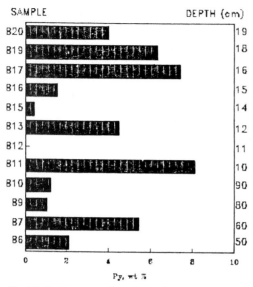

**Fig. 27.** Pyrite content in coal samples from section II determined by powder XRD.

The quantitative determinations were performed with external standards with compositions of: 3 wt% pyrite (Py) and 97 wt% coal (C); 5 wt% Py and 95 wt% C; 7 wt% Py and 93 wt% C; and 10 wt% Py and 90 wt%. These standards were selected because the preliminary semiquantitative analyses showed that Py in the samples was below 8–10 wt%. Thus the effect of absorption can be neglected. The samples were ground, homogenized and the powders were pressed by a glass slide in a standard sample holder.

The results showed that the pyrite content varies considerably in the range up to 10 wt% (Figs 26 and 27). It can be seen that pyrite varies largely on a centimetre scale, suggesting site-specific environmental changes occurring during sulphide deposition.

## Geochemistry

Thirty-seven trace elements were determined in the coal samples from the Pernik Basin. The average concentrations in the coal ash of As, Cu, Co, Ni, Zn, Pb and V are from two to 5.6 times higher than the Clarke values of Yudovich *et al.* (1985). The concentrations in the coals of As, Cu, Co, Ni, Zn, V, Ti, Cr, Mn, Mo, Ce, Tb, Yb, Ln, Hf, U and Th are 1.5 to 14.4 times the world averages given by Valcoevic (1983) (Table 2).

The organic or inorganic affinity of the trace elements was established by analysing their distribution with respect to the ash content (Fig. 28). Four groups of elements were separated:

(1) Elements with a high organic affinity: Ni, Ge, Sb, Mo, Lu, W, Co, Tb, La, U and B. The correlation coefficients with the ash content are negative with values below $-0.5$. Element concentrations decrease with increasing ash content and are highest in the coals ($A^d$ up to 30%) and very low in the humic shales ($A^d > 80\%$) (Fig. 28: Ni, Co and U). This can be explained by a decrease in organic matter content with which the elements from this group are connected.

(2) Elements whose organic affinity is higher than their inorganic affinity: Be, Mn, Zr, V, Cr, Ag, Ba, Zn, Yb, Sc, Hf, Sm, Ce, Eu, As, Pb, Cu and Th. The correlation coefficients with the ash content are negative, below $-0.5$. The varying correlation coefficients of the different elements indicate different distributions with respect to the ash content. The concentration of elements with the lowest correlation coefficients (for instance Ce, $r_0 = -0.46$) decreases slightly with an increase in the ash content (Fig. 28). With the higher correlation coefficients (for example, Cu, $r_0 = -0.30$), the element content decreases strongly in the coaly material ($A^d$ 30–50%) (Fig. 28). The concentrations of elements with comparatively high correlation coefficients with ash (for example, Sm, $r_0 = -0.18$) is highest in the coaly material (Fig. 28).

(3) Elements with inorganic affinity higher than their organic affinity: Ga, Sn, Ti, Sr, Y and Cs. The elements have positive correlation coefficients with the ash content, with values up to $+0.5$. The correlation coefficients of elements increase with an increase in the ash content and in the shale partings ($A^d$ 50–80%). The humic shale content decreases slightly with a lower correlation coefficient (for instance, Cs, $r_0 = +0.23$) (Fig. 28). The contents of elements with a higher correlation coefficient (for instance, Sn, $r_0 = +0.42$) increase slightly with an increase in the ash content (Fig. 28).

(4) Elements with a high inorganic affinity: Rb and Ta. The correlation coefficients with the ash content are from $+0.5$ to $+1.0$. The concentrations of the elements increases with an increase in the ash and is highest in humic shale (Fig. 28; Rb).

**Table 2.** *Trace elements concentrations in coal ash and coals*

| Elements | Average concentration (ppm) | | Subbituminous coal ash* | World average (Valcoevic 1983) |
|---|---|---|---|---|
| | Coal ash | Coals | | |
| B† | 569 ± 130 | 68 ± 11 | 560 | 75 |
| Be | 7.6 ± 1.5 | 1.8 ± 0.4 | 11 | 3 |
| Sc | 18.7 ± 2.9 | 4.5 ± 1.0 | 15 | 5 |
| Ti | 4785 ± 392 | 1162.7 ± 101 | 2600 | 500 |
| V | 250 ± 48 | 60.8 ± 11 | 120 | 25 |
| Cr | 106 ± 22 | 25.8 ± 4 | 70 | 10 |
| Mn | 755 ± 132 | 183 ± 34 | 510 | 50 |
| Co | 92.4 ± 15 | 22.4 ± 4.1 | 20 | 5 |
| Ni | 270 ± 41 | 65.7 ± 11 | 51 | 15 |
| Cu | 237 ± 43 | 60 ± 10.2 | 48 | 15 |
| Zn | 301.1 ± 47 | 75.4 ± 10.4 | 100 | 50 |
| Ga | 24.3 ± 4.4 | 5.9 ± 1.0 | 36 | ND |
| Ge | 7.0 ± 1.2 | 1.7 ± 0.3 | 9 | ND |
| As | 338.7 ± 49 | 82.3 ± 13.1 | 60 | 5 |
| Rb | 77.2 ± 10 | 18.8 ± 2.1 | 46 | 10 |
| Sr | 219.6 ± 35 | 53.4 ± 10.2 | 1100 | 500 |
| Y | 24.1 ± 3.6 | 5.9 ± 0.9 | 37 | ND |
| Zr | 113.6 ± 19 | 27.6 ± 4.3 | 160 | ND |
| Mo | 19.7 ± 15 | 4.8 ± 1.0 | 13 | 3 |
| Ag | 0.3 ± 0.05 | 0.08 ± 0.02 | 1 | ND |
| Sn | 2.7 ± 0.3 | 0.65 ± 0.08 | 4.1 | ND |
| Sb | 12.9 ± 1.3 | 2.9 ± 0.4 | 5–10 | 8 |
| Cs | 5.1 ± 0.6 | 1.2 ± 0.2 | 5–15 | 1.5 |
| Ba | 722.7 ± 66 | 175.6 ± 12.5 | 890 | 500 |
| La | 37.2 ± 5.1 | 9.0 ± 1.1 | 3–80 | 10 |
| Ce | 59.4 ± 2.3 | 14.4 ± 1.8 | 100–200 | 0.4 |
| Sm | 6.3 ± 0.6 | 1.5 ± 0.3 | ND | 1.6 |
| Eu | 2.0 ± 0.4 | 0.49 ± 0.1 | ND | 0.7 |
| Tb | 2.6 ± 0.3 | 0.61 ± 0.1 | ND | 0.3 |
| Yb | 2.5 ± 0.4 | 0.51 ± 0.08 | 5 | 0.5 |
| Lu | 2.1 ± 0.3 | 0.49 ± 0.1 | ND | 0.07 |
| Hf | 4.4 ± 0.6 | 1.1 ± 0.15 | 1–3 | 0.6 |
| Ta‡ | 0.94 ± 0.4 | 0.16 ± 0.03 | ND | 0.2 |
| W | 3.6 ± 0.5 | 0.83 ± 0.08 | 20–40 | 2.5 |
| Pb | 101 ± 21 | 24.5 ± 5.1 | 53 | 25 |
| Th | 13.3 ± 1.8 | 3.2 ± 0.5 | 22.8 | 2 |
| U | 13.7 ± 1.7 | 3.3 ± 0.4 | 10–20 | 1 |

* Yudovich *et al.* (1985).
† Eskenazy *et al.* (1994).
‡ Eskenazy (1990).
ND, No data.

Four geochemical associations are established in the coals from the Pernik Basin (Fig. 29):

(I)   Be–Ge–Zr–Y–Yb–Tb–U–Sr;
(II)  Mn–Sc–Th–Sm–Hf–Ce–Lu–La–Eu–Ni–
      Co–V–Ba–As–Sb–W–Mo–Cr;
(III) Pb–Cu–Ag–Zn; and
(IV)  Ga–Ti–Pb.

Figure 30 shows the chondrite-normalized distribution pattern of the average content of rare earth elements. The investigated coals are rich in light rare earth elements.. A negative anomaly of Sm and Eu concentrations is seen.

Such an anomaly is typical only for Eu and has also been established in other Bulgarian coals (Eskenazy 1987; Eskenazy & Brinkin 1991). A positive anomaly for Tb and Lu concentrations is also established in the Pernik coals (Fig. 30).

The shale-normalized patterns are given in Fig. 31. They show a distinct relative increment in the heavy rare earth elements. A negative anomaly of Sm is seen in contrast with other Bulgarian coals (Eskenazy 1987; Eskenazy & Brinkin 1991). A positive anomaly of Lu concentration is also seen in the shale-normalized pattens (Fig. 31).

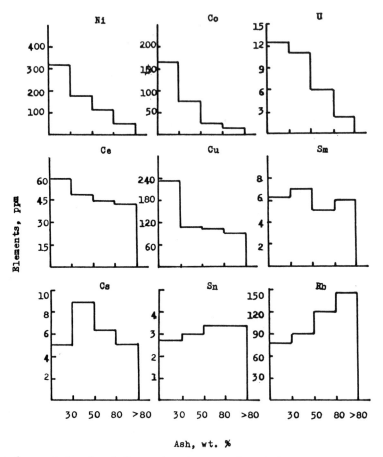

**Fig. 28.** Plot of concentrations (ppm) of some elements versus ash content.

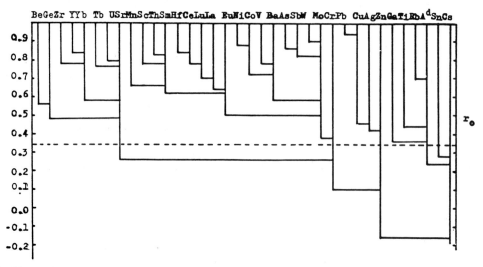

**Fig. 29.** Dendrogram of cluster analysis for trace elements in samples from Pernik coal basin.

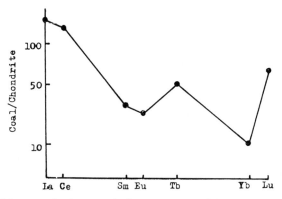

Fig. 30. Average chondrite-normalized rare earth element patterns of the coals.

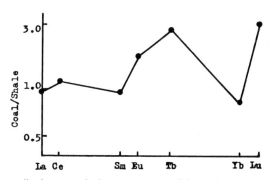

Fig. 31. Average shale-normalized rare earth element patterns of the coals.

## Conclusions

These mineralogical studies on Pernik subbituminous coals show that the minerals found can be grouped on the basis of their abundance as: major (pyrite, kaolinite, siderite and quartz); minor (calcite, dolomite, ankerite, plagioclase and gypsum); and accessory (bassanite, anhydrite, jarosite, melanterite, copiapite, szomolnoquite, aluminocopiapite, ammoniojarosite, Mg-copiapite and halotrichite).

Four mineralization stages are recognized: pyrite–kaolinite–illite, pyrite, pyrite–siderite, and sulphate.

The pyrite forms found in Pernik subbituminous coal are: framboidal, represented by spherical aggregates of euhedral pyrite crystals which can be of either organic or inorganic origin; euhedral, represented by well-shaped pyrite crystals with pentagonal dodecahedral, octahedral or cubic crystal forms; massive, represented by pyrite particles with a wide variety of shapes and sizes; anhedral, the shape of which depends on the shape of the plant debris in which they were deposited; and infiltrational, deposited in fractures and cleats of the coal beds.

The XRD quantitative determinations showed that the pyrite content varies considerably up to 8 wt% (mean values about 3–4 wt%), which suggests site-specific environmental changes occurring during sulphide deposition.

The high concentrations of Co, Ni, Cu, Zn, Pb and As in the coal from Pernik Basin is a result of Cretaceous andesites supplying material to the ancient peat bog. The andesites are rich in sulphide mineralization, which is the source of these elements.

The negative anomaly of Sm concentrations (Figs 30 and 31) is a result of the low concentrations of this element in the Pernik coals compared with the world average given by Valcovic (1983) (Table 2).

Thanks are due to L. Filizova for helpful discussions and to M. Tumangelova for assistance in the preparation of the paper.

# References

CHERNYAVSKA, S. 1970. Sporopolenovi zoni v nyakoi starotertsierni vaglenosni sedimenti v Bulgaria. *Bulgarian Academy of Sciences, Bulletin of the Geological Institute – series stratigraphy and lithology*, **19**, 79–98 [in Bulgarian].

ESKENAZY, G. 1987. Rare earth elements in a sampled coal from the Pirin deposit, Bulgaria. *International Journal of Coal Geology*, **7**, 301–314.

——1990, Geochemistry of tantalum in Bulgarian coals. *International Journal of Coal Geology*, **15**, 137–149.

—— & BRINKIN, K. 1991. Trace elements in the Karlovo coal deposit. *Comptes Rendus Acad. Bulg. Sci.*, **44**(11), 67–70.

——, DELIBALTOVA, D. & MINCHEVA, E. 1994. Geochemistry of boron in Bulgarian coals. *International Journal of Coal Geology*, **6**, 251–276.

KAMENOV, B. 1964 Vurhu stratigrafiyata i vuglenosnostta na Paleogena ot Pernishkiya basein. *Izv. NIGI*, **I**, 233–246 [in Bulgarian].

KONSTANTINOVA, V. 1964. Petrographic and genetic characteristics of the old Tertiary basins in South-West Bulgaria. *Bulletin de L'Institute Scientifique de Rècherches geologiques*, **1**, 317–330 [in Bulgarian, English Summary].

KORTENSKI, J. 1990. Sindiagenetichna mineralizaciya vuv vuglistata ot Pernishkiya basein. *Annals of the Higher Institute of Mining and Geology*, **V. XXXVI**, part I, 159–168 [in Bulgarian]

——1991. Genetic classification of the supergenic, sedimentogenic and metamorphogenic minerals in coals set against the example of Bulgarian basins. *Comptes Rendus Acad. Bulg. Sci.*, **44**(10), 77–80.

——1995. Mineralite i tehnite formi na prisustvie v Bulgarskite vuglishta. *Reviews of the Bulgarian Geological Society*, in press [in Bulgarian].

—— & KOSTOVA, I. 1995. Occurrence and morphology of pyrite in some Bulgarian coals. *International Journal of Coal Geology* (in press).

QUEROL, X., ALASTNEY, A., CHINCHON, J. S., FERNANDEZ-TURIEL, J. L. & LOPEZ-SOLER, A. 1993. Determination of pyritic sulphur and organic matter contents in Spanish subbituminous coals by X-ray powder diffraction. *International Journal of Coal Geology*, **22**, 279–293.

——, CHINCHON, J. S. & LOPEZ-SOLER, 1989. Iron sulphide precipitation sequence in Albian coals from Maestrazgo Basin, NE Spain. *International Journal of Coal Geology*, **11**, 171–189.

SISKOV, G., VALCEVA, S. & SALLABASHEVA, V. 1982. Klassifikaciya bolgarskih uglei po stepni uglefikacii. *Chemical Communications of the Bulgarian Academy of Science*, **15**, 721–726 [in Russian].

VALCOEVIC, V. 1983. *Trace Elements in Coal*, Vol. I. CRC Press, Boca Raton.

VALTCHEVA, S. & MINCHEV, D. 1977. Petrologichni izsledvania na vuglishta ot Pernishkia basein. *Annals de l'University de Sofia, Faculty de Geologie et Geographie, Livre 1 Geologie*, 225–233 [in Bulgarian].

WIESE, R. G. JR & FYFE, W. S. 1986. Occurrence of iron sulfides in Ohio coals. *International Journal of Coal Geology*, **6**, 251–276.

——, POWELL, M. A. & FYFE, W. S. 1987. Spontaneous formation of hydrated iron sulphates on laboratory samples of pyrite- and marcasite-bearing coals. *Chemical Geology*, **63**, 29–38.

YUDOVICH, Y. E., KETVIS, M. P. & MERZ, A. B. 1985 *Elementi-primesi v Iskopaemikh Uglistakh*. Nauka, Moscow [in Russian].

# Subsurface correlation of Carboniferous coal seams and inter-seam sediments using palynology: application to exploration for coalbed methane

DUNCAN McLEAN[1] & IAIN MURRAY[2]

[1] *Industrial Palynology Unit, Department of Earth Sciences, University of Sheffield, Mappin Street, Sheffield S1 3JD, UK*

[2] *Conoco (UK) Ltd, Rubislaw House, Anderson Drive, Aberdeen AB2 4AZ, UK*

**Abstract:** Palynology has a long history of application in the hydrocarbon exploration and development industry. Palynological methodologies developed for the study of traditional hydrocarbon reservoirs are directly transferable to exploration for coalbed methane. The principal difference in approach relates to the fact that coal seams usually contain abundant palynomorphs, whereas conventional reservoirs (sandstones and carbonates) do not. Work since the 1930s in the onshore British coalfields and work in the offshore southern North Sea Carboniferous basin illustrate that the high-resolution correlation of coal seams is achievable using palynology. The methods lend themselves to studies of coalbed methane reservoir connectivity and correlation. Other applications, such as the identification of coring points and stratigraphic targets, the monitoring of vertical and horizontal drilling and the study of coal seam maturation by spore colour analyses represent standard palynological applications.

This paper describes the principles and techniques of palynostratigraphic correlation of coal seams and their potential for application to exploration for coalbed methane (CBM). The principles are illustrated by a case study from the Murdoch Gas Field, offshore UK, in which coal seam palynology is used to correlate several exploration well sections and subsequently to monitor the drilling of development wells, including horizontal penetrations. This illustrates the benefits of palynology available to CBM exploration programmes.

Coalbed methane exploration in Britain is concerned primarily with the coal-bearing strata of the Westphalian, Late Carboniferous (Ayers *et al.* 1993; Creedy 1993). These strata typically consist of fluvial-dominated lower to upper alluvial plain facies associations, characterized by rapid lateral and vertical variations in lithology (Guion & Fielding 1988). Basin-wide correlation is possible where laterally persistent horizons, such as coal seams (e.g. Smith *et al.* 1967), marine bands (Calver 1969; Ramsbottom *et al.* 1978), tuffs and tonsteins (Richardson & Francis 1970; Spears & Kanaris-Sitiriou 1979) or *Estheria* bands (Jones 1980) are recognizable. The acquisition of electric log data (Archard & Trice 1990; Hurst *et al.* 1990; Leeder *et al.* 1990) and seismic data (Hatherly *et al.* 1989; Fraser & Gawthorpe 1990; Evans *et al.* 1992) offer additional correlation methods. Palaeontological correlation utilizes goniatites and other marine fauna (Calver 1968a; 1968b; 1973;

Paproth *et al.* 1983), non-marine bivalves (Calver 1956; 1969), arthropods (Paproth *et al.* 1983; Pollard 1969; Anderson 1970; Calver 1973), plants (Stockmans 1962; Cleal 1991), palynomorphs (Smith & Butterworth 1967; Clayton *et al.* 1977) or, less commonly, conodonts (Higgins 1975; 1985) and foraminifera (Paproth *et al.* 1983; Pastiels 1956). Additionally, local correlations may use the techniques of within-seam seismic (Hatherly & Holt 1984; Doyle 1987).

During subsurface exploration for CBM, macropalaeontological techniques are of limited use in correlation as cores are rarely cut and macrofossils are destroyed during drilling. Microfossils are commonly preserved in the fragmentary ditch cuttings produced by drilling and so represent the principal groups useful for biostratigraphic correlation.

Palynology is the study and use of acid-resistant, organic-walled microfossils (palynomorphs). These palynomorphs are dominantly the reproductive spores (miospores and megaspores *sensu* Guennel 1952) produced by terrestrial pteridosperm, sphenopsid and lycopod vegetation (Bennie & Kidston, 1886; Smith 1962; Scott 1979). Of significance in the application of palynology to stratal correlation are the great abundance, resistance to destruction and extraordinary scope of dispersal, particularly by water, of miospores (Richardson & McGregor 1986). Assemblages in coals are autochthonous, composed of miospores from the local mire vegetation. Vertical and lateral variations in

*From* Gayer, R. & Harris, I. (eds) 1996, *Coalbed Methane and Coal Geology,*
Geological Society Special Publication No 109, pp 315–324.

coal seam assemblages indicate palaeo-ecological responses to local edaphic factors (Smith 1957; 1962; 1964; 1968; Marshall & Smith 1965). These different, vertically stacked assemblages seen in a coal seam are termed phases. Smith & Butterworth (1967) reviewed the causes of the variable development of phases in different seams and the implications of this feature for seam characterization and correlation. They documented changes in phases related to seam splitting and the application of this to the correlation of seams to their various splits. Assemblages of dispersed miospores may be preserved in both terrestrial and marine sediments. Except in coals, each assemblage as a whole tends to reflect the combined vegetation of various parts of the basin, whereas individual components of the assemblage may vary both quantitatively and qualitatively, according to sedimentary processes and local concentrations of the parent vegetation (Richardson & McGregor 1986).

## Historical review of palynological coal correlation methods

Thiessen & Voorhees (1922) of the US Bureau of Mines were the first to describe the utility of megaspores and miospores for the correlation of coal seams. Thiessen & Staud (1923) and Thiessen & Wilson (1924) considered that individual seams contain characteristic species by which they may be identified. In Britain, Raistrick (1935; 1939) and subsequent workers also recognized the correlative potential of the stratigraphic distribution of certain miospore taxa through a series of coal seams. Smith & Butterworth (1967; and references cited therein) provide a comprehensive synthesis of palynological data relating to all the coalfields of Britain. Similar work was carried out in Europe (e.g. Potonié & Kremp 1956; Dybová & Jachowicz 1957) and North America (e.g. Schopf 1936; 1938; Kosanke 1950; 1954; Guennel 1952; 1958; Peppers 1970).

Raistrick (1937) foresaw the application of recognizing quantitative changes in miospore assemblages through a vertical seam profile to coal seam identification and correlation. Such changes are related to the temporal development of the original mire palaeo-ecology and were documented and interpreted by Smith (1957; 1962; 1964) and Marshall and Smith (1965).

Argillaceous inter-seam sediments commonly contain abundant well-preserved, allochthonous palynomorphs (Chaloner 1958; Neves 1958). The preservation of palynomorphs in non-coal lithologies is controlled by a complex of distributional, depositional and palaeo-biogeographical factors. Non-coal palynology is useful in the correlation of coal seams. Palaeosols and seam roofs yield slightly different palynological assemblages (Habib 1966), which can add to the successful discrimination of coal assemblages which otherwise might possess very similar palynological characters (Lele & Chandra 1975).

Of the published palynological divisions which have been introduced for the Westphalian (Fig. 1), that of Smith & Butterworth (1967) has been most extensively applied. They used palynological data exclusively from coal seams from the British coalfields and produced a scheme of 11 biozones. Loboziak et al. (1976) and Coquel et al. (1976) synthesized data from coal basins of Western Europe and proposed more widely applicable zonations for the Westphalian and Stephanian. Data from all lithologies from the whole of the Carboniferous of northwestern Europe were synthesized by Clayton et al. (1977) to produce a palynostratigraphic classification applicable in a regional context.

Coal seam correlation is usually based on the quantitative and qualitative comparison of miospore assemblages. Samples are either taken from a channel cut through the entire seam or from discrete points representing specific parts of the seam. Channel samples are preferred as this

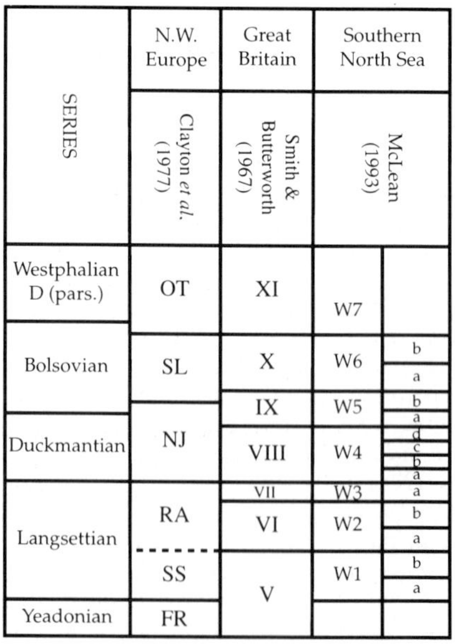

| SERIES | N.W. Europe — Clayton et al. (1977) | Great Britain — Smith & Butterworth (1967) | Southern North Sea — McLean (1993) | |
|---|---|---|---|---|
| Westphalian D (pars.) | OT | XI | W7 | |
| Bolsovian | SL | X | W6 | b / a |
| | | IX | W5 | b / a |
| Duckmantian | NJ | VIII | W4 | d / c / b / a |
| Langsettian | RA | VII | W3 | a |
| | | VI | W2 | b / a |
| | SS | V | W1 | b / a |
| Yeadonian | FR | | | |

**Fig. 1.** Comparison of palynostratigraphic classifications for the Westphalian of onshore and offshore UK. Note the increase in biostratigraphic resolution with decreasing geographical applicability.

sampling method can be consistently applied at outcrop, to core and, most importantly, to drill cuttings. Both qualitative and statistical methods of correlation rely on two assumptions. (1) Miospores were distributed uniformly across a coal mire at the time of accumulation such that, at any locality a statistical average of all types of miospores would be found. (2) There is sufficient difference in the composition of the flora of successive coal mires to produce miospore assemblages which are recognizably different (Raistrick 1937). If it is assumed that each coal seam contains a distinct miospore assemblage that does not differ geographically, the problem of correlation then consists essentially of evaluating sampling errors in the statistical sense (Gray & Guennel 1961). Tomlinson (1957) showed that such errors become statistically low if a large enough sample of an assemblage population (c. 500 specimens) is recorded. Statistical seam assemblage correlation can be achieved by the $\chi^2$ (Gray & Guennel 1961; Stone 1969), minimum distance analysis (De Jekhowsky 1963) and discriminant function analysis (Smith & Butterworth 1967). General practice is to make correlations based on abundances of specimens identified to the specific level (Smith & Butterworth 1967), or of genera and species combined (Guennel 1958); experience indicates that emphasis should be placed on particular taxa rather than total assemblages.

## Application of palynological correlation to CBM exploration

Techniques for the subsurface correlation of coal-bearing strata developed for the coal and hydrocarbon industries are immediately applicable to exploration for CBM as the coal reservoirs are palynologically highly fossiliferous. Biostratigraphic correlation of reservoirs can therefore be achieved directly without recourse to the correlation of inter-reservoir strata. Rightmire & Choate (1986) indicate that gas from inter-seam sandstones is important in recharging methane in coals. Palynological correlation and mapping of these inter-seam sediments, which can be achieved in conjunction with seam correlation, may also be important in assessing methane reserves. Individual seams in coal measures are generally not visible as individual reflectors on seismic (Van Riel 1965; Ruter & Schepers 1978) and the absence of regional seismic reflectors in coal-bearing sediments may be related to rapid lateral and vertical facies variations (Kingston & Steeves 1979). Therefore, palynostratigraphy can provide a correlative tool where other techniques

are limited. Palynology can also be used to correlate seams across faults and wash-outs. Such correlation provides a framework for coal thickness and CBM reserve estimation. The palynological correlation of seams in the coal basins of Nova Scotia by Hacquebard (1986) and Hacquebard et al. (1989) was applied to a CBM volumetric assessment by Grant & Moir (1992). Hacquebard (1986) provided palynological correlation of well sections in the Sydney Coalfield. He identified several palynologically distinct and laterally consistent stratigraphic intervals, each containing between four and 10 seams. The case study from the Murdoch Gas Field described later in this paper identifies palynologically distinct stratigraphic intervals which contain between five and six seams (Fig. 5). It is apparent that palynostratigraphic resolution increases with reduction in geographical scope reflecting the gradual lateral change of coal seam palaeo-flora as reflected in the palynological assemblages.

Routine palynological techniques used during drilling operations are suitable for CBM exploration. Direct applications include the identification of coring points and stratigraphic total depths, the monitoring of vertical and horizontal drilling and the study of coal seam maturation by spore colour analyses (Staplin 1969). Air or foam (air/fluid mix) drilling is commonly applied in CBM exploration (Rowecamp & Johnson 1993). This returns limited amounts of small-sized cuttings compared with the amounts produced by conventional mud or water drilling. The high concentrations of palynomorphs in coals means that the relatively impoverished samples still provide adequate data. New methods of laboratory preparation of palynological samples using microwave digestion (Jones 1994; Ellin & McLean 1994) allow rapid analysis of siliciclastic samples to monitor drilling progress and enhance MWD and LWD interpretation. Microwave digestion allows coal samples to be prepared in under 10 minutes (R. Jones pers. comm. 1995) and so palynological analysis can keep pace with rapidly drilled CBM exploration wells. The absence of oil-, polymer- or water-based muds in CBM wells is beneficial as such drilling media may present difficulties during palynological preparation (Desezar & Poulsen 1994). When high-speed air or foam drilling is used, circulation lag times and corresponding stratigraphic positions can be difficult to calculate; turbulent mixing of material may occur (e.g. of coal from several seams) and the end of circulation results in a mixed bottom hole slurry which will provide little biostratigraphic resolution. Evaluating reports of coal from cable tool drilling is also problematic. Caving and pulverization of coaly material in rapidly drilled wells can make logging difficult (Howie 1977). The downhole logging speed can be too fast for the accurate delineation of coaly intervals on density logs (Hacquebard 1986). This makes the identification of suitable coal sampling points difficult. Further, the failure of logging tools to identify thin coals biases sampling towards thicker seams.

## Case study: seam correlation in the Murdoch Gas Field, southern North Sea Carboniferous basin

The following case study relates to the palynological correlation of coal seams around a conventional hydrocarbon reservoir. The correlation methodology remains applicable to all palynomorph-rich, Carboniferous coal-bearing strata.

The Murdoch Gas Field is located in Block 22 of Quadrant 44 in the Silver Pit area of the southern North Sea (Fig. 2). The field is an early Duckmantian (Westphalian B) gas accumulation trapped within a faulted, northwest to southeast trending anticline, which is bounded to the northeast by a major reverse fault. The succession consists of fluvial sediments deposited on a poorly drained, low relief, low gradient alluvial plain cut by channels flowing generally southwards (Bailey et al. 1993). The 'Murdoch reservoir sandstone' (= pars Caister Coal Formation of Cameron 1993) consists of stacked, tabular, cross-bedded, pebbly sandstones and

conglomerates (Green & Slatt 1992). The reservoir is sealed by Duckmantian mudstones and by mudstones of the Rotliegend Group Silver Pit Formation (Green & Slatt 1992; Bailey et al. 1993; Ritchie & Pratsides 1993). Westphalian coals provide the source of the gas (Cornford 1984; Bailey et al. 1993). A marine, high-gamma claystone underlies the reservoir sandstone. Macropalaeontology and palynology (Turner et al. 1991; McLean unpublished data) identify this as the horizon of the *Anthracoceras vanderbeckei* Marine Band, which defines the base of the Duckmantian. Another high-gamma mudstone unit overlies the reservoir sandstone. This contains assemblages of the non-marine bivalve *Anthracosia ovum* and is interpreted as a laterally extensive lacustrine facies (McLean 1993).

Following the discovery of the Murdoch Gas accumulation by Conoco (UK) Ltd in 1984 (Gunn et al. 1993), palynostratigraphic monitoring of the exploration phase in the field was carried out. The distribution of palynomorphs (Fig. 3) was interpretted in terms of the onshore classifications of Smith & Butterworth (1967) and Clayton et al. (1977). Before the start of the development phase of drilling, a refined biostratigraphic framework was required to maximize the resolution of monitoring of well penetrations. This refinement concentrated on palynological assemblages from coal seams and differentiation of the Vanderbeckei Marine Band and *Anthracosia ovum* lacustrine bed (Figs 3–5). Detailed quantitative analysis resulted in the recognition of three palynologically distinctive

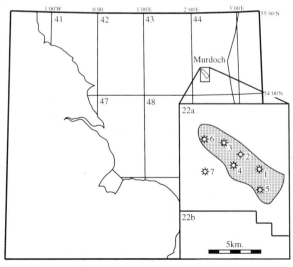

**Fig. 2.** Location of the Murdoch Gas Field, Quadrant 44, Block 22 and position of wells.

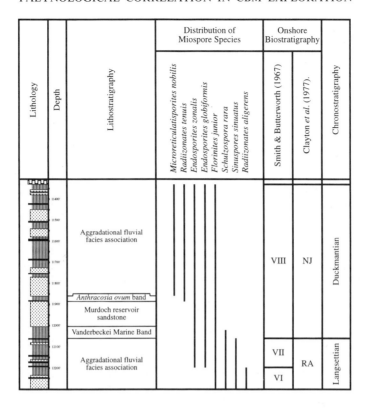

**Fig. 3.** Subdivision of Carboniferous sediments in the Murdoch Gas Field showing biostratigraphic resolution available by comparison with onshore biozonations. The generalized lithological section is from well 44/22-1. Lithological symbols as in Fig. 5.

seams. The remaining seams could not be characterized individually, but were divided into palynologically similar groups of seams (Figs 4a and 5). The distinctive seams and groups of seams were recognizable in all of the exploration wells, allowing detailed correlation of the reservoir sequence (Fig. 5).

Application of the coal correlation scheme to the monitoring of five development penetrations confirmed the on-going penetration of several critical horizons. The fifth development well included a substantial subhorizontal penetration through the reservoir sandstone. To monitor progress it was necessary to differentiate palynologically the claystone units above, within and below the reservoir. Each of these claystone units is the product of different sedimentological environments (lacustrine above; fluvial overbank within; and marine below the reservoir) and have distinct palynofacies (total organic residue) and palynological characters (Fig. 4b). Despite problems encountered with caving of material from the overlying lacustrine claystone into the reservoir and sub-reservoir section

of the well, the palynological recognition of the claystone units proved successful and was used to confirm the interpretation of MWD gamma readings as the well penetrated beneath the reservoir and through a fault (Gunn et al. 1993).

This case study illustrates that palynology can be used to correlate coal seams in the subsurface at geographical scales applicable to the traditional hydrocarbon industry. Lateral continuity of palynological assemblages in individual seams indicates that similar edaphic conditions existed on the coal mires over large areas. Inevitably, edaphic conditions change laterally, resulting in a corresponding change in vegetation and miospore assemblages, but such changes do not affect the viability of the correlation technique at the kilometre scale. Much more closely spaced well sections are produced in CBM exploration (Hewitt 1984) and the implication is that lateral edaphic variations should not adversely affect the feasibility of palynological correlation. High resolution palynostratigraphic correlations are best made on such localized groups of wells or

**a.**

**CHARACTERISTIC MIOSPORE TAXA**

| COAL HORIZON | *Ahrensisporites guerickei* | *Endosporites globiformis* | *Camptotriletes* spp. | *Cirratriradites saturni* | *Densosporites* spp. | *Vestispora* spp. | *Raistrickia fulva* | *Sinuspores "infans"* | *Radiizonates striatus* | *Schulzospora rara* | *Hymenospora "murdochensis"* | *Radiizonates aligerens* | *Radiizonates* cf. *aligerens* | *Radiizonates difformis* | *Grumosisporites* spp. |
|---|---|---|---|---|---|---|---|---|---|---|---|---|---|---|---|
| Group 4 coals | o | ○ | o | ○ | o | o | | | | | | | | | |
| Seam D | o | ○ | | ○ | o | o | o | | | | | | | | |
| Group 3 coals | ○ | ○ | | ○ | ● | o | o | o | o | | | | | | |
| Seam C | | ○ | | | ■ | ○ | | o | o | | | | | | |
| Seam B | | | | | ● | | | o | o | ○ | o | | | | |
| Group 2 coals | | | | | o | o | | o | | o | o | o | | | |
| Seam A | | | | | o | o | o | o | | o | | | o | | |
| Group 1 coals | | | ○ | | o | ○ | o | o | | o | | ○ | o | ○ | ○ |

**b.**

**CHARACTERISTIC MIOSPORE TAXA**

| MUDSTONE HORIZON | *Florinites* spp. | *Potonieisporites elegans.* | *Auroraspora* spp. | *Verrucosisporites* spp. | *Spelaeotriletes* spp. | *Kraeuselisporites* spp. | *Crassispora kosankei.* | *Laevigatosporites* spp. | *Sinuspores "infans"* | *Schulzospora rara* |
|---|---|---|---|---|---|---|---|---|---|---|
| *Anthracosia ovum* lacustrine bed | ■ | ● | ● | ○ | o | ○ | o | o | o | |
| Intra-reservoir mudstones | o | o | | ○ | | o | ● | ● | o | |
| Vanderbeckei Marine band | ■ | ■ | ○ | o | ○ | ○ | o | o | | o |

| ABUNDANCE KEY | | | |
|---|---|---|---|
| o  0.5-1% | ○  1-5% | ●  5-10% | ■  10-20% |

**Fig. 4.** Palynological characteristics of (**a**) distinctive seams and groups of seams and (**b**) mudstone horizons in the Murdoch Gas Field. Informally described taxa are shown in inverted commas.

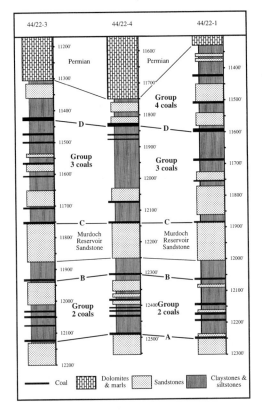

**Fig. 5.** Correlation of palynologically distinctive seams and groups of seams in the Murdoch Gas Field wells 44/22-1, 44/22-3 and 44/22-4.

exploration since the 1930s. Similar techniques have been applied to the correlation of coal seams and inter-seam sediments in the offshore southern North Sea Carboniferous basin. These methods are immediately applicable to the stratigraphic requirements of CBM exploration and development in Britain. Palynological correlation of coal seams will enhance reservoir description and evaluation. The success of CBM exploitation will be increased by the careful analysis of localized geological features which are below the resolution of three-dimensional seismic surveys, but within the resolution of palynostratigraphy. The techniques are applicable to ditch cuttings and core samples from air or foam drilled boreholes, as well as from those drilled with oil- or water- based muds. Rapid preparation and analysis of ditch cuttings samples using microwave solution and oxidation techniques can be used to monitor drilling progress and to identify coring points and stratigraphic total depths and can be used to enhance the accuracy of interpretations based on MWD and LWD methods.

The authors thank B. MacCarthy of Kinetica Ltd for invaluable discussion. Conoco (UK) Ltd, Arco British Ltd and Total Oil Marine plc are thanked for permission to publish data from the Murdoch Field. R. Neves (University of Sheffield) and N. Turner (British Geological Survey) are thanked for their contributions to the palynostratigraphy of the field. I. M. Fulton and another, anonymous, reviewer are thanked for constructive criticism of the manuscript.

sections. The geographical extension of such correlation can only be realistically achieved by close comparison of nearest neighbour wells or sections stepwise across a coal basin.

In the Westphalian, the length of time between successive coal seams is usually smaller than that required for the evolution of vegetation. In addition, successive seams are likely to have had the same edaphic conditions, so that the palynological assemblages are likely to be similar. In the case study, this problem has been addressed by the recognition and correlation of groups of palynologically similar seams. With the smaller distances between wells in the CBM industry and the possibility of some core material, more detailed correlations may be achievable.

## Conclusions

Qualitative and quantitative palynological methods of correlating Carboniferous coal seams in the subsurface have been applied to onshore coal

## References

ANDERSON, F. W. 1970. Carboniferous ostracoda—the genus *Carbonita* Strand. *Bulletin of the Geological Survey of Great Britain*, **32**, 69–121.

ARCHARD, G. & TRICE, R. 1990. A preliminary investigation into the spectral radiation of the Upper Carboniferous Marine Bands and its stratigraphic application. *Newsletter on Stratigraphy*, **21**, 167–173.

AYERS, W. B., TISDALE, R. M., LITZINGER, L. A. & STEIDL, P. F. 1993. Coalbed methane potential of Carboniferous strata in Great Britain [abstract]. *In: Proceedings of the 1993 International CBM Symposium, University of Alabama, Tuscaloosa*, 1–14.

BAILEY, J. B., ARBIN, P., DAFFINOTTI, O., GIBSON, P. W. & RITCHIE, J. S. 1993. Permo-Carboniferous plays of the Silver Pit Basin. *In:* PARKER, J. R. (ed.) *Petroleum Geology of Northwest Europe: Proceedings of the 4th Conference.* Geological Society, London, 707–715.

BENNIE, J. & KIDSTON, R. 1886. On the occurrence of spores in the Carboniferous Formation of Scotland. *Royal Physical Society of Edinburgh Proceedings*, **9**, 82–117.

CALVER, M. A. 1956. Die stratigraphische Verbreitung der nichtmarinen Muscheln in den penninischen Kohlenfeldern Englands. *Zeitschrift Deutsche Geologische Gesellschaft*, **107**, 26–39.

—— 1968a. Distribution of Westphalian marine faunas in northern England and adjoining areas. *Proceedings of the Yorkshire Geological Society*, **37**, 1–37.

—— 1968b. Coal Measures invertebrate faunas. *In*: MURCHISON, D. G. & WESTOLL, T. S. (eds) *Coal and Coal-bearing Strata*. Oliver & Boyd, Edinburgh and London, 147–177.

—— 1969. Westphalian of Britain. *Compte Rendu, 6ème Congrès International de Stratigraphie et de Géologie du Carbonifère, Sheffield 1967*, **1**, 233–245.

—— 1973. Marine fauna of the Westphalian B and C of the British Pennine coalfield. *Compte Rendu 7ème Congrès Internationale de Stratigraphie et de Géologie du Carbonifère, Krefeld 1971*, **2**, 253–266.

CAMERON, T. D. J. 1993. Carboniferous and Devonian (Southern North Sea). *In*: KNOX, R. W. O'B. & CORDEY, W. G. (eds) *Lithostratigraphic Nomenclature of the UK North Sea*, Vol. 5. British Geological Survey, Nottingham, 1–93.

CHALONER, W. G. 1958. The Carboniferous upland flora. *Geological Magazine*, **95**, 261–262.

CLAYTON, G., COQUEL, R., DOUBINGER, J., GUEINN, K. J., LOBOZIAK, S., OWENS, B. & STREEL, M. 1977. Carboniferous miospores of Western Europe: illustration and zonation. *Mededelingen Rijks Geologische Dienst*, **29**, 1–71.

CLEAL, C. J. 1991. Carboniferous and Permian biostratigraphy. *In*: CLEAL, C. J. (ed.) *Plant Fossils in Geological Investigations. The Palaeozoic*. Ellis Horwood, Chichester, 182–215.

COQUEL, R., DOUBINGER, J. & LOBOZIAK, S. 1976. Les microspores-guides du Westphalien à l'Autunien de Europe occidentale. *Revue de Micropaléontologie*, **18**, 200–212.

CORNFORD, C. 1984. Source rocks and hydrocarbons of the North Sea. *In*: GLENNIE, K., (ed.) *Introduction to the Petroleum Geology of the North Sea*. Blackwell, Oxford, 171–205.

CREEDY, D. P. 1993. An introduction to the geology of CBM in Britain [abstract]. *In*: *Proceedings of the Symposium Coalbed Methane '93*, Geological Society, London, 5–7.

DE JEKHOWSKY, B. 1963. La méthode des distances minimales, nouveau procédé quantitatif de corrélation stratigraphique; exemple d'application en palynologie. *Revue de l'Institut Français du Pétrole*, **18**, 629–653.

DESEZAR, Y. & POULSEN, N. E. 1994. On palynological preparation technique. *American Association of Stratigraphic Palynologists Newsletter*, **27**(Aug), 12–13.

DOYLE, J. F. 1987. In-seam seismic—a tool for the coal industry. *In*: *Geological Society of Australia, Geology and Coal Mining Conference Proceedings*, 251–259.

DYBOVÁ, S. & JACHOWICZ, A. 1957. Microspores of the Upper Silesian coal measures. *Prace Instytut Geologiczne*, **23**, 1–328.

ELLIN, S. J. & McLEAN, D. (1994) The use of microwave heating in hydrofluoric acid digestions for palynological preparations. *Palynology*, **18**, 23–31.

EVANS, D. J., MENEILLY, A. & BROWN, G. 1992. Seismic facies analysis of Westphalian sequences of the southern North Sea. *Marine and Petroleum Geology*, **9**, 578–589.

FRASER, A. J. & GAWTHORPE, R. L. 1990. Tectono-stratigraphic development and hydrocarbon habitat of the Carboniferous in northern Britain. *In*: HARDMAN, R. F. P. & BROOKS, J. (eds) *Tectonic Events Responsible for Britain's Oil and Gas Reserves*. Geological Society, London, *Special Publication*, **55**, 49–86.

GRANT, A. C. & MOIR, P. N. 1992. *Observations on Coalbed Methane Potential, Prince Edward Island*. Geological Survey of Canada, Paper, **92–1E**, 269–278.

GRAY, H. H. & GUENNEL, G. K. 1961. Elementary statistics applied to palynologic identification of coal beds. *Micropaleontology*, **7**, 101–106.

GREEN, C. & SLATT, R. M. 1992. Complex braided stream depositional model for the Murdoch Field Block 44/22 UK Southern North Sea [abstract]. *In*: *Proceedings of the Conference Braided Rivers: Form Process and Economic Applications*. Geological Society, London, 16.

GUENNEL, G. K. 1952. Fossil spores of the Alleghenian Coals of Indiana. *Indiana Department of Conservation Geological Survey Report of Progress*, **4**, 1–40.

—— 1958. Miospore analysis of the Pottsville coals. *Department of Conservation and Geological Survey of Indiana Bulletin*, **13**, 1–101.

GUION, P. D. & FIELDING, C. R. 1988. Westphalian A and B sedimentation in the Pennine Basin. *In*: BESLEY, B. M. & KELLING, C. R. (eds) *Sedimentation in a Synorogenic Basin Complex—the Carboniferous of Northwest Europe*. Blackie, Glasgow & London, 153–177.

GUNN, R., HWANG, N. J., MURRAY, I. & REZIGH, A. 1993. Teamwork brings success in Quad 44—the Murdoch Field development [abstract]. *Newsletter of the Petroleum Exploration Society of Great Britain*, February 1993, 4–7.

HABIB, D. 1966. Distribution of spore and pollen assemblages in the Lower Kittanning Coal of western Pennsylvania. *Palaeontology*, **9**, 629–666.

HACQUEBARD, P. A. 1986. The Gulf of St. Lawrence Carboniferous basin: the largest coalfield of Eastern Canada. *Bulletin of Canadian Institution of Mining and Metallurgy*, **79**, 67–78.

——, GILLIS, K. S. & BROMLEY, D. S. 1989. *Re-evaluation of the Coal Resources of Western Cape Breton Island*. Nova Scotia Department of Mines and Energy, Paper, **89–3**, 1–47.

HATHERLY, P. J. & HOLT, G. E. 1984. In-seam seismic surveys in the Hunter Valley [abstract]. *In*: *Geoscience in the Development of Natural Resources, 7th Australian Geological Convention, Sydney*, 221.

——, EVANS, B., REICH, S. & POOLE, G. R. 1989. 3D seismic reflection surveying for detailed coal seam mapping. *In: Advances in Studies of the Sydney Basin, 23rd Newcastle Symposium Proceedings*, 85–92.

HEWITT, J. L. 1984. Geologic overview, coal and coalbed methane resources of the Warrior basin—Alabama and Mississippi. *In:* RIGHTMIRE, C. T., EDDY, G. E. & KIRR, J. N. (eds) *Coalbed Methane Resources of the United States.* American Association of Petroleum Geologists, Studies in Geology Series, **17**, 73–104.

HIGGINS, A. C. 1975. Conodont zonation of the late Viséan–early Westphalian strata of the south and central Pennines of northern England. *Bulletin of the Geological Survey of Great Britain*, **53**, 1–90.

——1985. The Carboniferous System: part 2—conodonts of the Silesian Subsystem from Great Britain and Ireland. *In:* HIGGINS, A. C. & AUSTIN, R. L. (eds) *A Stratigraphical Index of Conodonts.* British Micropalaeontological Society Series, Ellis Horwood, Chichester, 210–227.

HOWIE, R. D. 1977. *Geological Studies and Evaluation of MacDougall Core Hole 1A, Western Prince Edward Island.* Geological Survey of Canada, Paper, **77–20**, 1–26.

HURST, A., LOVELL, M. A. & MORTON, A. C. (eds) 1990. *Geological Application of Wireline Logs.* Geological Society, London, Special Publication, **48**.

JONES, J. M. 1980. Carboniferous Westphalian (Coal Measures) rocks. *In:* ROBINSON, D. A. (ed.) *Geology of Northeast England.* Natural History Society of Northumberland and Durham, Special Publication, 23–36.

JONES, R. A. 1994. The application of microwave technology to the oxidation of kerogen for use in palynology. *Review of Palaeobotany and Palynology*, **80**, 333–338.

KINGSTON, P. W. E. & STEEVES, B. A. 1979. *Shallow Seismic Data from the Eastern Part of the New Brunswick Platform: Stratigraphic and Structural Implications.* New Brunswick Department of Natural Resources, Open File Report, **79–14**, 1–31.

KOSANKE, R. M. 1950. Pennsylvanian spores of Illinois and their use in correlation. *Illinois State Geological Survey Bulletin*, **74**, 1–128.

——1954. Correlation of coals and spore analysis. *University of Missouri School of Mines and Metallurgy Bulletin*, Technical Series, **85**, 11–16.

LEEDER, M. R., RAISWELL, R., AL-BIATTY, H., MCMAHON, A. & HARDMAN, M. 1990. Carboniferous stratigraphy, sedimentation and correlation of well 48/3-3 in the southern North Sea Basin: integrated use of palynology, natural gamma/sonic logs and carbon/sulphur geochemistry. *Journal of the Geological Society, London*, **147**, 287–300.

LELE, K. M. & CHANDRA, S. 1975. Associations of mio- and megafloras in the roof shales of some Barakar coal seams, South Karanpura Coalfield, Bihar. *Palaeobotanist*, **24**, 254–260.

LOBOZIAK, S., COQUEL, R. & JACHOWICZ, A. 1976. Stratigraphie du Westphalien d'Europe occidentale et de Pologne à la lumiere des etudes palynologiques (microspores). *Annales de la Societé de Géologie du Nord*, **96**, 157–172.

MARSHALL, A. E. & SMITH, A. H. V. 1965. Assemblages of miospores from some Upper Carboniferous coals and their associated sediments in the Yorkshire coalfield. *Palaeontology*, **7**, 656–673.

NEVES, R. 1958. Upper Carboniferous plant spore assemblages from the *Gastrioceras subcrenatum* horizon, North Staffordshire. *Geological Magazine*, **95**, 1–19.

PAPROTH, E., DUSAR, M., & 11 others 1983. Bio- and lithostratigraphic subdivision of the Silesian of Belgium, a review. *Annales de la Société Géologique de Belgique*, **106**, 241–283.

PASTIELS, A. 1956. Contribution à l'étude des foraminifères du Namurien et du Westphalien de la Belgique. *Publication de l'Association de l'Etude Paléontologique et Stratigraphique Houiller*, **27**, 1–32.

PEPPERS, R. A. 1970. Correlation and palynology of coals in the Carbondale and Spoon Formations (Pennsylvanian) of the northeastern part of the Illinois Basin. *Illinois State Geological Survey Bulletin*, **93**, 1–173.

POLLARD, J. E. 1969. Three ostracod-mussel bands in the Coal Measures (Westphalian) of Northumberland and Durham. *Proceedings of the Yorkshire Geological Society*, **37**, 239–276.

POTONIÉ, R. & KREMP, G. O. W. 1956. Die sporae dispersae des Ruhrkarbons, ihre Morphographie und Stratigraphie mit Ausblicken auf Arten andere Gebeite und Zeitabschnitte: Teil II. *Palaeontographica B*, **99**, 85–191.

RAISTRICK, A. 1935. The correlation of coal-seams by microspore content. I. The seams of Northumberland. *Transactions of the Institution of Mining Engineers*, **88**, 142–153; 259–264.

——1937. The correlation of coal seams by microspore content. *Colliery Engineering*, **14**, 299–302.

——1939. The correlation of coal seams by microspore content: II. The Trencherborne Seam, Lancashire and the Busty Seam, Durham. *Transactions of the Institution of Mining Engineers, London*, **97**, 425–437; **98**, 95–99; 171–175.

RAMSBOTTOM, W. H. C., CALVER, M. A., EAGAR, R. M. C., HODSON, F., HOLLIDAY, D. W., STUBBLEFIELD, C. J. & WILSON, R. B. 1978. *A Correlation of Silesian Rocks in the British Isles.* Geological Society of London, Special Report, **10**, 1–81.

RICHARDSON, G. & FRANCIS, E. H. 1970. Fragmental clayrock (FCR) in coal-bearing sequences in Scotland and north-east England. *Proceedings of the Yorkshire Geological Society*, **38**, 229–259.

RICHARDSON, J. B. & MCGREGOR, D. C. 1986. Silurian and Devonian spore zones of the Old Red Sandstone continent and adjacent regions. *Bulletin of the Geological Survey of Canada*, **364**, 1–79.

RIGHTMIRE, C. T. & CHOATE, R. 1986. Coal-bed methane and tight gas sands interrelationships. *In*: SPENCER, C. W. & MAST, R. F. (eds) *Geology of Tight Gas Reservoirs*. American Association of Petroleum Geologists, Studies in Geology, **24**, 87–110.

RITCHIE, J. S. & PRATSIDES, P. 1993. The Caister Fields, Block 44/23a, UK North Sea. *In*: PARKER, J. R. (ed.) *Petroleum Geology of Northwest Europe: Proceedings of the 4th Conference*. Geological Society, London, 759–769.

ROWECAMP, K. J. & JOHNSON, P. 1993. Air drilling in CBM reservoirs [abstract]. *In*: *Proceedings of the Symposium CBM '93*. Geological Society, London, 14.

RUTER, H. & SCHEPERS, R. 1978. Investigation of the seismic response of cyclically layered Carboniferous rocks by means of synthetic seismograms. *Geophysical Prospecting*, **26**, 29–47.

SCHOPF, J. M. 1936. Spores characteristic of Illinois Coal No. 6. *Transactions of the Illinois State Academy of Sciences*, **28**, 173–176.

——1938. Spores from the Herrin (No. 6) Coal bed of Illinois. *Illinois Geological Survey Report of Investigations*, **50**, 1–55.

SCOTT, A. C. 1979. The ecology of Coal Measure floras from northern Britain. *Proceedings of the Geologists' Association*, **90**, 97–116.

SMITH, A. H. V. 1957. The sequence of microspore assemblages associated with the occurrence of crassidurite in coal seams of Yorkshire. *Geological Magazine*, **94**, 345–363.

——1962. The palaeoecology of Carboniferous peats based on the miospores and petrography of bituminous coals. *Proceedings of the Yorkshire Geological Society*, **33**, 423–474.

——1964. Zur petrologie und Palynologie der Kohlenflöze des Karbons und ihrer Begleitschichten. *Fortschritten in der Geologie von Rheinland und Westfalen*, **12**, 285–302.

——1968. Seam profiles and seam characteristics. *In*: MURCHISON, D. G. & WESTOLL, T. S. (eds) *Coal and Coal-bearing Strata*. Oliver & Boyd, Edinburgh & London, 31–40.

—— & BUTTERWORTH, M. A. 1967. Miospores in the coal seams of the Carboniferous of Great Britain. *Special Papers in Palaeontology*, **1**, 1–324.

SMITH, E. G., RHYS, G. H. & EDEN, R. A. 1967. *Geology of the Country Around Chesterfield, Matlock and Mansfield*. Memoir of the Geological Survey of Great Britain, HMSO, London, 1–430.

SPEARS, D. A. & KANARIS-SITIRIOU, R. 1979. A geochemical and mineralogical investigation of some British and other European tonsteins. *Sedimentology*, **26**, 407–425.

STAPLIN, F. L. 1969. Sedimentary organic matter, organic metamorphism and oil and gas occurrence. *Bulletin of Canadian Petroleum Geology*, **17**, 47–66.

STOCKMANS, F. 1962. Paléobotanique et stratigraphie. *Compte Rendu, 4ème Congrès Internationale de Stratigraphie et de Géologie du Carbonifère, Heerlen 1958*, **3**, 657–682.

STONE, J. F. 1969. Palynology of the Eddleman Coal (Pennsylvanian) of North-Central Texas. *Bureau of Economic Geology, University of Texas at Austin, Report of Investigations*, **64**, 1–55.

THIESSEN, R. & STAUD, J. N. 1923. Correlation of coal beds in the Monangahela formation of Ohio, Pennsylvania and West Virginia. *Coal-Mining Investigations*, **9**, 1–64.

—— & VOORHEES, A. W. 1922. A microscopic study of the Freeport Bed, Pennsylvania. *Carnegie Institute of Technology Bulletin*, **2**, 1–75.

—— & WILSON, F. E. 1924. Correlation of coal beds of the Allegheny Formation of Western Pennsylvania and Eastern Ohio. *Coal-Mining Investigations*, **109**, 1–56.

TOMLINSON, R. C. 1957. Coal Measures miocrospore analyses: a statistical investigation into sampling procedures and some other factors. *Geological Survey of Great Britain, Bulletin*, **12**, 18–26.

TURNER, N., MCLEAN, D., NEVES, R., MASON, M. & WATSON, H. K. 1991. The Vanderbeckei Marine Band of the Murdoch Field, Block 44/22a, Southern North Sea: biostratigraphy and geology [abstract]. *In*: *Programme of the Joint Meeting of the Geological Society Coal Geology Group and the Yorkshire Geological Society, Carboniferous of the Southern North Sea Basin and the Onshore Flanks, Sheffield, September 1991*, 5.

VAN RIEL, W. J. 1965. Synthetic seismograms applied to the seismic investigation of a coal basin. *Geophysical Prospecting*, **13**, 105–121.

# Coal geology, chemical and petrographical characteristics, and implications for coalbed methane development of subbituminous coals from the Sorgun and Suluova Eocene basins, Turkey

A. I. KARAYIGIT, E. ERIS & E. CICIOGLU

*Hacettepe University, Department of Geological Engineering,*
*06532, Beytepe, Ankara, Turkey*

**Abstract:** The Sorgun and Suluova basins contain a thick and laterally extensive Lower Eocene coal seam. The stratigraphy of both basins comprises basement rocks, the Yozgat Granitoid of Pre-Eocene age in the Sorgun basin and Jurassic–Cretaceous limestones in the Suluova basin, overlain unconformably by the coal-bearing Lower Eocene Celtek Formation and by transgressive units of Middle–Upper Eocene age. In the Sorgun basin the Artova Ophiolitic Complex tectonically overlies the transgressive units. Neogene and Quaternary deposits cover all the older units unconformably in both basins. The single coal seam is located at the base of the Celtek Formation and reaches a thickness of 6 m in the open-cast mine at Sorgun and 8 m in boreholes in the Suluova basin. Some coal is produced by opencast mining, but most is produced by underground mining methods.

A total of 75 coal samples was collected from both basins and analysed chemically and petrographically. The coals are characterized by a relatively low moisture content (0.7–9.3 wt% on an air-dried basis) and variable ash yields (2.4–67.8 wt% on an air-dried basis) and huminite contents (28.4–89.6 vol%). Measurements of the mean random reflectance of ulminite (mainly eu-ulminite A) vary between 0.46 and 0.60% (0.53% average). These reflectance values and the chemical analyses show that the rank is subbituminous A or transitional to high-volatile bituminous C coal according to the ASTM classification.

The coalbed methane potential of both basins is at present unknown. No borehole has been drilled to explore for this resource and no gas capacity value has been obtained for the coal seam and bituminous shales. The methane explosions which have occurred during coal production in the underground mines imply that both basins may have been an important potential for coalbed methane.

Coal is the main raw material for energy provision in Turkey. The total Turkish lignite reserve is about 8.05 Gt and about 50 Mt of coal were produced in 1994. Extensive coal geological studies and chemical analyses have been carried out by the General Directorate of Mineral Resource and Exploration (MTA) and Turkish Coal Enterprise (TKI). However, detailed geological investigations of many coal basins in Turkey have not yet been completed. In addition, petrographical studies of coal and coalbed methane exploration are limited in Turkey.

The Eocene rocks of Turkey consist of marine sediments and submarine volcanic rocks with sedimentary intercalations (Fig. 1). Small Eocene coal basins occur in the northern Anatolian region, mainly adjacent to the Northern Anatolian Fault. Among those that contain economic coals are the Bolu-Mengen, Yozgat-Sorgun and Amasya-Suluova basins. The total coal reserves of the Sorgun and Suluova basins presented in this study are about 45 Mt and approximately 2000–3000 t/day of coal is produced by the coal companies. Coal production in both basins is hindered by many mining problems—for example, methane explosions, overburden, groundwater and spontaneous combustion. Methane explosions are a more serious problem than the others, but detailed investigations have not yet been made in either basin. Unfortunately, according to our knowledge, 148 mine workers in the Yeni Celtek coal mine in Suluova and, on 15 March 1995, 40 mine workers in the Madsan coal mine in Sorgun were killed as a result of methane explosions.

Limited regional geological and coal geology studies have been carried out by Blumenthal (1937), Lahn (1940), Barutoglu (1953), Brelie (1953), Pekmezciler (1953), Ketin (1955), Agrali (1965; 1966), Ozdemir & Bekmezci (1983) and Gumussu (1984) in the two basins. The first and most important study of the coal geology of the Sorgun basin was carried out by Agrali (1965; 1966), who proposed that the Sorgun coals may have accumulated in a lagoonal environment. In the Suluova basin, Blumenthal (1937) was the first to determine a Lower Eocene age for the coal-bearing Celtek Formation. Gumussu (1984), in common with

*From* Gayer, R. & Harris, I. (eds) 1996, *Coalbed Methane and Coal Geology,*
Geological Society Special Publication No 109, pp 325–338.

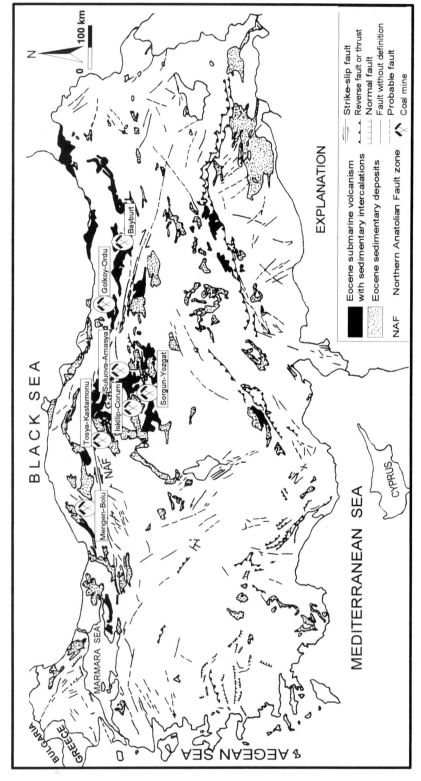

**Fig. 1.** Distribution of Eocene rocks in Turkey and location of the Sorgun and Suluova basins (modified after Bingol 1989).

some previous workers, thought that two different coal seams existed, one at the base of the Lower Eocene Celtek Formation and the other at the base of the Middle Eocene Armutlu Formation. In this study, only a single coal seam was identified. The overall objectives of this study are to determine the geological setting of the Sorgun and Suluova basins, to characterize the geochemistry and petrography of the included coals, and to interpret the results in terms of the coalbed methane potential in the two basins.

## Coal geology

Revised geological maps, generalized stratigraphic sequences and cross-sections for the Sorgun and Suluova coal basins are presented in

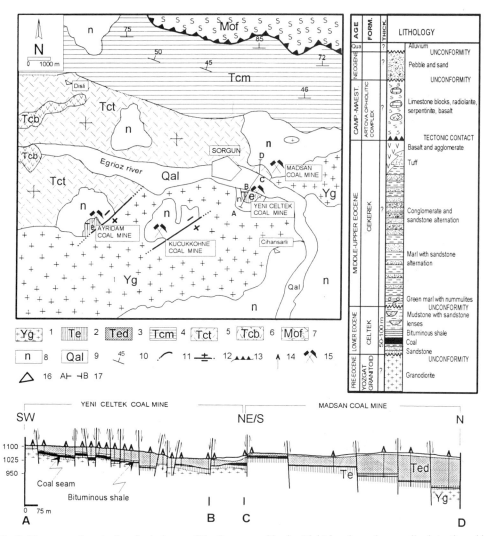

**Fig. 2.** Upper panel: revised geological map of the Sorgun coal basin. Right-hand panel: generalized stratigraphic sequence. Lower panel: geological cross-section through the Yeni Celtek to Madsan coal mines. Locations of the coal mines are also presented on the geological map. Key for map and cross-section: (1) Yozgat Granitoid; (2) Celtek Formation; (3) Undifferentiated Eocene rocks; (4) Cekerek Formation–marl; (5) Cekerek Formation–tuff; (6) Cekerek Formation–basalt and agglomerate; (7) Artova Ophiolitic Complex; (8) Neogene deposits; (9) alluvium; (10) bedding; (11) boundary; (12) normal fault; (13) thrust fault; (14) borehole; (15) coal mine; (16) village; and (17) line of cross-section.

Figs 2 and 3 respectively. The locations of coal mines are also presented on these maps.

## Basement

The basement of the Sorgun basin is the Yozgat Granitoid (Fig. 2), which crops out as light and dark brown massive blocks. The age of this granitoid was inferred to be between Late Mesozoic and Early Eocene, according to Lahn (1940), but pre-Eocene according to Erler *et al.* (1991). The basement of the Suluova basin con-sists of Jurassic–Cretaceous limestones (Fig. 3). They are grey, thin-bedded (20–50 cm), highly folded and show karstification.

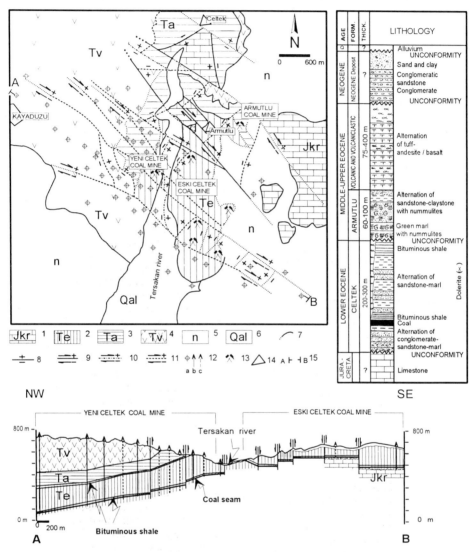

**Fig. 3.** Upper panel: revised geological map of the Suluova coal basin. Right-hand panel: generalized stratigraphic sequence. Lower panel: geological cross-section through the Yeni Celtek to Eski Celtek coal mines (c). Locations of the coal mines are also presented on the geological map. Key for map and cross-section: (1) Jurassic–Cretaceous limestone; (2) Celtek Formation; (3) Armutlu Formation; (4) volcanic and volcaniclastic rocks; (5) Neogene deposits; (6) alluvium; (7) boundary; (8) normal fault; (9) translation strike-slip fault; (10) covered translation strike-slip fault; (11) translation strike-slip fault determined from underground maps; (12) boreholes (a, borehole on the geological map; b, borehole on the cross-section; c, borehole transported on the cross-section); (13) coal mine; (14) village; and (15) line of cross-section.

## Celtek formation

This coal-bearing formation in the Sorgun basin has been identified by Karayigit & Cicioglu (1994) and Cicioglu (1995). These workers noted that this formation consists, from the base upwards, of grey sandstone, a coal seam, bituminous shales and mudstones with sandstone lenses (Fig. 2). The thickness of the formation from borehole data is between 50 and 100 m. The Celtek Formation in the Suluova basin has been studied in some detail by Karayigit & Eris (1994). These workers showed that this formation consists of, from bottom to top, alternations of conglomerate, sandstone and marl, a coal seam, bituminous shales, alternations of sandstone and marl, and bituminous shales. The thickness of the formation, from borehole data, varies between 200 and 300 m.

Macroscopic seam sections and some sample locations for the two basins are given in Fig. 4. The Sorgun coals are mainly bright and banded. Some clayey coals and coaly claystones are developed within the coal seam; they are especially common in the coal seam in the Kucukkohne underground mine. Kucukkohne coals appear dull, but thin (<1 cm), bright bands can be traced within the coal seam. The coal thickness is about 6 m in the Yeni Celtek opencast mine and 14–16 m in some boreholes from the Sorgun basin.

The coals of the Suluova basin have a similar appearance, but thin limestones with included brown and light brown claystones are formed within the coal seam. Some vertebrate bones, turtle fossils and teeth are occasionally found in the coal seam and bituminous shale. Figure 4 shows that, with the exception of the Armutlu underground coal mine, because of the great thickness of the coal seam in the Yeni Celtek coal mine, only the upper part of the coal seam is mined. In the Eski Celtek coal mine only the lower part of the coal seam was exploited . The total coal thickness in this basin reaches 8 m in some boreholes.

In the Sorgun basin, 1 : 500 or 1 : 1000 scale mine maps show many normal faults with small slips. They are not shown on the geological map because they are unconformably overlain by Neogene deposits and alluvium. The coal seam and overburden units are downthrown by faults towards the Madsan coal mine, which has fewer normal faults than the other coal mines in this area. In the Suluova coal basin there are many strike-slip faults that have similar strike directions to the Northern Anatolian Fault, i.e. NW–SE.

Some palynological investigations of coal samples from both basins were performed and

**Table 1.** *Palynomorphs identified in the Sorgun and Suluova coals*

| |
| --- |
| *Leiotriletes* sp. |
| *Cicatricosisporites dorogensis* (R. Pot and Gell.) Kedves |
| *Inaperturopollenites hiatus* (R. Pot.) Th. and Pf. |
| *Triatriopollenites excelsus* (R. Pot.) Th. and Pf. |
| *Triatriopollenites rurobituitus* Pf. in Th. and Pf. |
| *Triporopollenites ef. megagranifer* (R. Pot.) Th. and Pf. |
| *Subtriporopollenites anulatus* Pf. and Th. in Th. and Pf. |
| *Tricalporopollenites pseudocingulum* (R. Pot.) Th. and Pf. |
| *Tricalporopollenites cingulum* (R. Pot.) Th. and Pf. |
| *Tetracolporopollenites sp.* |

Table 1 shows the spore pollens that were identified. They indicate broadly an Eocene age. Nakoman (1966) identified the age of the Sorgun coals as Lower or Middle Eocene from palynological studies. However, green marls with nummulites of middle Eocene age were located at the base of the transgression units and constitute a fossiliferous marker zone. As this marl lies above the coal, the age of coals for the two basins can be identified as Early Eocene.

It was suggested by Ketin (1955) and Agrali (1965; 1966) that the Sorgun coals accumulated in a lagoonal facies, on the basis that the green marls with nummulites were deposited in a marine environment. However, in this study, micropalaeontological investigations were made on samples collected from bituminous shales in the coal-bearing Celtek Formation and green marls with nummulites in the transgressive units from both basins; no marine fossil has been found. Marine fossils, especially dinoflagellates, were only found in the green marls of the Cekerek Formation in the Sorgun basin and the Armutlu Formation in the Suluova basin. Additionally, the bituminous shales of the Celtek Formation include abundant amorphous kerogens, possibly formed in freshwater environments. These results show that the bituminous shales and the coal of the Celtek Formation probably accumulated in a lacustrine environment with little or no marine influence. The tectonic setting was such that freshwater lakes developed in fault-bounded basins, in which peat-forming mires formed. These mires were subsequently covered by a marine transgression.

## Transgressive units

These units comprise the Cekerek Formation in the Sorgun basin and the Armutlu Formation and volcanic–volcaniclastic rocks in the Suluova basin. The Cekerek Formation was defined

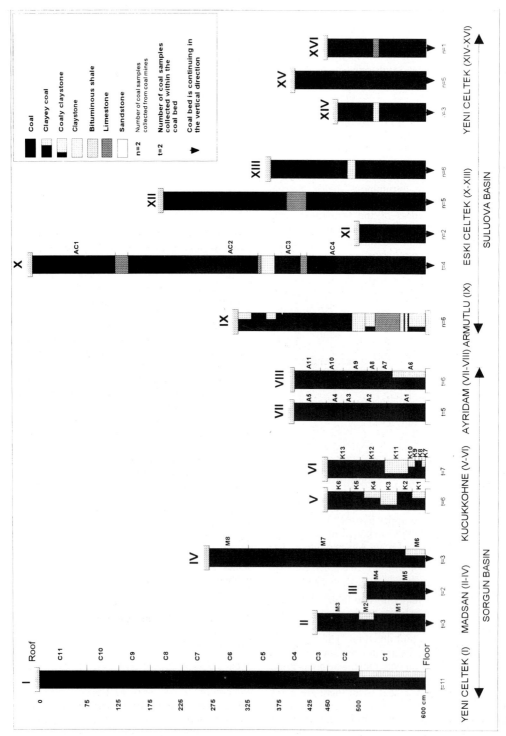

**Fig. 4.** Macroscopic seam sections and some sample locations within the coal seam for the Sorgun and Suluova coal basins.

**Table 2.** *Fossils determined in the green marls with nummulites in the Sorgun basin*

---

*Nummulites* sp.
*Discocyclina* sp.
*Rotalidae*
*Eorupertia magna le Calves*
*Miliolidae*
*Bryozoa*
*Ditrupa* sp.
*Nummulites* cf. *beaumonti d'Archiae*
*Assilina* sp.

---

initially by Ozcan *et al.* (1980) at the northern and eastern side of the study area. This formation extends into the northern part of the Sorgun basin (Fig. 2) and consists of a basal green marl containing nummulites and rare ostrea fossils, with, from the base upwards, intercalations of marl and thin (10–20 cm) sandstones, conglomerates and sandstone alternations, tuffs, basalts and agglomerates. A typical reddish coloured alteration zone occurs between the coal-bearing Celtek Formation and the green marl with nummulites of the Cekerek Formation. The Armutlu Formation was first identified by Blumenthal (1937) in the Suluova basin. The base of this formation contains a similar succession to the Sorgun basin. The upper part of this formation in the Suluova basin is different from the Sorgun basin and consists of alternations of sandstone and claystone with nummulites. Volcanic and volcaniclastic rocks, common around Kayaduzu village (Fig. 3), contain alternations of bedded tuff and andesite/basalt. The Celtek Formation is cut by a small doleritic intrusion to the south of Armutlu village (Fig. 3).

The thickness of the Cekerek Formation in the Sorgun basin has not been determined because of cover units, but, in the Suluova basin, borehole data indicate that the Armutlu Formation and the volcanic–volcaniclastic rocks are between 60–100 and 75–400 m respectively. Fauna from the green marls with nummulites in the Sorgun basin (Table 2) indicate a Middle Eocene age, the same as for the Armutlu Formation in the Suluova basin (Blumenthal 1937; Gumussu 1984). The age of the volcanic–volcaniclastic rocks in the two basins is Middle-Upper Eocene.

## Artova ophiolitic complex

This complex was identified by Ozcan *et al.* (1980). It rests tectonically on the Cekerek Formation only in the Sorgun basin. The ophiolitic complex, containing blocks of limestone and serpentinite, has a matrix of basalt, radiolarite and chert.

## Neogene and Quaternary deposits

These units unconformably overlie the older units of the two basins. The Neogene deposits consist of unconsolidated pebbles and sands in the Sorgun basin, and conglomerates, conglomeratic sandstones, unconsolidated sands and clays in the Suluova basin. Unconsolidated alluvium represents the Quaternary in both basins.

## Methods

Forty-three channel coal samples were collected from the Sorgun basin and 32 from the Suluova basin. Some of these are represented on the seam sections in Fig. 4. All the samples were prepared for analysis according to the ASTM (1991) procedure. Proximate analyses of all 75 samples and ultimate analyses of 16 selected samples were performed in accordance with ASTM (1991) procedure. Maceral analysis was made on polished coal briquettes under incident and fluorescence light. The same briquettes were used for the measurement of random ulminite reflectance, using 50 points on every briquette. ICCP (1963; 1971), Stach *et al.* (1982) and ASTM (1991) maceral identification methods were adopted. Maceral and mineral matter abundances are reported as volume percentages of the whole coal and are based on a minimum of 500 points counted on every petrographic briquette. During petrographic analyses, a Leitz MPV II type microscope, 50× objective, mercury lamp (CS 100 W-2), double B12 filter and one B38 heating–absorbing filter, K510 barrier filter and point counter were used. For ulminite reflectance measurements, saphire (0.551% *R*) and glass (1.23% *R*) standards were used for calibration of the microscope.

## Analytical results and discussion

### Coal chemistry

Moisture content, ash yield and total sulphur contents on an air-dried basis, volatile matter content on a dry, ash-free basis and calorific value on both an air-dried basis and a moisture mineral matter-free basis for coals from both basins are presented in Fig. 5. In addition, coal seam samples from the Yeni Celtek opencast mine in the Sorgun basin are used to evaluate vertical variations in chemistry and petrography within the coal seam (Fig. 6).

The moisture content of the coals varies from 0.7 to 9.3 wt% (Fig. 5a). The vertical variation of the moisture content shows no significant variation (Fig. 6a). Ash yields (air-dried basis) in both basins show a wide range from 2.4 to 67.8 wt% (Fig. 5b), with an average of 19.9 wt%. The average ash yield of the Sorgun and Suluova coals is 18.3 and 22.0 wt%, respectively. The base of the coal seam has a higher ash yield due

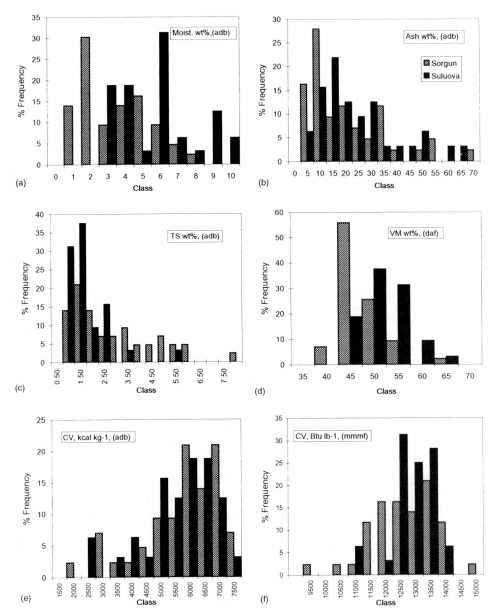

**Fig. 5.** Histograms of (a) moisture content, (b) ash yield, (c) total sulphur (TS) contents on an air-dried basis (adb), (d) volatile matter (VM) content on a dry, ash-free basis (daf) and (e) and (f) calorific value (CV) on both adb and moist, mineral matter-free (mmmf) basis for the Sorgun and Suluova coal basins.

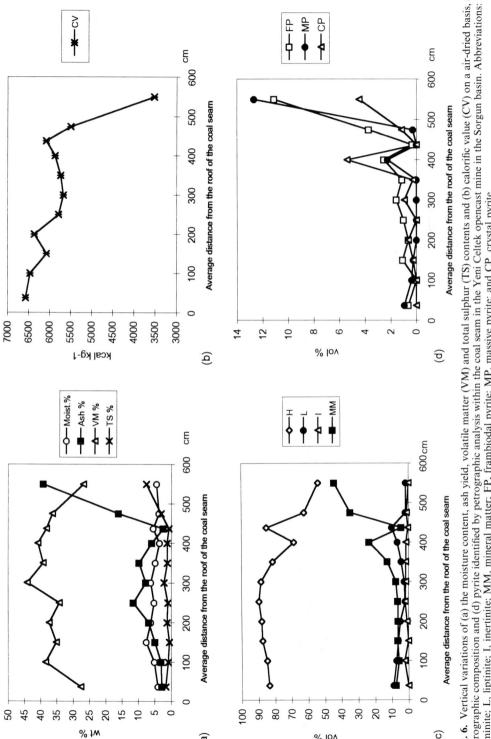

**Fig. 6.** Vertical variations of (a) the moisture content, ash yield, volatile matter (VM) and total sulphur (TS) contents and (b) calorific value (CV) on a air-dried basis, (c) petrographic composition and (d) pyrite identified by petrographic analysis within the coal seam in the Yeni Çeltek opencast mine in the Sorgun basin. Abbreviations: H, huminite; L, liptinite; I, inertinite; MM, mineral matter; FP, framboidal pyrite; MP, massive pyrite; and CP, crystal pyrite.

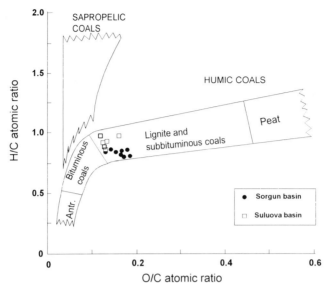

**Fig. 7.** Location of the O/C and H/C atomic ratios of the coals on Van Krevelen diagram. Reproduced with permission from Killops & Killops (1993).

to clay admixtures (Fig. 6a). However, apart from these clay-rich coals, the coal seam has a generally low ash yield.

The total sulphur contents (air-dried basis) of the coals is between 0.6 and 7.6 wt% (Fig. 5c). However, 35% of the Sorgun coals and 69% of the Suluova coals have total sulphur contents between 0.5 and 2.0 wt%, making them relatively low total sulphur coals. The vertical variation of total sulphur content, like the ash yield, shows a large increase at the base of the coal seam (Fig. 6a). This increase comes from the clay-rich coals. Additional work is required to explain why the total sulphur content is so high in the clayey coals. The sulphur content appears not to have been influenced by the marine transgression, in contrast with the results of Casagrande *et al.* (1977) for marine influences on sulphur in coal.

The volatile matter content (dry, ash-free basis) of the coals shows a broad range from 38.5 to 62.3 wt%, with that of the Suluova coals being relatively higher than that of the Sorgun coals (Fig. 5d). The Suluova coals also have relatively higher ash yields (Fig. 5b), suggesting that the volatile matter content of the Suluova coals may have been increased by the carbonate within the seams (see Fig. 4). In addition, the free-swelling index (FSI) of the Sorgun coals (there is currently no FSI information on the Suluova coals) is between 0 and 1 (Cicioglu 1995). This indicates that the Sorgun coals have a non-agglomerating or weak agglomerating character.

The calorific values (air-dried basis), as with the ash yield and total suphur content, show a broad range between 1689 and 7480 kcal kg$^{-1}$ (Fig. 5e). The calorific values (moisture, mineral matter-free) for the two basins are between 9441 and 14 550 Btu lb$^{-1}$. The calorific value of 91% of the Sorgun coals and of 94% of the Suluova coals is between 11 500 and 14 000 Btu lb$^{-1}$. The calorific value decreases at the base of the coal seam (Fig. 6b) because of the high ash yield.

Ultimate analyses of 16 samples from both basins gave the following results: carbon, 70.83–77.50 wt%; hydrogen, 4.75–6.19 wt%; nitrogen, 1.46–2.54 wt%; and oxygen by difference, 13.41–20.07 wt%. The H/C and O/C atomic ratios from these results are plotted on a Van Krevelen diagram (Fig. 7). All the samples plot at the subbituminous–bituminous coal boundary. The H/C atomic ratios of the Suluova coals are slightly higher than the Sorgun coals.

*Petrographic composition*

The petrographic composition of the coals was determined by maceral analysis. The coals are characterized by a variable content of huminite group macerals (28.4–89.6 vol%), relatively low liptinite group macerals (0–15.2 vol%) and only minor amounts of inertinite group macerals (0–3.6 vol%). Mineral matter has a wide range between 4.1 and 69.6 vol%. The petrographic data are also presented in the form of triangular diagrams on a mineral-matter free basis for the two basins (Fig. 8a and 8b), from which it can be seen that the maceral composition of the coals of the two basins are in general similar. The huminite content is higher than the other two maceral groups in both basins, but the Suluova coals have a slightly higher liptinite group maceral content. This explains the higher hydrogen content of the Suluova coals (Fig. 7).

The huminite group consists of textinite (0–24.4 vol%), ulminite (1.1 46.1 vol%), attrinite (1.0–41.6 vol%), densinite (0–7.1 vol%), gelinite (0.6–41.8 vol%) and corpohuminite (0.5–32.5 vol%) macerals. In general, the most important macerals in the huminite group are thus ulminite, attrinite, gelinite and corpohuminite. Sporinite, cutinite, liptodetrinite and alginite of the liptinite group have been determined in

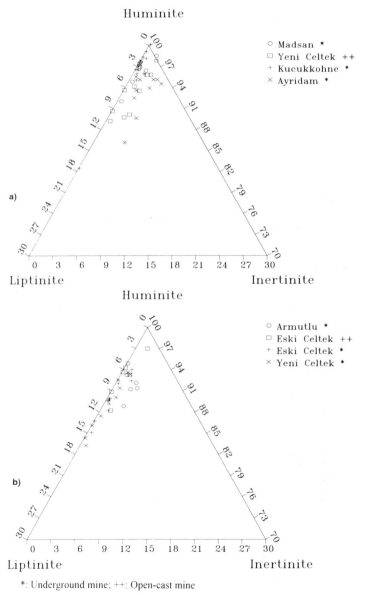

*: Underground mine; ++: Open-cast mine

**Fig. 8.** Ternary diagrams showing petrographical composition on a mineral matter-free basis in the coal mines in (a) the Sorgun basin and (b) the Suluova basin.

the coal samples. The sporinite (0–3.5 vol%) shows a generally yellow and sometimes reddish-orange colour in fluorescent light. Cutinite (0–8.5 vol%) is more common than sporinite and shows orange and reddish orange colours in fluorescent light. Some samples contain cutinite clarite, where cutinite occurs within the humi-nites, indicating a high water level in the peat-forming environment (Stach *et al.*, 1982). Lipto-detrinite (0–11.1 vol%) generally shows a yellow colour in fluorescence light. A few samples, especially from the Suluova coals, have virtually no or only minor amounts of alginites. The inertinite group macerals are sclerotinite (0–2.8 vol%) and inertodetrinite (0–2.5 vol%). The large amounts of attrinite macerals, minor amounts of inertinite macerals and the presence of cutinite clarite show that the peat developed in a reducing environment with high water levels.

During the maceral analyses three different groups of pyrite were generally recognized. These are framboidal, massive and crystalline pyrite. Framboidal pyrite normally forms spherical clusters between 10–15 $\mu$m in diameter. Crystal-line pyrite consists of individual euhedral-shaped pyrite grains 1–5 $\mu$m in diameter. Massive pyrite generally shows pyritized cell textures. Framboi-dal and massive pyrite contents show wide ranges from 0 to 11.1 vol% and from 0 to 29.1 vol%, respectively. The content of crystalline pyrite ranges from 0 to 5.4 vol%. The mineral matter content, excluding pyrite, ranges from 4.0 to 67.6 vol% and consists of clay minerals plus quartz and syngenetic carbonate minerals (including siderite in the Suluova coals).

The vertical variations in the maceral groups, mineral matter content and pyrite in the Yeni Celtek coal mine from the Sorgun basin are presented in Fig. 6c and 6d. The maceral composition and mineral matter content, in a similar manner to the proximate analyses, show significant changes at the base of the coal seam and these, as mentioned earlier, are the result of clay-rich coal. The huminite group especially decreases, whereas the mineral matter content increases (Fig. 6c). Framboidal, massive pyrite and, to a lesser extent, crystalline pyrite, in a similar manner to the total sulphur content, clearly increase at the base (Fig. 6d). The remaining coals show a similar composition.

## Reflectance measurements and rank

The results of the random reflectance measure-ments of ulminite (mainly eu-ulminite A) are shown in Table 3. The mean random ulminite reflectance values in the two basins vary between 0.46 and 0.60% (0.53% average). There are small differences between the mean ulminite reflectance values of the samples taken from the same seam in different coal mines. These differences in the two basins may be related to the different gelification of maceral varieties of the ulminite group. The average reflectance value from the Madsan coal mine in the Sorgun basin, where a coalbed methane explo-sion occurred, is similar to the coals from the other mines. Relatively higher reflectance values have only been measured in the Ayridam coal mine, but no coalbed methane has been detected. In the Suluova coal basin coalbed methane is known to occur in the Yeni Celtek coal mine, where slightly higher reflectance values have been measured.

Taking into account some of the coal rank parameters, such as the ulminite reflectance value

**Table 3.** *Results of the mean random reflectance measurements of ulminite (mainly eu-ulminite A) for every coal sample and comparison between basins*

| Coal mine | Range (mean) % *Rr* | Range (mean) Standard deviation |
|---|---|---|
| Sorgun basin | | |
| Masdan (n = 8)* | 0.52–0.57 (0.54) | 0.03–0.05 (0.03) |
| Yeni Celtek (n = 11)† | 0.46–0.51 (0.49) | 0.02–0.04 (0.03) |
| Kucukkuohre (n = 13)* | 0.49–0.55 (0.53) | 0.02–0.04 (0.03) |
| Ayridam (n = 11)* | 0.54–0.60 (0.57) | 0.02–0.04 (0.03) |
| Suluova basin | | |
| Armutlu (n = 6)* | 0.50–0.54 (0.51) | 0.03–0.04 (0.04) |
| Eski Celtek (n = 6)† | 0.50–0.55 (0.52) | 0.02–0.03 (0.03) |
| Eski Celtek (n = 11)† | 0.50–0.54 (0.52) | 0.01–0.04 (0.03) |
| Yeni Celtek (n = 9)* | 0.50–0.59 (0.54) | 0.02–0.04 (0.02) |

%*Rr*, Mean random reflectance of ulminite on every coal sample.
* Underground mine.
† Opencast mine.

(0.46–0.60% range and 0.53% average), the calorific value between 11 500 and 14 000 Btu lb$^{-1}$ (moisture, mineral matter-free), the non-agglomerating or weakly agglomerating character of the coals and the H/C and O/C atomic ratios, these coals, according to the ASTM classification, are either subbituminous A coals or transitional to high-volatile C bituminous coal in the two basins.

## Implications for coalbed methane development

During coalification, coal releases large amounts of gas and coal can be a source for hydrocarbons. Methane is typically a predominant substance in coalbed gas, but a variety of other substances are also liberated during mining or reservoir production, including $CO_2$, $H_2O$, $N_2$, $C_2H_6$, $C_3H_8$, and condensates, and, in some instances, even oil (Levine 1994). Bituminous rank coals are ideally suited for coalbed methane generation and retention (Dawson & Kalkreuth 1994). However, the coals described in this paper are subbituminous in rank and the presence of methane during coal production occurs only rarely. At present, the coalbed methane resource potential of either basin is unknown. No borehole has been drilled to explore for this resource and no gas capacity value has been reported for the coal seam or overlying bituminous shales. The occurrence of methane explosions during coal production, especially in the Madsan mine in the Sorgun basin and the Yeni Celtek mine in the Suluova basin, implies that both basins may have an important potential for coalbed methane production. Some of the factors which indicate a good potential for coalbed methane in both basins are: (1) thick, impermeable bituminous shales (up to 40–50 m in some boreholes) immediately overlying the coal seam and clayey floor rocks; (2) the average coal thickness of 6 m which reaches up to 16 m in some boreholes in the Sorgun basin and 8 m in the Suluova basin, and their large areal extent; (3) relatively thick overburden above the coal seam—for example, coal production is at about 150 m depth in the Madsan coal mine in the Sorgun basin and at 300 m in the Yeni Celtek mine in the Suluova basin; (4) coal quality, such as the relatively low moisture contents of coals (0.7–9.3 wt%, air-dried basis), relatively low average ash yield (19.9 wt%, air-dried basis) and higher huminite contents (28.4–89.6 vol%) than liptinite and inertinite group macerals; and (5) a relatively stable tectonic structure of the coalfields and the sealing of any fault zones by calcium carbonate.

Without undertaking additional investigations, such as desorption tests, of coal samples obtained from relatively deep underground mines and boreholes, no accurate assessment of coalbed methane resources can be made. In this paper, the geological setting of the coal-bearing formation, the chemical characteristics, petrographic composition and rank of the coals were described and the potential for coalbed methane development presented. Additional work is required to examine the coalbed methane resource potential.

## Conclusions

Geological studies and petrographic analyses show that the coal-bearing Celtek Formation is Early Eocene in age and it is believed to have been deposited in a lacustrine environment before a marine transgression by units of Middle–Late Eocene age. There are a number of small normal faults in the Sorgun basin and strike-slip faults striking parallel to the Northern Anatolian Fault in the Suluova basin.

The results of the chemical and petrographic analyses show that the coals in both basins are similar. The coals are characterized by a relatively low moisture content (0.7–9.3 wt%, air-dried basis), variable ash yields (2.4–67.8 wt%, air-dried basis) and huminite contents (28.4–89.6 vol%). Measurements of the mean random reflectance of ulminite (mainly eu-ulminite A) vary between 0.46 and 0.60% (0.53% average). These reflectance values and the chemical analyses show that the rank, according to the ASTM classification, is a subbituminous A transitional to high-volatile bituminous C coal.

The presence of methane in some coal mines in the two basins implies that the coals have the potential to contain a coalbed methane resource. The geology of the two basins suggests that the conditions are suitable for the extensive development of coalbed methane within the basins. Further work is required to test this resource potential.

We acknowledge the Turkish Scientific and Research Council (TUBITAK) for supporting the Sorgun investigations (Project No. TBAG/YBAG113) and Yeni Celtek Coal Company for providing logistical support during the field studies; F. Akgun for palynological analyses; S. Orcen, A. Hakyemez and N. Bozdogan for palaeontological analyses; and A. Culfaz for ultimate analyses. We also thank R. Gayer and M. Whateley, who critically reviewed the manuscript.

# References

AGRALI, B. 1965. *Geological Mapping of Yozgat-Sorgun Coal Basin at the Scale of 1/10000 and Determination of Coal Reserve using the Exploration Boreholes Drilled in 1965–66 years.* MTA Report, No. 3895 [in Turkish].

——, 1966. *A Report About the Investigation of Lignite Occurrence in Buzacioglu-Calatli (Yozgat) Region and the Geological Map at the Scale of 1/25000.* MTA Report No. 3914 [in Turkish].

ASTM 1991. *Annual Book of ASTM Standards, Gaseous Fuels; Coal and Coke.* ASTM, Philadelphia.

BARUTOGLU, O. H. 1953. *A Report on the Celtek Coal Potential and Exploration,* MTA Special Report [in Turkish].

BINGOL, E. 1989. *Geological Map of Turkey at the Scale of 1 : 2000000.* MTA Publication.

BLUMENTHAL, M. M. 1937. *Coal Geology of Merzifon and Suluova (Amasya).* MTA Report No. 7063 [in Turkish].

BRELIE, G. 1953. *A Suggestion for Two Exploration Boreholes Drilled in the Celtek Coal Basin.* MTA Report No. 2091 [in Turkish].

CASAGRANDE, D., SIEFERT, L., BERSHINISKI, C. & SUTTON, N. 1977. Sulfur in peat forming systems of Okefenokee Swamp and Florida Everglades: Origins of sulfur in coals. *Geochimica et Cosmochimica Acta,* **41**, 161–167.

CICIOGLU, E. 1995. *Investigation of chemical and petrographical characteristics of Sorgun (Yozgat) coals.* MSc Thesis, Hacettepe University, Ankara [in Turkish].

DAWSON, F. M. & KALKREUTH, W. D. 1994. Coal rank, distribution, and coalbed methane potential of the lower Cretaceous Luscar Group, Bow River to Blackstone River, Central Alberta Foothills. *Geological Survey of Canada Bulletin,* **473**.

ERLER, A., AKIMAN, O., UNAN, C., DALKILIC, F., DALKILIC, B., GEVEN, A. & ONEN, P. 1991. Petrology and geochemistry of magmatic rocks of the Kirsehir Massive in the Kaman (Kirsehir) and Yozgat regions. *TUBiTAK, Nature, Journal of Engineering and Environmental Sciences,* **15**, 76–100 [in Turkish].

GUMUSSU, M. 1984. *Geology of Celtek (Amasya) coal basin and evaluation of coal potential.* PhD Thesis, Ankara University, Ankara [in Turkish].

ICCP 1963 and 1971. *International Lexikon für Kohlenpetrologie.* Centre National de la Recherche Scientifique 15, Quiai-Anatole-France, Paris.

KARAYIGIT, A. I. & CICIOGLU, E. 1994. Geology, depositional environment and coal petrology of Sorgun Eocene coals, Yozgat-Turkey *In: International Conference & Short Course on Coalbed Methane and Coal Geology, University of Wales Cardiff, UK, 12–16 September 1994.*

—— & ERIS, E. 1994. Geological setting and petrology of Celtek Eocene coals, (Amasya) & the influence of the North Anatolian Fault and volcanism on the coal rank. *In: International Conference & Short Course on Coalbed Methane and Coal Geology, University of Wales Cardiff, UK, 12–16 September 1994.*

KETIN, I. 1955. Geology of Yozgat region and tectonics of Central Anatolian Massive. *TJK Bulletin.* **6:1**, 1–28 [in Turkish].

KILLOPS, S. D. & KILLOPS, V. J. 1993. *An Introduction to organic geochemistry.* Longman Geochemistry Series.

LAHN, E. 1940. *Report about the region between Kizilirmak and Yesilirmak.* MTA Special Report, No. 1026 [in Turkish].

LEVINE, J. R. 1994. Coal petrology in exploring for coal seam gas. *In: International Conference, & Short Course on Coalbed Methane and Coal Geology, University of Wales Cardiff, UK, 12–16 September 1994.*

NAKOMAN, E. 1966. Palynological investigation of Eocene Sorgun lignite. *MTA Bulletin,* **67**, 69–89 [in Turkish].

OZCAN, A., ERKAN, A., KESKIN, A., ORAL, O., OZER, S., SUMENGEN, M. & TEKELI, O. 1980. *Geology of the Area Between North Anatolian Fault and Kirsehir Massive.* MTA Report No. 6722 [in Turkish].

OZDEMIR, I. & BEKMEZCI, F. 1983. *Geology of Celtek lignite basin and coal exploration using boreholes data,* MTA Report No. 7396 [in Turkish].

PEKMEZCILER, S. 1953. *An Exploration Report on the Celtek lignite.* MTA Special Report [in Turkish].

STACH, E., MACKOWSKY, M.-TH., TEICHMULLER, M., TAYLOR, G. H., CHANDRA, D. & TEICHMULLER, R. 1982. *Stach's Textbook of Coal Petrology.* Gebruder Borntraeger, Berlin.

# Index